W0080122

Mental Health in Historical Perspective

Series Editors
Catharine Coleborne,
School of Humanities and Social Science
University of Newcastle
Callaghan, NSW, Australia

Matthew Smith
Centre for the Social History of Health and Healthcare
University of Strathclyde
Glasgow, UK

Covering all historical periods and geographical contexts, the series explores how mental illness has been understood, experienced, diagnosed, treated and contested. It will publish works that engage actively with contemporary debates related to mental health and, as such, will be of interest not only to historians, but also mental health professionals, patients and policy makers. With its focus on mental health, rather than just psychiatry, the series will endeavour to provide more patient-centred histories. Although this has long been an aim of health historians, it has not been realised, and this series aims to change that.

The scope of the series is kept as broad as possible to attract good quality proposals about all aspects of the history of mental health from all periods. The series emphasises interdisciplinary approaches to the field of study, and encourages short titles, longer works, collections, and titles which stretch the boundaries of academic publishing in new ways.

Jane Freebody

Work and Occupation in French and English Mental Hospitals, c.1918–1939

palgrave
macmillan

Jane Freebody
Oxford Brookes University
Oxford, UK

ISSN 2634-6036 ISSN 2634-6044 (electronic)
Mental Health in Historical Perspective
ISBN 978-3-031-13104-2 ISBN 978-3-031-13105-9 (eBook)
https://doi.org/10.1007/978-3-031-13105-9

Cover illustration: Colin Waters / Alamy Stock Photo

This Palgrave Macmillan imprint is published by the registered company Springer Nature Switzerland AG.
The registered company address is: Gewerbestrasse 11, 6330 Cham, Switzerland

To my mother, Yvonne, whose fortitude and stoicism have been an inspiration

Preface and Acknowledgements

This book is about the use of occupation by psychiatrists to treat mental illness. The value attached to work and occupation as therapeutic tools has varied over time and in different places, but being active has, since ancient times, been seen as essential to mental health and wellbeing. This was never more apparent than during the Covid-19 lock-downs, when many of us struggled with the imposition of free time. To combat that feeling of "rudderlessness" (around time not spent working at the kitchen table) we took up new hobbies, re-organised our living spaces, began a fitness regime, and mastered new technologies to keep in touch with friends and colleagues.

As Karl Marx explained, "purposeful activity" is necessary for the "full realisation" of our humanity. Humans need to be stimulated, challenged, engaged. We need the boost to our self-esteem that comes from solving problems and achieving our objectives. From a personal perspective, I know that to feel fulfilled I need to have an absorbing project, a goal to achieve, or a task to complete. This book has been one of those projects. It has been both challenging and rewarding.

I first became interested in the therapeutic power of occupation at a symposium in 2013, organised by Professor Waltraud Ernst of Oxford Brookes University. The theme of the symposium was "Therapy and Empowerment, Coercion and Punishment", and as the title suggests, its focus was the tension between the different ways occupation could be used in institutions for the mentally ill, not all of them therapeutic. The topic drew me in, and I decided to conduct a comparative study on how occupation was used in mental institutions in France and England for my

PhD, for which I received funding from the Wellcome Trust (award number 108615/Z/15/Z).

The book originated from the research conducted for my PhD, awarded in February 2020, a month before the first lock-down began. The book would not have happened without Wellcome's financial support for my research. I am extremely grateful to Wellcome for this support, and for their enriching programme for doctoral scholars that provided opportunities to network, learn new skills and think about life after the PhD.

I am indebted to and warmly thank my PhD supervisors, Waltraud Ernst, Glen O'Hara and Viviane Quirke, and Tom Crook, Post-Graduate Research Tutor, at Oxford Brookes University. Their expertise, advice and support were invaluable. They continued to believe in me and kept me on the right track. I would also like to acknowledge and thank my PhD examiners, Hilary Marland, Clare Hickman and Katherine Watson, whose insightful comments about my work have been extremely helpful during the planning stages of this book.

The various archives where I conducted my research also deserve special thanks. Bethlem Museum of the Mind, Oxfordshire Health Archives and Special Collections at Oxford Brookes University Library in the UK, and the Archives de Paris, Archives de la Sarthe and Bibliothèque Henri Ey in France all went out of their way to assist me. The editorial team, notably Lucy and Eliana, at Palgrave Macmillan have also been very supportive, leading me through the production process and always on hand to answer questions. The peer reviewers they selected provided me with hugely valuable constructive criticism.

My friends and family all deserve medals for their support and forbearance—they thought it was all over when I completed my PhD, but then along came the book! My love and gratitude go to Kevin, Sophie and Imogen, and to Rufus the dog, who maintained a dignified silence from the sofa when "walkies" were late.

Oxford, UK
June 2022

Jane Freebody

A Note on Terminology

Although some of the language and terminology used during the early twentieth century can sound offensive to twenty-first-century ears, I have used the language of contemporaries. Terms such as "mental deficiency" and "feebleminded", for example, are used in the book for the sake of maintaining authenticity. Today, we might speak of "service-users" rather than patients, and we would not refer to "males and females" when discussing patients, or to "paupers" when referring to the poor, but this was the language used at the time.

I have used "institution" as a generic term to include asylums, hospitals and colonies. The terms "hospital" and "asylum" are used interchangeably. Differentiation is complicated by the fact that in England, asylums became known as mental hospitals in 1926 (according to the recommendations of the Macmillan Commission of 1924-6), while in France they remained asylums (or "asiles") until 1938, when they became known as psychiatric hospitals ("hôpitaux psychiatriques"). A "colony" was the name given to a specialist institution for incurable cases.

The terms curable, incurable and chronic, used in contemporary texts, also need explanation. Curable cases of mental disorder are those that were believed to stand a chance of recovery, according to contemporary diagnoses. It was generally believed that to be curable, patients needed to begin treatment within a year of the onset of symptoms, preferably within three months. Patients at this early stage of mental disorder were known as "acute" cases. Incurable cases were those who had either been mentally disordered from birth (such as the so-called mentally deficient), or who had suffered a life-changing injury or disease (such as syphilis). Chronic cases were those whose conditions, although potentially curable if treated earlier, had worsened to the point of incurability.

Changing nosology during the interwar period made comparisons of different types of mental disorder difficult. Psychiatrists of the same nationality used different classification systems, which changed over time. A comparison of French and English disease categories was impeded for the same reasons. That said, specific diagnoses did not appear (from the records) to influence what type of occupation was allocated to patients, or whether they were occupied at all. More relevant was whether the patient was considered acute or convalescent, curable or incurable, and turbulent or calm, and whether they were physically fit.

Contents

Abbreviations

ADP	Archives de Paris
ADS	Archives de la Sarthe
AOT	American-style occupational therapy
AU	Author Unknown
BMM	Bethlem Museum of the Mind
DPM	Diploma in Psychological Medicine
GPI	General Paralysis of the Insane
IT	Industrial Therapy
MACA	Mental After Care Association
MAT	More Active Therapy
OHA	Oxfordshire Health Archives
PSW	Psychiatric Social Worker
(R)MPA	(Royal) Medico-Psychological Association (the society became "Royal" in 1926)

List of Figures

List of Tables

CHAPTER 1

Introduction

"It is my day for visiting the workshops, laundry and farm, and I will ask the reader to accompany me. Leaving the airing court, I turn down past the female ward blocks and the female hospital and enter the first workshop. There are four workshops in this block: the coir-picking shop, the tailor's shop, the bootmaker's shop, and the painter's shop. … There are some dozen male patients working in the coir-picking shop, picking the coir, or cocoa-nut fibre, with which most of the mattresses used by the patients are stuffed. It is unpleasant, unhealthy work, reminiscent of oakum-picking to those who have been in jail or worked as "casuals" in workhouses, and patients with weak chests or a tendency to bronchitis should not be employed at it, as the dust given off causes considerable bronchial irritation. But it is very useful work from the point of view of the asylum authorities, for it saves them much expense. In the tailor's shop some half a dozen patients are now employed under the superintendence of the asylum tailor, who is also a part-time attendant. In the bootmaker's shop only one patient is at present employed, for not many lunatics can be trusted with sharp tools.

"From the workshops we cross over to the laundry, which provides one of the most important and useful employments to which patients are put, women equally with men. The laundry is the stepping-stone to liberty for more patients than any other workshop. For only the best and most trustworthy patients are employed there, and few decline to take the job, though it is not a particularly healthy one, because they know that in many

J. Freebody, *Work and Occupation in French and English Mental Hospitals, c.1918–1939*, Mental Health in Historical Perspective,
https://doi.org/10.1007/978-3-031-13105-9_1

cases it is the half-way house to freedom. There are a score or more employed there this morning under the charge of the laundryman attendant, and I stop to chat with three or four, whom I am sending up for discharge at the next meeting of the Board, to satisfy myself as to their mental progress. From the laundry, I cross over to the boiler-house and thence to the engineer's shop, where three or four of the more intelligent patients are working. Two of these are also on my list for the next discharge, one of whom, an old man of about sixty, has been in the asylum for twenty-three years, and from all accounts has been fit for discharge for many years past, had anyone ever taken the trouble to interest himself sufficiently in his case.

"As I pass down the road on my way to the farm … I encounter a string of patients garbed in white overalls, who are wheeling boxes on barrows under the charge of an attendant. This is the "closet-barrow gang", and numbers twelve in all, and it has been at work, with an interval for breakfast, some four and a half hours. Theirs is the most unpleasant and unhealthy work of all. … The work of emptying the asylum closets must, of course, be done by somebody, and it is much cheaper to employ asylum labour than to get it done outside. Emptying earth-closets is a class of labour which, though unpleasant, is common enough in various parts of the country [and in France], and the particular type of patient employed in this instance is certainly not likely to suffer from undue fastidiousness. Were there no alternative to the earth-closet system there would certainly be no harm in employing healthy lunatics to empty the earth-closets, if they were not averse to the job, provided also that they were well fed, well clothed, and properly compensated, and that every care was taken to make the work as little exhausting and unhealthy as possible. As a matter of fact … none of these conditions were complied with."[1]

The same year that Montagu Lomax's gloomy account of how patients were occupied at Lancashire's Prestwich Asylum was published, the American psychiatrist Dr Adolf Meyer (1866–1950) gave a lecture on "The Philosophy of Occupation Therapy" at the fifth annual meeting of the National Society for the Promotion of Occupational Therapy in Baltimore, USA. Lomax's description, based on his experiences as a medical officer between 1917 and 1919, is borne out by remarks made by Meyer about the occupation of patients in English asylums during the late nineteenth century. Meyer observed that work in the "industrial shops and work in the

[1] Montagu Lomax, *Experiences of an Asylum Doctor: With suggestions for asylum and lunacy law reform* (London: G.Allen and Unwin Ltd, 1921), 104–107.

laundry and kitchen and on the wards [were] very largely planned to relieve the employees."[2] Meyer was also critical of occupation in American institutions. At the hospital in Kankakee, Illinois to which Meyer was appointed in 1893, "there was little in the atmosphere to foster interest in occupation" and where work was organised "merely from the point of view of utility".[3] His words could equally apply to the situation in French asylums. Meyer's lecture then turned to the new conception of occupation, developed in the USA just before the outbreak of World War I, that involved a "blending of work and pleasure—all made possible by a wide supplementing of centralisation by individualisation and a kind of de-centralisation." This new approach, which included a range of more interesting craft activities, such leather work, basketry and book-binding, generated amongst patients, "a pleasure in achievement … and a happy *appreciation of time*".[4]

The two sets of observations, made by Lomax (1860–1933) and Meyer in 1921, raise the question of what had happened to the individualised, therapeutic work programmes that had formed the cornerstone of moral treatment in early nineteenth-century France and England? Why did the nature of work allocated to patients change in the late nineteenth and early twentieth centuries, while asylums continued to operate within the framework of moral treatment? This study will explore what happened within psychiatry, and wider society, to foster a type of patient work that appeared to benefit the institution as much, if not more than the patient. It will identify the combination of medical, economic and social factors in the wake of World War I that caused a re-appraisal of psychiatry, patient work and occupation in England, and amongst a small group of Parisian psychiatrists. Was the approach outlined by Meyer in his lecture really new, or was it merely a re-fashioning of moral treatment? How helpful was it to those hoping to re-join the labour force after leaving hospital? Analysis of factors such as the fiscal crises following World War I and the Great Depression, changing attitudes towards work and welfare, the influence of contemporary notions of class and gender on work, and the changing nature of industry and working practices will help to explain the differences in approach to patient occupation that developed in France and England between 1918 and 1939. Fundamental to this discussion are the divergent professional

[2] Adolf Meyer, "The Philosophy of Occupation Therapy," in *The Collected Papers of Adolf Meyer: Vol IV Mental Hygiene*, ed. Eunice Winters (Baltimore: The Johns Hopkins Press, 1952), 87.

[3] Ibid.

[4] Ibid., 87–88. (Italics in the original).

trajectories of French and English psychiatry and disparate views concerning the origin, curability and treatment of mental disorder.

Selection of Countries and Institutions

The historical parallels in the development of the French and English asylum systems were marked. Theories regarding the moral treatment of mental illness developed by the English William Tuke (1732–1822) and the French Philippe Pinel (1745–1826) emerged contemporaneously in the early nineteenth century and became the model of treatment to which physicians on both sides of the Channel aspired. Work formed a key element of moral treatment and the type of work given to patients, such as farm work, work in the workshops, kitchen and laundry, was similar in French and English asylums. Fast-forward one hundred years, and patient occupation had taken on a different character in the two countries. This divergence in approach, after a century of close alignment, prompted the selection of France and England for study during the interwar period. French and English psychiatrists (or alienists as they were called) had collaborated during the nineteenth century, visiting each other's institutions and reporting on developments in psychiatry in each other's countries. Despite this professional collaboration and shared history, work therapy developed differently in each country after World War I, an anomaly that warranted further investigation.

The four metropolitan institutions selected for study all specialised in the treatment of acute-stage, presumed curable cases of mental illness, while the two provincial institutions cared for a mixed clientele of curable and incurable cases. The selection facilitates a comparison of the different approaches to occupation in French and English institutions, between institutions admitting patients at the onset of their symptoms and those only accepting certified patients, and between metropolitan and rural institutions. Ste Anne's in central Paris included both the Asile Clinique (established in 1867) and the Henri Rousselle Hospital (1922). In London, the two hospitals specialising in acute, curable cases were the Maudsley Hospital, which opened for civilian cases in 1923, and Bethlem Royal Hospital, which had been in existence since the thirteenth century. The Henri Rousselle and Maudsley hospitals were similar in that they were both established for the specific purpose of treating mild, incipient cases of mental disorder that did not warrant certification. They were the only public mental institutions in France and England where the poorest members of society could be admitted voluntarily, before their symptoms became entrenched.[5] The institutions, founded by Édouard Toulouse (1865–1947) and Henry Maudsley (1835–1918), both became important centres of psychiatric research and provided models for the mental hospitals of the future, offering a blend of in-patient and out-patient care.

Prior to the opening of the Maudsley, the only other English mental hospital that specialised in acute, curable cases, and admitted voluntary as well as certified patients, was the Bethlem Royal Hospital. Bethlem was a registered hospital and therefore not subject to the same jurisdiction as county and borough mental hospitals; it admitted poor patients (although not paupers) voluntarily on a charitable basis. Bethlem had operated a policy of restricting admissions to patients in "a presumably curable condition" since the mid-nineteenth century.[6] This acceptance of curable patients on a voluntary basis, coupled with its long history, made Bethlem an interesting institution to compare with its near neighbour and rival, the modern Maudsley Hospital.[7] Despite their similar admissions policies, the treatment offered at the two institutions differed significantly. The various factors that contributed to these differences are analysed, such as Bethlem's links with the specialist neurological National Hospital for Nervous Diseases in Queen Square, the pervasiveness of a traditional approach to treatment, and the social class of patients. Treatment at the Maudsley, the early plans for which were based on Emil Kraepelin's clinic in Munich, drew on the latest psychiatric thinking. Occupational therapy was introduced as soon as the Maudsley opened in 1923, while it was only provided at Bethlem from 1932. Both Bethlem and the Maudsley had a medical school, but Bethlem's foundered in the mid-1920s in the face of competition from the Maudsley school. Bethlem's move from south London to Kent in 1930, and a lack of high-profile research sealed the fate of the medical school, which ceased in 1937.[8]

As well as the "deserving poor", who were not expected to pay fees, Bethlem also admitted an increasing number of private patients who could

[5] At all other public mental hospitals, admission was via the often-lengthy process of certification. This situation changed in England in 1930 with the passing of the Mental Treatment Act which permitted voluntary admission to all public mental hospitals.

[6] Jonathan Andrews, Asa Briggs, Roy Porter, Penny Tucker and Keir Waddington, *The History of Bethlem* (Abingdon and New York: Routledge, 1997), 649.

[7] Under the National Health Service, the Bethlem and Maudsley hospitals joined forces to become a Joint Hospital in 1948.

[8] Andrews *et al.*, *History of Bethlem*, 573.

afford to contribute towards the costs of their care, but not the exorbitant fees of a private asylum. The occupations of these middle-class patients, for whom manual work was an anathema, facilitated a comparison with those of the private patients of the Asile de la Sarthe, which accepted both pauper and paying patients. Occupation was considered important at Bethlem, but the focus was on sport and recreation rather than work. The Asile de la Sarthe's pauper patients were obliged to work, while private patients were exempt. The French asylum lacked the extensive programme of leisure activities provided at Bethlem and did not offer occupational therapy, but a small proportion of the paying patients engaged in some form of work. The variation in the way patients were occupied in the different classes (pauper, poor or working-class and middle-class) in France and England offered interesting points of comparison.

When it was established in 1867, the Asile Clinique catered for a mixed clientele of acute, curable and incurable cases. Since its opening, there had been calls for it to become a hospital exclusively for acute, curable patients. These plans were resurrected in 1918, although the transformation was not fully completed until 1927.[9] As a hospital for acute cases, the Asile Clinique was expected to provide the most up-to-date psychiatric treatment by doctors at the peak of their profession.[10] Most patients still had to undergo the process of certification before they could be admitted to the Asile Clinique, thereby delaying the commencement of treatment. Others were brought directly to the Admissions Office by their families (an arrangement made possible by special legislation passed in 1876) which meant that treatment could begin earlier.[11] Discharge rates after 1928 indicate that many patients were believed to have recovered. The active treatment delivered at the Asile Clinique was heavily influenced by neurology, a characteristic shared with Bethlem. Psychiatrists at the Henri Rousselle Hospital, and some of the junior doctors at the adjacent Faculty Clinic, adopted a more holistic approach, treating patients psychologically as well as biologically. These differences raise the question of the extent to which occupation was used therapeutically in the various institutions, and whether certain

[9] Such plans were originally put forward in the 1870s and periodically resurfaced, including during the 1910s, but were not acted upon.

[10] Louis Dausset, Rapport Général au nom de la 3e Commission sur les propositions budgétaires pour le Service des Aliénés (budget de 1919), Conseil Général de la Seine, 1918, 20. ADP/D.10K3/27/20.

[11] Patricia E. Prestwich, "Family strategies and medical power: 'voluntary' committal in a Parisian asylum, 1876–1914," *Journal of Social History* 27, no. 4 (1994): 799.

approaches attached greater importance to therapeutic occupation. The differences in approach of divisions within the same complex (Ste Anne's) offered interesting territory for exploration and demonstrate the importance of the treatment preferences of individual psychiatrists.

The two provincial institutions, the Littlemore Hospital in Oxford and the Asile de la Sarthe in Le Mans, both catered for a mixed clientele of curable and incurable cases. Both institutions were situated in or near provincial towns that served the surrounding rural populations of the county of Oxfordshire, England, and the department of La Sarthe, France. Established in 1846 and 1828 respectively, both institutions had been managed according to the principles of moral therapy during their early incarnations. Both institutions were able to provide patients with a plentiful supply of farm work (considered so important by the moral therapists) on the land surrounding the asylums. By the end of the nineteenth century much of the agricultural land near the Asile de la Sarthe had been built on, while the farm at the Littlemore remained a significant aspect of asylum life throughout the interwar period. This raises interesting questions regarding each asylum's priorities in terms of providing therapeutic work for patients on the one hand and offsetting costs on the other. The financial importance to each institution of growing fresh produce for consumption by patients and staff, or for sale, are assessed in Chap. 5.

A crucial difference between the Littlemore and the Asile de la Sarthe was the Littlemore's access to the medical students who studied at Oxford University and the Radcliffe Infirmary. Le Mans did not have a university (the University of Maine was founded in 1977) which limited the Asile de la Sarthe's ability to recruit interns. A shortage of medical staff compromised the ability of the chief medical officer to deliver treatment. The medical superintendent at the Littlemore enjoyed the assistance of medical students and junior medical staff. It was far more feasible for the Littlemore superintendent to deliver active treatment, such as psychotherapy or malaria therapy, than it was for the Asile de la Sarthe's chief medical officer to deliver any form of treatment. The English Mental Treatment Act of 1930 permitted institutions outside London, like the Littlemore, to admit patients voluntarily, without the need for certification. This meant they could begin treatment at an earlier, potentially curable, stage of their illness. This legislation instigated a gradual change in the proportion of curable patients at the Littlemore, setting it apart from the Asile de la Sarthe where no such changes took place. The approaches to treatment taken by the medical superintendents of the Littlemore and the chief medical officers of the Asile de la Sarthe were markedly different. These differences, examined in Chap. 6, were highlighted by the psychiatrists' attitudes towards the occupation of patients.

Analysis of patient work and occupation in each of the institutions studied demonstrates how occupation was associated with different psychiatric traditions and approaches to mental disorder. The presumed curability of the patient and stage of their illness, a patient's class and gender, the institution's rural or urban location, the availability and quality of staff, and the treatment preferences of individual psychiatrists were also important factors. The selected institutions were not necessarily representative of psychiatric care in their respective nations, but their use of occupation highlights differences between "modern" and "traditional" institutions and signposts the future development of mental hospitals in France and England. A comparison of occupation in these institutions advances and adds a new dimension to themes identified by the existing historiography of patient work and occupational therapy and throws new light on approaches to psychiatry during the interwar period.

Sources

The annual reports of the six institutions formed the main source material for this study. The nature of the reports varied, with some including administrative, "moral" and medical reports and others written solely by the medical superintendent, but they incorporated similar information and statistics on the workings of each institution, its patients and staff, treatment regimes, facilities and financial situation. Whilst these reports provided the details essential for conducting an investigation into patient work and occupation, it must be remembered that they were written to present a certain image to the outside world. As Kathryn McKay notes, the annual reports of medical superintendents in British Columbia, Canada, included carefully balanced accounts of patient labour that overcame the inherent "tension between exploitation and therapy".[12] It has been important to try and "read between the lines" of the annual reports

[12] Kathryn McKay, "From Blasting Powder to Tomato Pickles: Patient work at the provincial mental hospitals in British Columbia, Canada, 1855–1920." In *Work, Psychiatry and Society, c.1750–2015*, ed. Waltraud Ernst (Manchester: Manchester University Press, 2016), 102.

to piece together what life was really like in these institutions, and to use other sources to put them into context.

The reports of the Board of Control (England) and the Commission de Surveillance (France), the bodies that oversaw the provision of institutional care for the mentally disordered, were consulted for their reports on individual institutions and to ascertain their stance (and therefore that of government) on patient work and occupation. In the case of the French Commission de Surveillance, the contents of the reports tended to focus on practical matters, such as sanitation and building works. The Board of Control reports were more policy-driven, with sections dedicated to "lunacy" (or mental illness) and "mental deficiency" (or intellectual impairment), as well as inspection reports on individual institutions.[13] The Board of Control, which reported to the Ministry of Health after the latter's formation in 1919, formed a link between the mental hospital system and the state and acted as a conduit for government policy towards the mentally disordered.[14] It could not enforce compliance but highlighted the standards expected of institutions. In France, asylum directors and chief medical officers reported to the General Council (*Conseil Général*) of the department, an elected body responsible to the Prefect who, operating under the Ministry of the Interior, represented the state at local level.

The professional psychiatric journals of each country, the *Journal of Mental Science* (established in 1855) and the *Annales Médico-psychologiques* (founded in 1843), were consulted to gain a sense of the important contemporary issues within psychiatry and to ascertain the level of interest amongst psychiatrists in therapeutic occupation. These two publications were associated with the Medico-Psychological Association (the MPA became the Royal Medico-Psychological Association, or RMPA, in 1926) in England, and the Société Médico-psychologique in France. As such, they were the "official" voice of French and English psychiatry. Articles advocating new approaches to patient work written by French psychiatrists also appeared in *L'Aliéniste français*, which was produced by the *Association amicale des Médecins des établissements publics d'aliénés de France*, an association for doctors working in public asylums founded in 1907, and *L'Hygiène Mentale*, the journal of the French League of Mental Hygiene, edited by Édouard Toulouse, co-founder of the French league. *L'Hygiène Mentale* was therefore supportive of the principles of mental hygiene,

[13] The Board of Control replaced the Lunacy Commission in 1913.

[14] Previously, the Board had reported to the Home Office.

which included the re-education of patients through occupation, while *L'Aliéniste français* was in favour of asylum reform. Other key sources included the reports from government enquiries, such as the Macmillan Report of 1926, and legislation, such as the Mental Treatment Act of 1930 and the French Ministerial Circulars of 1937 and 1938. These give further insight into government thinking on the use of occupation within the overall contexts of institutional management and psychiatric treatment, revealing occasional differences between central policy and local practice.

Themes and Historiography

The prescription of work for patients originated in the context of moral treatment and the emergence of the asylum system. Its prescription was influenced by a range of factors that varied across time and space. The contemporary economic climate; the nature of industry; the provision of welfare; notions of class and gender; psychiatric ideology; and the professionalisation of psychiatry, mental nursing and occupational therapy, all had a role to play. These topics have their own specialist histories, but few (with the obvious exception of occupational therapy) are linked to institutional patient work. This study draws these various influences together to enrich our understanding of the rationale for patient work and occupation. It highlights the tendency of psychiatry to encroach upon areas of everyday social life and to "medicalise" the concept of employment. In so doing, it reveals our changing and often ambivalent attitudes towards work, its perception as a moral good, as a means of subsistence, as a source of identity and personal satisfaction, and as the basis of national wealth and competitiveness. Conversely, this study also draws attention to ways of thinking about unemployment, as a moral failing or as a social problem to be solved, and about the responsibility of the state towards those without work or income. These issues were particularly relevant during the political and economic uncertainty of the interwar years.

Work and Working Practices

The devastation of the economies of Europe by World War I and by the Great Depression of the 1930s, have been well documented by economic historians.[15] The direct and indirect effects of this devastation on the practice of patient work and occupation in institutions have attracted less attention. Both the war and the economic precarity of the 1930s had a significant effect on the mental health of individuals, on how they were treated and occupied in institutions, on the budgets available for healthcare and on the work that might be available outside those institutions to discharged patients. As well the premature death of ten million people, massive migration and the redrawing of national borders, the war caused major disruptions to markets and industry. After the conflict, industry had to re-adjust to peace-time conditions. In France, industries in the north-east destroyed during the conflict had to be reconstructed. It was estimated in 1919 that the war had left France short of three million men from its labour force, necessitating the recruitment of immigrant workers.[16] In Britain, the government had to decide how to redeploy some five million servicemen, and a similar number of civilians (including *c.*700,000 women) who had been employed in war-related work.[17] The war had encouraged women to leave sectors traditionally associated with female labour, such as textiles, clothing manufacture or domestic service and to move into clerical work, shopwork and factory work.[18] Whether such changes were reflected by the work provided in asylums is an area addressed in Chap. 9.

During the process of recovery from these war-induced challenges, the profile of industry was changing. Coal-powered steam engines were gradually being replaced by electric motors in manufacturing, and electricity and diesel threatened the use of locomotives in rail transport.[19] The new sectors of electrical engineering, rayon production and automobile manufacture were poised for growth after the war. There remained, however,

[15] For example: Stephen Broadberry and Kevin O'Rourke, eds, *The Cambridge Economic History of Modern Europe, Vol II: 1870 to the Present* (Cambridge: Cambridge University Press, 2010).

[16] James F. McMillan, *Twentieth-Century France: Politics and Society 1898–1991* (London: Arnold, 1992), 79.

[17] Julian Greaves, "The First World War and Its Aftermath," in *Twentieth-Century Britain: Economic, Cultural and Social Change*, eds Francesca Carnevali and Julie-Marie Strange (Harlow and New York: Pearson Education, 2007), 132.

[18] Arthur J. McIvor, *A History of Work in Britain, 1880–1950* (Basingstoke: Palgrave, 2001), 38–9.

[19] Erik Buyst and Piotr Franaszek, "Sectoral Developments, 1914–1945," in *The Cambridge Economic History of Modern Europe*, eds Stephen Broadberry and Kevin O'Rourke (Cambridge: Cambridge University Press, 2010), 224.

pockets of traditional, artisanal occupations, that coexisted alongside the newer industries, and the extent and nature of industrial development varied regionally, in both France and Britain.[20] The percentage of the work force employed in industry remained stable between 1913 and 1930 at *c*.46% in Britain and *c*.33% in France.[21] The agricultural sector continued to shrink in both France and Britain, although in France the proportion of workers remaining on the land was much higher. In Britain, one of the least agricultural countries of Europe, just 10% of people were employed in agriculture in 1913, and by 1930 this percentage had decreased to 6%. In France, 41% were employed in agriculture in 1913, and 36% in 1930.[22] The services sector, however, increased in both countries, with a modest rise in Britain from 45% in 1913 to 48% in 1930, and in France from 26% in 1913 to 31% in 1930.[23] This sector comprised shipping; rail and road transport; financial services; distribution; education, medical and social services; and religious and domestic services.[24] These factors had implications for what might constitute rehabilitative work within asylums and raise questions over the ability of asylums to provide work that prepared individuals for the labour market outside. Was agriculture, for example, an area to encourage given the declining numbers employed in the sector?

As Geoffrey Searle has highlighted, attempts were being made by the British to maximise productivity before the outbreak of World War I, amid concerns regarding "national efficiency" and maintaining Britain's reputation as "the workshop of the world".[25] The war accelerated these efforts and stimulated innovation in working methods. Research into how to minimise "industrial fatigue" and maximise output by workers increased, as the war-time requirement for long hours and intense, sustained effort took their toll on the health of munitions workers.[26] The war, as Roger Cooter and Steve Sturdy have maintained, acted as a catalyst for the development of scientific management techniques to maximise productivity and efficiency.[27] The new methods, developed in the

[20] Maxine Berg and Pat Hudson, "Rehabilitating the Industrial Revolution," *Economic History Review* 45 no.1 (1992): 24–50.

[21] Buyst and Franaszek, "Sectoral Developments," 210.

[22] Ibid.

[23] Ibid.

[24] Ibid., 228.

[25] G. R. Searle, "The Politics of National Efficiency and War, 1900–1918," in *A Companion to Early Twentieth-Century Britain*, ed. Chris Wrigley (Oxford: Blackwells, 2003), 56.

[26] Steffan Blayney, "Industrial Fatigue and the Productive Body: the science of work in Britain, 1900–1918," *Social History of Medicine* 32 no. 2 (2019): 310–28.

[27] Steve Sturdy and Roger Cooter, "Science, Scientific Management and the Transformation of Medicine in Britain *c*.1870–1950," *History of Science*, 36 no. 1 (1998): 421–46.

USA and often known collectively as Taylorism, involved the breaking down and subdivision of tasks involved in the production process. This resulted in a de-skilling of the workforce; tighter controls over workers; a loss of worker discretion and autonomy; closer links between effort and earnings; and more stringent monitoring of performance.[28] The routinised and fragmented working practices, analysed by Anson Rabinbach, were not only unpopular with workers but could be perceived as harmful to their mental health.[29] According to the Mental Hygiene Movement's preventative agenda, work should be satisfying and well-suited to the aptitudes of the individual worker. This raises the question over whether work for patients was aimed at preparing them for the labour market, whilst exposing them to potentially harmful working practices, or at protecting their mental health by providing satisfying occupations that failed to equip them for the modern workplace. Equally damaging to mental health was the anxiety, loss of self-esteem and sense of hopelessness generated by being unemployed.[30] Arguably, recently discharged patients might be more prone to unemployment if they did not leave the institution with a marketable skill.

Welfare and Unemployment

Much has been written on changing attitudes towards unemployment and poverty, and on the beginnings of state welfare provision in both France and England.[31] Poverty and joblessness, regarded as a moral failure and a matter of individual fault for most of the nineteenth century, gradually

[28] McIvor, *History of* Work, 54–5.

[29] See: Anson Rabinbach, *The Human Motor: Energy, Fatigue and the Origins of Modernity* (Berkeley: University of California Press, 1992).

[30] Richard Warner, *Recovery from Schizophrenia: Psychiatry and Political Economy* (London and New York: Routledge and Kegan Paul, 1985), 50–1.

[31] Examples include: Bernard Harris, *The Origins of the British Welfare State: Social Welfare in England and Wales, 1800–1945* (Basingstoke: Palgrave Macmillan, 2004); Edward Royle, *Modern Britain: A Social History 1750–2011*, Third ed. (London and New York: Bloomsbury, 2012); Lynn Hollen Lees, *The Solidarities of Strangers: The English Poor Laws and the People, 1700–1948* (Cambridge: Cambridge University Press, 1998); John Burnett, *Idle Hands: The Experience of Unemployment, 1790–1990* (London and New York: Routledge, 1994); Paul V. Dutton, *Origins of the French Welfare State: The Struggle for Social Reform in France, 1914–1947* (Cambridge: Cambridge University Press, 2002); Philip Nord, "The Welfare State in France, 1870–1914," *French Historical Studies* 18, no. 3 (1994); Susan Pederson, *Family, Dependence, and the Origins of the Welfare State: Britain and France, 1914–1945* (Cambridge: Cambridge University Press, 1993).

came to be seen as a social problem in the late nineteenth and early twentieth centuries. Measures to tackle the root causes of poverty, such as old age, sickness and unemployment, were introduced in England in the decade before World War I, but not until the late 1920s in France. The existing literature does not explore the relationship between these changes taking place in society and their effect on patient care and how patients were occupied in mental institutions, a lacuna addressed by this study. The existence of the twenty-year time lag between the introduction of significant welfare measures in England and France raises questions about the impact of these measures on the way patients were prepared for life outside English and French institutions. Did the existence of welfare measures in post-war England enable psychiatrists to focus more on using occupation therapeutically, rather than as a means of ensuring patients' employability, than in France? Late nineteenth-century English psychiatric texts made frequent references to the "creation of useful members of society" who could earn their own living after discharge, as Sarah Chaney observes, but this type of rhetoric was far less common after World War I.[32] Was the development of self-sufficiency amongst patients no longer a priority for mental hospitals, despite the economic uncertainty of the period?

The problem of unemployment grew very rapidly after World War I, as Bernard Harris observes.[33] Between 1921 and 1938 in Britain, the average number of unemployed people never dropped below one million and remained above two million during the first half of the 1930s, peaking in 1932 as the Great Depression took its toll.[34] In France, the effects of the Depression lasted longer than in Britain, causing widespread hardship between 1931 and 1938.[35] High levels of unemployment during the Depression put increasing strain on existing welfare provision, as well as on health budgets. In England, the poor law buckled under the strain. The Boards of Guardians were abolished, and the 1929 Local Government

[32] Sarah Chaney, "Useful members of society, or motiveless malingerers? Occupation and malingering in British asylum psychiatry, 1870–1914," in *Work, Psychiatry and Society, c.1750–2015*, ed. Waltraud Ernst (Manchester: Manchester University Press, 2016), 282.

[33] Harris, *Origins of the British Welfare* State, 198.

[34] Ibid.

[35] McMillan, *Twentieth-Century* France, 101.

Act saw responsibility for the poor transferred to the local authorities, who were obliged to set up Public Assistance Committees.[36] This study examines how such pressures outside mental institutions affected the occupation of patients inside them. As Richard Warner notes, in the USA, increased levels of admissions for schizophrenia were observed during economic slumps, and recovery rates were significantly lower during the Great Depression.[37]

The Emergence of the Asylum System and Moral Treatment

The history of Western psychiatry has tended to focus on the late-eighteenth and nineteenth centuries, a period characterised by the rise of the asylum and by new ways of treating the insane according to the humanitarian principles of the Enlightenment.[38] "Moral treatment" became the method to which most asylum doctors aspired, and its

[36] Harris, *Origins of the British Welfare State*, 202–3.

[37] Warner, *Recovery from Schizophrenia*, 269.

[38] Institutional accounts: Anne Digby, *Madness, Morality and Medicine: A Study of the York Retreat, 1796–1914* (Cambridge: Cambridge University Press, 1985); Steven Cherry, *Mental Health Care in Modern England: The Norfolk Lunatic Asylum/St. Andrew's Hospital, 1810–1998* (Woodbridge: Boydell, 2003); D. Gittins, *Madness in Its Place: Narratives of Severalls Hospital, 1913–1997* (London and New York: Routledge, 1998); Andrews *et al.*, *The History of Bethlem*; John Crammer, *Asylum History: Buckinghamshire County Pauper Lunatic Asylum—St John's* (London: Gaskell, 1990); Charlotte MacKenzie, *Psychiatry for the Rich: A History of Ticehurst Private Asylum, 1792–1917* (London: Routledge, 1992); Works on moral therapy: Roy Porter, *Mind-Forg'd Manacles: A History of Madness in England from the Restoration to the Regency* (London: Penguin, 1990); Roy Porter, "Was There a Moral Therapy in Eighteenth-Century Psychiatry?," *Lychnos* (1981): 12–26; Digby, "Moral Treatment at the Retreat, 1796–1846," in *The Anatomy of Madness, Vol II: Institutions and Society*, eds W.F. Bynum, Roy Porter and Michael Shepherd (London and New York: Tavistock), 52–72, Alexander Walk, "Some Aspects of the 'Moral Treatment' of the Insane up to 1854," *Journal of Mental Science*, no. 421 (1954): 807–37; Eric Carlson and Norman Dain, "The Psychotherapy that was Moral Treatment," *The American Journal of Psychiatry*, no. 117 (1960): 519–24; Donald Gerard, "Chiarugi and Pinel Considered: Soul's Brain/ Person's Mind," *Journal of the History of the Behavioral Sciences* 33, no. 4 (1997): 381–403; Louis Charland, "Benevolent Theory: Moral Therapy at the York Retreat," *History of Psychiatry*, 18 no. 1 (2007): 61–80; Jane Freebody, "The Role of Work in Late Eighteenth- and Early Nineteenth-Century Treatises on Moral Treatment in France, Tuscany and Britain," in *Work, Psychiatry and Society, c.1750–2015*, ed. Waltraud Ernst (Manchester: Manchester University Press, 2016), 31–54.

principles provided the administrative and medical framework for the emerging asylum systems in both France and England. The therapy concentrated on the rational and emotional, rather than the organic causes of insanity, aiming to build up patients' self-esteem and self-restraint, thereby equipping them with sufficient self-discipline to master their condition.[39] Integral to moral treatment was giving patients some form of work, such as assisting the attendants with their duties or working in the fields or gardens.[40] Patient work was, according to Andrew Scull (1993), a "major cornerstone" of nineteenth-century moral treatment, while Leonard Smith (2007) describes it as "a rationalised central element of therapy and rehabilitation in public lunatic asylums".[41]

The merits of moral treatment, and thus of patient work, have been the topic of vigorous debate among historians. Psychiatric practitioners writing the history of their profession in the 1930s, 1940s and 1950s, extolled the virtues of moral treatment as the harbinger of a new compassionate era that witnessed a transformation of the profession of psychiatry from "cruelty and barbarism to organised, institutional humanitarianism, and from ignorance, religion and superstition to modern medical science".[42] These progressivist, internalist accounts became known as Whiggish histories.[43] They were challenged by Michel Foucault in his *Histoire de la Folie* (1961), in which he questioned the alleged benign nature of moral treatment, and rejected the notion of liberal humanism in nineteenth-century psychiatry, medicine and penal reform. In his view, moral treatment merely replaced

[39] Digby, "Moral Treatment at the Retreat, 1796–1846," 53.

[40] By way of an example, the work regime at the York Retreat, where moral therapy was pioneered by William Tuke (1732–1822), is outlined by Digby in: ibid., 52–72.

[41] Leonard Smith, *Lunatic Asylums in Georgian England 1750–1830* (London and New York: Routledge, 2007), 156; Andrew Scull, *The Most Solitary of Afflictions: Madness and Society in Britain, 1700–1900* (New Haven and London: Yale University Press, 1993), 102.

[42] Roy Porter and Mark Micale, "Introduction: Reflections on Psychiatry and Its Histories," in *Discovering the History of Psychiatry*, eds Mark Micale and Roy Porter (Oxford and New York: Oxford University Press, 1994), 6.

[43] See: Ernst Mayr, "When is historiography Whiggish?" *Journal of the History of Ideas* 5 no.2 (1990): 301–09.

physical abuse with mental abuse. Seen through Foucault's lens, the asylum was a means of moulding inmates into willing, acquiescent workers, ready to contribute to the state's economic requirements. Foucault characterised work within the asylum as "a constraining power superior to all forms of physical coercion" due to the regularity of the hours, the need to pay attention, and the obligation to produce results.[44] Work carried out by inmates was "deprived of any productive value", imposed as a "moral rule, a limitation of liberty, a submission to order, an engagement of responsibility" designed to rid the mind of "of all exercises of the imagination" and supplant "delirious illusions".[45] Foucault's critical account, together with those of Erving Goffman, R.D. Laing and Thomas Szasz, prompted a series of revisionist works on the history of psychiatry, including those by Klaus Doerner, David Rothman, Robert Castel and Andrew Scull.[46] According to Scull, writing in the late 1970s, the asylum system was designed to "repair damaged human capital", to instil bourgeois values of self-control, self-sufficiency and productivity through a programme of "moral therapy", and to "warehouse" those unable to support themselves.[47]

A new generation of scholars turned away from revisionist arguments focusing on negative aspects of social control to embrace what Joseph Melling has termed a late-Whiggish position on the nineteenth-century

[44] Michel Foucault, *Madness and Civilization: A History of Insanity in the Age of Reason*, trans. Richard Howard (New York: Random House, 1965), 247.

[45] Ibid., 248.

[46] Erving Goffman, *Asylums: Essays on the Social Situation of Mental Patients and Other Inmates* (Garden City, NY: Anchor, 1961); R.D. Laing, *The Divided Self: A Study of Sanity and Madness* (London: Tavistock, 1960); Thomas Szasz, *The Myth of Mental Illness: Foundations of a Theory of Personal Conduct* (New York: Hoeber-Harper, 1961); Klaus Doerner, *Madmen and the Bourgeoisie: A Social History of Insanity and Psychiatry* [1969], trans. Joachim Neugroschel and Jean Steinberg (Oxford: Blackwell, 1981); David Rothman, *The Discovery of the Asylum: Social Order and Disorder in the New Republic* (Boston: Little Brown, 1971); Robert Castel, *L'Ordre psychiatrique: L'âge d'or de l'aliénisme* (Paris: Minuit, 1976).

For a full discussion of the "Great Revision" see: Porter and Micale, "Introduction: Reflections on Psychiatry and Its Histories," 6–14.

[47] Andrew T. Scull, *Museums of Madness: The Social Organization of Insanity in Nineteenth-Century England* (London: Allen Lane, 1979), 94.

asylum, suggesting its function was more benign.[48] Leonard Smith, Akihito Suzuki and David Wright, contributors to Melling and Forsythe's edited volume (1999), present the asylum as a therapeutic establishment that responded to community needs.[49] In their study of the Welsh asylum at Denbigh, Pamela Michael and David Hirst assert that "the operational philosophy of the asylum was based on the 'rule of kindness'".[50] These more recent studies, based on the detailed study of institutional practices, indicate that the pendulum has swung back towards a more compassionate interpretation of the asylum. The present study of patient occupation helps to clarify the extent to which the mental hospital fulfilled a political aim of turning pauper patients into productive workers, and whether the *primary* rationale for the prescription of patient work was to offset institutional running costs, or to provide therapy for the benefit of the patient.

Psychiatry and Ideology

Other areas of the history of psychiatry that have attracted the attention of historians include the period between *c.*1870 and 1914,[51] the war

[48] Joseph Melling, "Accommodating Madness: New Research in the Social History of Insanity and Institutions," in *Insanity, Institutions and Society, 1800–1914, A social history of madness in comparative perspective,* eds Joseph Melling and Bill Forsythe (London and New York: Routledge, 1999), 14.

[49] Ibid., 22.

[50] Pamela Michael and David Hirst, "Establishing the 'Rule of Kindness': The Foundation of the North Wales Lunatic Asylum, Denbigh," in *Insanity, Institutions and Society, 1800–1914: A social history of madness in comparative perspective,* eds Joseph Melling and Bill Forsythe (London and New York: Routledge, 1999), 171.

[51] For example: Robert Ellis, *London and its Asylums, 1888–1914* (Cham: Palgrave Macmillan, 2020); Stef Eastoe, *Idiocy, Imbecility and Insanity in Victorian Society: Caterham Asylum, 1867–1911* (Cham: Palgrave Macmillan, 2020); Joseph Melling and Bill Forsythe, eds, *Insanity, Institutions and Society, 1800–1914: A social history of madness in comparative perspective* (London and New York: Routledge, 1999); Jean-Christophe Coffin, *La Transmission de la Folie 1850–1914* (Paris: L'Harmattan, 2003).

neuroses experienced by soldiers during World War I,[52] the history of psychoanalysis and the teaching of Sigmund Freud,[53] and aspects of psychiatry after 1945, such as the "pharmaceutical revolution" of the 1950s and 1960s.[54] In England, the process of de-institutionalisation after 1959, and in France the reforms taking place within psychiatry in the aftermath of World War II, have been particular foci for recent scholarship.[55] The inter-war period has received less attention, with the exception of studies of the Mental Hygiene Movement; the effect of eugenic ideology on psychiatry;

[52] Peter Leese, *Shell Shock: Traumatic Neurosis and the British Soldiers of the First World War* (Basingstoke: Palgrave Macmillan, 2002); Tracey Loughran, "Shell Shock, Trauma, and the First World War: The Making of a Diagnosis and Its Histories," *Journal of the History of Medicine & Allied Sciences* 67, no. 1 (2012): 94–119; Fiona Reid, *Broken Men: Shell Shock, Treatment and Recovery in Britain, 1914–1930* (London: Continuum, 2010); Hervé Guillemain and Stéphanie Tison, *Du Front à l'Asile, 1914–1918* (Paris: Alma éditeur, 2013); Marc Roudebush, "A Battle of Nerves: Hysteria and Its Treatments in France During World War I," in *Traumatic Pasts: History, Psychiatry and Trauma in the Modern Age, 1870–1930*, eds Paul Lerner and Mark S. Micale (Cambridge: Cambridge University Press, 2001), 253–79; Paul Lerner, *Hysterical Men: War, Psychiatry and the Politics of Trauma in Germany, 1890–1930* (Ithaca, NY: Cornell University Press, 2003); Laurent Tatu and Julien Bogousslavsky, *La Folie Au Front: La Grande Bataille des Névroses de Guerre (1914–1918)* (Paris: Éditions Imago, 2012); Marie Derrien, *'La Tête En Capilotade': Les Soldats De La Grande Guerre Internés Dans Les Hôpitaux Psychiatriques Français (1914–1980)*, Thèse (Lyon: Université de Lyon 2, 2015); Anne Creton, *Psychiatrie Française et Première Guerre Mondiale: Évolution des Idées et Situation au Sein des Établissements de Soins du Nord-Pas-De-Calais*, Thèse (Lille: Université de Lille, 2015).

[53] Frances M. Moran, *The Paradoxical Legacy of Sigmund Freud* (Abingdon and New York: Routledge, 2018); Eli Zaretsky, *Political Freud: A History* (New York: Columbia University Press, 2015); Mikkel Borch-Jacobsen, Sonu Shamdasani, *The Freud Files: an inquiry into the history of psychoanalysis* (Cambridge: Cambridge University Press, 2012); Reuben Fine, *A History of Psychoanalysis* (New York: Continuum Publishing, 1990).

[54] Such as Joanna Moncrieff, "An Investigation into the Precedents of Modern Drug Treatment in Psychiatry," *History of Psychiatry*, 10 no. 4 (1999): 475–90; David Healy, "The History of British Psychopharmacology," in *150 Years of British Psychiatry Vol II: The Aftermath*, eds Hugh Freeman and German E. Berrios (London and New Jersey: Athlone, 1996), 61–88.

[55] For example: John Welshman and Jan Walmsley, eds, *Community Care in Perspective: Care, Control and Citizenship* (Basingstoke and New York: Palgrave Macmillan, 2006); Isabelle von Bueltzingsloewen, *L'Hécatombe des Fous: La Famine dans les hôpitaux psychiatriques français sous l'occupation* (Paris: Éditions Flammarion, 2007); Nicholas Henckes, "Réformer et Soigner. L'émergence de la psychothérapie institutionelle en France, 1944–55," *Psychiatries dans l'histoire*, ed. Jacques Arveiller (Caen: PUC, 2008), 277–88.

and the introduction of shock treatments and psycho-surgery in the late 1930s.[56] The interwar period in France has been dismissed by some historians as "immobile" or "frozen", as Isabelle von Bueltzingsloewen observes.[57] This notion is challenged by the present study which demonstrates that, while by no means nationwide, new approaches to psychiatric care were being developed in France (albeit limited to Paris) and England during the interwar period, and it was during the interwar years that occupational therapy emerged in England. The present study's focus on patient work and occupation highlights what Mark Micale has described as the competition between "two explanatory models" of mental disorder that

[56] Jonathan Toms, *Mental Hygiene and Psychiatry in Modern Britain* (Basingstoke: Palgrave Macmillan, 2013); David Freis, *Psycho-Politics between the Wars: Psychiatry and Society in Germany, Austria and Switzerland* (Cham: Palgrave Macmillan, 2019); Anne-Laure Simonnot, *Hygiénisme et Eugénisme au XXe Siècle: à travers la psychiatrie française* (Paris: Éditions Seli Arslan, 1999); Jean-Bernard Wojciechowski, *Hygiène Mentale et Hygiène Sociale: contribution à l'histoire de l'hygiénisme, tomes I et II* (Paris and Montreal: Éditions L'Harmattan, 1997); Mathew Thomson, "Mental Hygiene in Britain during the First Half of the Twentieth Century," in *International Relations in Psychiatry: Britain, Germany and the United States to World War II*, eds Volker Roelcke, Paul Weindling and Louise Westwood (Rochester: University of Rochester Press, 2010), 134–55; Isabelle von Bueltzingsloewen, "Réalité et Perspectives de la Médicalisation de la Folie dans la France de l'Entre-Deux-Guerres," *Genèses* 1, no. 82 (2011); Mark B. Adams, ed., *The Wellborn Science: Eugenics in Germany, France, Brazil and Russia* (Oxford and New York: Oxford University Press, 1990); Mathew Thomson, "Disability, Psychiatry and Eugenics," in *The Oxford Handbook of the History of Eugenics*, eds Alison Bashford and Philippa Levine, (Oxford and New York: Oxford University Press, 2010), 116–133; Deborah Blythe Doroshow, "Performing a Cure for Schizophrenia: Insulin Coma Therapy on the Wards," *Journal of the History of Medicine and Allied Sciences* 62, no. 2 (2007): 213–43; Niall McCrae, "A Violent Thunderstorm: Cardiazol Treatment in British Mental Hospitals," *History of Psychiatry* 17, no. 1 (2006): 67–90; G. E. Berrios, "Early Electroconvulsive Therapy in Britain, France and Germany: A Conceptual History," in *150 Years of British Psychiatry, Vol II: The Aftermath*, eds Hugh Freeman and German E. Berrios (London: Athlone Press, 1996), 3–15; Jack Pressman, *Last Resort: Psychosurgery and the Limits of Medicine* (New York: Cambridge University Press, 1998); Joanna Moncrieff, "An Investigation into the Precedents of Modern Drug Treatment in Psychiatry," *History of Psychiatry* 10, no. 4 (1999): 475–90; Lucy King, "The Best Possible Means of Benefiting the Incurable: Walter Bruetsch and the Malaria Treatment of Paresis," *Annals of Clinical Psychiatry* 12, no. 4 (2004): 197–203; Andrew Scull, "Focal Sepsis and Psychosis: The Career of Thomas Chivers Graves (1883–1964)," in *150 Years of British Psychiatry, Volume II: The Aftermath*, eds Hugh Freeman and German E. Berrios (London and New Jersey: Athlone, 1996), 517–36.

[57] von Bueltzingsloewen, "Réalité et Perspectives," 54.

characterised psychiatry in the twentieth century.[58] One model is based on a psychosocial theory of insanity as an "illness of the mind or spirit" that emphasises external factors, such as family upbringing, personal history, and environmental factors, and the other on organic or somatic theories based on the view that mental disorder as "a cerebral disease with physical determinants such as heredity, foetal milieu, hormonal environment, and brain anatomy, physiology, and chemistry".[59] The two explanatory models are associated with different models of care. The somatic model is often associated with a custodial model of care known as "alienism", while the psychosocial model is indicative of the move towards modern "psychiatry" (a term that was rarely used before 1900) that involved the active treatment of patients.[60]

The adoption and rejection of these models took place at different times in different locations. Adoption of the psychosocial model in England was stimulated by the experiences of psychiatrists who treated soldiers suffering from war neuroses during World War I. The conflict had a profound effect on English psychiatrists' attitudes towards the causation and treatment of mental disorder. Although it would be naïve to assume that all English psychiatrists made this ideological shift, they were much more likely to accept that mental disorder had a psychological or social origin after the conflict than before the war, as Chris Feudtner has observed.[61] The fact that the war did not have the same impact on French attitudes towards mental disorder, and did not therefore prompt the move from a custodial to a psychosocial model of care, affords an interesting angle for comparison. This difference in approach is highlighted by the type of occupations assigned to patients, and the type of patients to whom it was assigned. In custodial institutions, such as those found in provincial France, only the calm, chronic and incurable patients and convalescent patients, who required little supervision, were given work around the hospital. At institutions where the psychosocial model had been adopted, such as the Maudsley Hospital, carefully supervised occupational therapy or work was prescribed to all patients, including those at the early, acute stage of their illness.

[58] Mark Micale, "The Psychiatric Body," in *Companion to Medicine in the Twentieth Century*, eds Roger Cooter and John Pickstone (London and New York: 2000), 323.

[59] Ibid.

[60] Aude Fauvel, "Aliénistes contre psychiatres. La médecine mentale en crise (1890–1914)," *Psychologie clinique* 17 (2004), 61–76.

[61] Chris Feudtner, "'Minds the Dead Have Ravished': Shell-shock, History and the Ecology of Disease Systems," *History of Science* 31 no. 4 (1993): 337–420.

Professionalisation of Psychiatry and Mental Nursing

The ideological shift from a physiological to a psychosocial approach is linked in the present study to levels of professionalisation in French and English psychiatry. It is argued that psychiatry followed very different professional trajectories in France and England. French psychiatry remained closely aligned with the more prestigious profession of neurology and did not develop as an independent discipline until after World War II. Psychiatry in England, on the other hand, developed independently of neurology, and the two disciplines diverged after World War I.[62] While there exists extensive literature on the professionalisation of medicine, there are fewer studies of the professionalisation of psychiatry.[63] Many tend to focus on the stages of development in training and professional organisations, rather than on the ideological shifts taking place within the profession, that occurred at different times in different places.[64] The present comparative study links the different stages of ideological and professional development in England and France, which are emphasised by the prescription of occupational therapy by psychiatrists.

The professionalisation of mental nursing is another key area addressed by the present study, since, in the absence of professional occupational therapists, the successful application of occupational therapy depended on the skills, competence and training of mental nurses. Peter Nolan traces the development of mental nursing in England from its origins in the eighteenth century to the end of the twentieth century. He highlights the goal of self-sufficiency through the use of patient labour in the nineteenth-century asylum, but pays scant attention to the role of mental nurses

[62] Hugh Freeman, "Psychiatry in Britain, *c.*1900," *History of Psychiatry* 21 no. 3 (2010): 12–24.

[63] See: Eliot Freidson, *Profession of Medicine: a study of the sociology of applied knowledge* (New York: Dodd Mead, 1970); Harold Perkin, *The Rise of Professional Society: England since 1880* (Abingdon: Routledge, 2002); Ivan Waddington, "The Movement towards the Professionalisation of Medicine," *British Medical Journal* 301 (1990): 688–90.

[64] John Crammer, "Training and Education in British Psychiatry, 1770–1970," in *150 Years of British Psychiatry, Vol II: The Aftermath*, eds Hugh Freeman and German E. Berrios (London: Athlone Press, 1996), 209–42; Thomas Bewley, *Madness to Mental Illness* (London: Royal College of Psychiatrists, 2008); Jan Goldstein, *Console and Classify: The French Psychiatric Profession in the Nineteenth Century* (Chicago and London: Chicago University Press, 2001); J. Guyotat, "La formation des psychiatres en France," *Raison Présente* 83 (1987): 69–81; Ian Dowbiggin, "French Psychiatry and the search for a professional identity: the Société Médico-Psychologique 1840–1870," *Bulletin of the History of Medicine* 63 no.3 (1989): 331–55.

supervising occupation in the 1920s and 1930s. Nolan acknowledges, however, that mental nurses felt threatened by the new profession of occupational therapy.[65] An edited volume by Anne Borsay and Pamela Dale examines the changing role of mental nurses in Britain, Ireland, Wales and Australia, addressing such issues as training, the move from asylum to community, gender, violence, recruitment and retention, but does not focus on the supervision of occupation.[66] The history of the mental nursing profession in France, which attracted considerable criticism from interwar French psychiatrists, is addressed by Alexandre Klein, Patrice Krzyzaniak and Benoît Majerus.[67] Their studies highlight the paucity of training and education received by nurses during the early twentieth century, but are not focused on the role of mental nurses in supervising occupation, a role for which contemporary psychiatrists found them lacking. The history of Dutch mental nursing profession is discussed in a monograph by Geertje Boschma.[68] Mental nursing in Holland attracted educated, middle-class women, for whom training was available from the 1870s.[69] These nurses were better equipped with the skills required for delivering occupational therapy, than their working-class, poorly educated colleagues in France, as this study highlights.

Patient Work and Occupation

The specific nature of the work and occupation prescribed for mental hospital patients has only recently attracted scholarly attention in its own right; more frequently it is described in studies focusing on individual asylums or on moral treatment (see footnote 37). Studies by historians

[65] Peter Nolan, *A History of Mental Nursing* (London: Chapman and Hall, 1993), 44, 118.

[66] Anne Borsay and Pamela Dale, eds, *Mental Health Nursing: The working lives of paid carers in the nineteenth and twentieth centuries* (Manchester: Manchester University Press, 2013).

[67] Alexandre Klein, "Théodore Simon (1873–1961). Itinéraire d'un psychiatre engage pour la professionnalisation des infirmiers et infirmières d'asile," *Association de recherche en soins infirmiers* 135 (2018): 91–106; Patrice Krzyzaniak, *Georges Daumézon (1912–1979): un camisard psychiatre et pedagogue: une contribution singulière aux sciences de l'éducation*, Thèse (Lille: Université Charles de Gaulle, 2017); Benoît Majerus. "Surveiller, Punir et Soigner? Pratiques psychiatriques en Europe de l'ouest du 19e siècle aux années 1950," *Histoire, Médecine et Santé* 7 (2015): 51–62.

[68] Geertje Boschma, *The Rise of Mental Health Nursing: A history of psychiatric care in Dutch asylums, 1890–1930* (Amsterdam: Amsterdam University Press, 2003).

[69] Ibid., 94.

that put patient work centre-stage include an edited volume by Waltraud Ernst (2016) and works by Véronique Fau-Vincenti (2014), J-P. Arveiller and Clément Bonnet (1991); Jennifer Laws (2011); Vicky Long (2013, 2006); and Geoffrey Reaume (2006).[70] These studies examine the relationship between the meaning of work *inside* the asylum and the socio-economic, cultural and political contexts *outside*. A number of similar themes emerge from the existing literature, which are taken forward by the present study, such as the relationship between economic exploitation and therapy inherent in patient work, whether work offered an effective means of rehabilitation, and the influence of the socio-economic and political climate on work within mental institutions. Few focus on the interwar period,[71] and none offer a direct comparison between patient occupation in one country and another.

[70] Waltraud Ernst, ed., *Work, Psychiatry and Society, c.1750–2015* (Manchester: Manchester University Press, 2016); Véronique Fau-Vincenti, "Valeur du Travail à la Troisième Section de l'Hôpital de Villejuif: Entre Thérapie et Instrument de Discipline," *Savoirs, Politiques et Pratiques de l'exécution des peines en France au XXe siecle* (12/09/2014), http://criminocorpus.revues.org/2788 (accessed 03/03/2018); Jennifer Laws, "Crackpots and Basket-Cases: A history of therapeutic work and occupation," *History of the Human Sciences* 24 no. 2 (2011): 65–81; Vicky Long, "'A Satisfactory Job Is the Best Psychotherapist': Employment and Mental Health 1939–1960," in *Mental Illness and Learning Disability since 1850: Finding a Place for Mental Disorder in the United Kingdom*, eds J. Melling and P. Dale (Abingdon: Routledge, 2006), 179–99; "Rethinking Post-War Mental Health Care: Industrial Therapy and the Chronic Mental Patient in Britain," *Social History of Medicine* 26 no. 4 (2013): 738–58; Geoffrey Reaume, "Patients at Work: Insane Asylum Inmates' Labour in Ontario, 1841–1900," in *Mental Health in Canadian Society: Historical Perspectives*, eds James Moran and David Wright (Montreal: McGill Queen's University Press, 2006), 69–116; J.-P. Arveiller and Clément Bonnet, *Au Travail, Les Activités Productives dans le Traitement et la Vie du Malade Mental* (Toulouse: ERES, 1991).

[71] Exceptions include: John Hall, "From work and occupational therapy: the policies of professionalisation in English mental hospitals from 1919 to 1959," in *Work, Psychiatry and Society*, ed. Ernst, 314–333; Monika Ankele, "The Patient's View of Work Therapy: The Mental Hospital Hamburg-Langenhorn during the Weimar Republic," in *Work, Psychiatry and Society*, ed. Ernst, 238–261; Yolanda Eraso, "'A burden to the state': The reception of the German 'Active Therapy' in an Argentinian 'Colony-Asylum' in the 1920s and 1930s," in *Transnational Psychiatries: Social and Cultural Histories of Psychiatry in Comparative Perspective, c.1800–2000*, eds Waltraud Ernst and Thomas Mueller (Newcastle: Cambridge Scholars Publishing, 2010), 51–79; Jane Freebody, "'The Root of All Evil is Inactivity': The Response of French Psychiatrists to New Approaches to Patient Work and Occupation, 1918–1939," in *Voices in the History of Madness: Personal and Professional Perspectives on Mental Health and Illness* eds Robert Ellis, Sarah Kendal and Steen J. Taylor (Cham: Palgrave Macmillan, 2021), 71–94.

A key theme addressed by the present study is the balance between justifications for patient work and occupation as a means of therapy, rehabilitation and offsetting institutional running costs. None of these justifications were mutually exclusive, but the priority accorded to each changed over time and varied between France and England. The view that patient work was a means of "repairing damaged human capital" designed to mould the mentally ill into productive citizens, expressed by Andrew Scull (1979), has been challenged by contemporary historians.[72] But questions remain over the justifications for patient work during the interwar period. Was it to prepare recovered patients for employment after discharge, ready to assist with the rebuilding of the economy after the devastation of World War I? Or was the main purpose to save on employment costs by using patient labour to perform the tasks necessary for the smooth-running of the institution? Or were work and occupation deemed curative and therefore primarily deployed to expedite the recovery of the patient? Linked to the theme of rehabilitation is that of "idleness". Were work and occupation introduced into the asylum regime to prevent "malingering" and to instil a work ethic? The present study demonstrates that how patient work and occupation were justified by psychiatrists varied regionally and across time, in response to a range of factors both internal and external to the asylum.

A moral justification for patient work, as a means of combating idleness, is noted by several authors in relation to the late nineteenth century. The Board of Control's annual reports from the late nineteenth and early twentieth century refer to the provision of "useful employment" for patients, reflecting an expectation that patients would contribute to the costs of their care. Sarah Chaney notes that patients refusing to engage with work around the asylum were "regarded with suspicion".[73] James Moran, in his study of patient work in New Jersey, maintains that work in the asylum was "synonymous with the mid-nineteenth century

[72]Joseph Melling, "Accommodating Madness: New Research in the Social History of Insanity and Institutions," in *Insanity, Institutions and Society, 1800–1914: A social history of madness in comparative perspective,* eds Joseph Melling and Forsythe (London and New York: Routledge, 1999), 5. See: Michel Foucault, *Madness and Civilisation: A History of Insanity in the Age of Reason,* trans. Richard Howard (New York: Random House, 1965); Andrew T. Scull, *Museums of Madness: The Social Organization of Insanity in Nineteenth-Century England* (London: Allen Lane, 1979).

[73]Chaney, "Useful Members of Society," 277.

middle-class ideal of productivity".[74] Towards the end of the nineteenth century, combating idleness was increasingly emphasised in asylums in British-held territories in South Asia, as Ernst notes, replacing the early rhetoric of patient work that was linked to moral therapy.[75] This focus on combating idleness persisted into the interwar period in Germany, where productivity was seen as the route to economic recovery and restoration of national pride after defeat in World War I. A desire to re-establish Germany as a "resurgent and aspiring nation", put greater emphasis on labour and resulted in a significant increase in the number of German psychiatric patients engaged in work, as Monika Ankele observes.[76]

Under the National Socialist regime, as Thomas Mueller reports, labour and the ability to work were placed "at the centre of decision-making".[77] Following a re-interpretation of psychiatrist Hermann Simon's "more active" work therapy during the late 1930s, those patients who failed to respond to his treatment regime were dismissed as "hopeless cases". In a tragic and extreme example of politics influencing psychiatry, medical staff were made responsible for selecting patients who were unable to work, and therefore deemed "unworthy of living", for euthanasia.[78] In Argentina, those unable to work were considered a "burden to the state" during the economic crisis of the 1930s that put extreme pressure on the Argentinian welfare budget.[79] As a result, Yolanda Eraso notes that Hermann Simon's "more active" work therapy was implemented in asylums and adapted to ensure the institution's material needs were met with a minimum of financial input from the state.[80] Whilst there is no suggestion that such extreme measures were adopted in England or France during the interwar period, the present study questions whether the idea of work as moral duty still had traction. France, it must be remembered, was occupied by the

[74] James Moran, "Travails of Madness: New Jersey, 1800–70," in *Work, Psychiatry and Society*, ed. Ernst, 78.

[75] Waltraud Ernst, "'Useful Both to Patients as well as to the State': Patient Work in Colonial Mental Hospitals in South Asia, c.1818–1948," in *Work, Psychiatry and Society*, ed. Ernst, 122.

[76] Ankele, "The Patient's View of Work Therapy," 241.

[77] Thomas Mueller, "Between therapeutic instrument and exploitation of labour force: Patient work in rural asylums in Wurttemburg, *c*.1810–1945," in *Work, Psychiatry and Society*, ed. Ernst, 230.

[78] Ibid., 230–2.

[79] Yolanda Eraso, "A Burden to the State," 52–3.

[80] Ibid., 78–9.

Germans in World War II and from 1940 conditions in mental institutions took a sinister turn, as Isabelle von Bueltzingsloewen outlines.[81]

Studies by Ernst, Reaume, McKay and Moran highlight the increasing economic importance of patient work, revealing that as asylum budgets became increasingly restricted towards the end of the nineteenth century, patient work offered a means of making substantial cost-savings. This became as important as the therapeutic benefit of work to the patient.[82] This was certainly the case in Ontario, Canada, as Geoffrey Reaume observes. "Moral therapy", according to Reaume, "when stripped of its therapeutic veneer, was in reality a public works programme run on the 'free' labour of people confined in insane asylums".[83] Between 1841 and 1900, he notes, patients undertook an ever-increasing array of work duties, from knitting to digging and from masonry to nursing. On the one hand, psychiatrists downplayed the extent to which patient labour constituted real productive work, while promoting the cost-savings achieved by it on the other.[84] Such contradictions were also evident British Columbia, Canada, as Kathryn McKay's account demonstrates. The emphasis placed on economy or therapy in superintendents' annual reports varied according to prevailing medical ideology and socio-economic conditions.[85] This point is highlighted by the comparative nature of the present study, in which English psychiatrists' attempts to downplay the economic contribution of patient work are contrasted with the French focus on its importance as a cost-saving measure.

[81] Isabelle von Beultzingsloewen, *L'Hécatombe des fous: La Famine dans les hôpitaux psychiatriques français sous l'occupation* (Paris: Éditions Flammarion, 2007).

[82] Ernst, "'Useful Both to the Patients as Well as to the State'," 117–41; Thomas Mueller, "Between Therapeutic Instrument and Exploitation of Labour Force," 220–37; James Moran, "Travails of Madness: New Jersey, 1800–1870," in *Work, Psychiatry and Society*, ed. Ernst, 77–96; Reaume, "Patients at Work," 69–116.

[83] Reaume, "Patients at Work," 90.

[84] Ibid.

[85] Kathryn McKay, "From Blasting Powder to Tomato Pickles: Patient work at the provincial mental hospitals in British Columbia, Canada, 1855–1920," in *Work, Psychiatry and Society*, ed. Ernst, 112–3.

Occupational Therapy

The development of the para-medical professions, including occupational therapy, is addressed by historian Gerald Larkin.[86] Larkin emphasises the influence of social, technical and cultural forces, as well as the impact of developments within medicine, that gave rise to the para-medical professions.[87] Jean-Philippe Guihard is more specific; he maintains that the decision to include or exclude occupational therapy in psychiatry's therapeutic arsenal rested with psychiatrists.[88] In other words, the development of the profession of occupational therapy was driven by psychiatric demand for the specialty. These views resonate with the arguments presented in this study; in locations where psychiatrists saw a value in prescribing occupational therapy (such as England) the profession developed, whereas in areas where psychiatrists preferred prescribing biological remedies for curable patients (such as France), it did not.

Most histories of occupational therapy (rather than patient work in institutions) have been written by practitioners of the profession. As with histories of psychiatry written by psychiatrists, it is important to consider whether these histories constitute "Whiggish" accounts, offering a "purely descriptive reporting of facts" rather than a critical analysis, or an interpretation that "studies the past with reference to the present".[89] As Scull warns us, histories portraying a "triumphal procession towards the rational and humane practices of today" often fail to take account of alternative perspectives.[90] Foucault's view of patient work as a means of disciplining the unruly and of enforcing compliance to behavioural norms—of replacing physical shackles with mental ones—can seem abhorrent to occupa-

[86] Gerald Larkin, "The Emergence of Para-Medical Professions," in *Companion Encyclopaedia of the History of Medicine*, eds W.F. Bynum and R. Porter (London and New York: Routledge, 1994): 1329–40.

[87] Ibid., 1331–3.

[88] Jean-Philippe Guihard, "Quelle évolution pour l'ergothérapie en psychiatrie?" Conference paper delivered at the Institute of Occupational Therapy Training, Rennes, 2000, http://jp.guihard.pagesperso-orange.fr/articles/evolution/evolution.html (accessed 03/02/2022).

[89] Herbert Butterfield, *The Whig Interpretation of History* (London: G. Bell, 1931), 310.

[90] Andrew T. Scull, "Moral Treatment Reconsidered: some sociological comments on an episode in the history of British psychiatry," *Psychological Medicine* 3 (1979): 421.

tional therapists. They can find it hard to believe that their profession could be rooted in coercion or in any practices that could be considered inhumane or unkind. As Laws highlights, Foucault is rarely cited in histories of occupational therapy written by practitioners.[91] The danger of accepting "taken-for-granted" versions of events, even if this "threatens the accepted historical narrative" is highlighted by Brid Dunne, Judith Pettigrew and Katie Robinson, who aim to encourage other occupational therapists to delve into the history of their profession.[92] The authors explain how to use historical research to gain a deeper critical understanding of the profession of occupational therapy, using primary and secondary source material.[93] This may be a tacit acknowledgement that many histories of occupational therapy lack *critical* engagement.

Few monographs have been dedicated to the history of occupational therapy, although works by Ann Wilcock, former president of the International Society of Occupational Scientists, and Catherine Paterson, director of occupational therapy at the Robert Gordon University, Aberdeen (1990–2002), are two important British exceptions. Wilcock's detailed, two-volume account of occupational therapy in the UK traces the roots of using occupation therapeutically back to the ancient Greeks, through the teaching of the moral therapists to the emergence of the profession in the twentieth century, and its early adoption by Dr D.K. Henderson and Dr Elizabeth Casson.[94] Paterson has performed a similar task for the profession in Scotland,[95] while Virginia Quiroga has documented the earliest years of the profession in the USA.[96] Quiroga emphasises the holistic approach taken by the early occupational therapists. Margaret Drake, an occupational therapist in the USA for 30 years,

[91] Laws, "Crackpots and Basket Cases," 66.

[92] Bríd Dunne, Judith Pettigrew and Katie Robinson, "Using historical documentary methods to explore the history of occupational therapy," *British Journal of Occupational Therapy* 79, no.6 (2016): 382.

[93] Ibid., 376–84.

[94] Ann A. Wilcock, *Occupation for Health: A Journey from Prescription to Self Health, Vols I & II* (London: British College of Occupational Therapists, 2002).

[95] Catherine F. Paterson, *Opportunities not Prescriptions: The Development of Occupational Therapy in Scotland 1900–1960* (Aberdeen: Aberdeen History of Medicine Publications, 2010).

[96] Virginia Quiroga, *Occupational Therapy: The First Thirty Years, 1900–1930* (Bethesda: American Occupational Therapy Association, 1995).

has written a fictional account of the experiences of an American occupational therapist coming to France with the US Army in World War I. It offers an insight into daily life at the Front, the difficulties encountered by female practitioners of fitting into a male-dominated Army hospital, as well as the treatment of shell-shocked soldiers.[97]

Many histories of occupational therapy written by practitioners appear in the professional journals,[98] or in chapters devoted to the profession's history in occupational therapy textbooks.[99] The annual Elizabeth Casson Memorial Lectures, given at the Annual Conference and Exhibition of the College of Occupational Therapists in England, the Eleanor Clarke Slagle Lectures in the USA and the Doris Sym Memorial Lecture in Scotland have frequently been devoted to historical reflection.[100] Professional

[97] Margaret Drake, *Reconstructing Soldiers: An Occupational Therapist in World War I* (Universe, 2013).

[98] Anglo-Saxon journals include the *British Journal of Occupational Therapy*, the *American Journal of Occupational Therapy*, the *Australian Occupational Therapy Journal* and the *Canadian Journal of Occupational Therapy*, while in France the main journals are *ErgOThérapies* and *Le Journal d'Ergothérapie.*

[99] Such as E.M. Macdonald, *Occupational Therapy in Rehabilitation,* Chapter 1 (London: Baillière, Tindall and Cox, 1961); Nicole P. Gillette, "Occupational Therapy in Mental Health: A Historical Perspective," in *Occupational Therapy,* Fourth ed., eds Helen S. Willard and Clare S. Spackman, 51–60 (Philadelphia, J.B. Lippincott & Co, 1963); C.F. Paterson, "A Short History of Occupational Therapy in Psychiatry," in *Occupational Therapy and Mental Health,* Third ed., ed. Jennifer Creek, 3–15 (Edinburgh and London: Churchill Livingstone, 2002); Gabriel Gable, "Naissance de l'ergothérapie," in *Nouveau Guide de Pratique en Ergothérapie: entre concepts et réalité,* ed. J.-M. Caire, 85–89 (Marseille, Solal: 2008); É. Quevillon, "Construire la science de l'activité humaine en France," in *L'activité humaine: un potentiel pour la santé?* eds M.-C. Morel-Bracq and E. Trouvé, 323–330 (Paris: De Boeck-Solal, 2015); Isabelle Pibarot, "Une Histoire du soin par activité," in *Ergothérapie en psychiatrie: De la souffrance à la réadaptation,* Second ed., ed. H. Hernandez (Paris: De Boeck Supérieure, 2016), 28–36.

[100] Elizabeth Casson (1881–1954) was the doctor who founded the first school of occupational therapy in England in 1930; Eleanor Clarke Slagle (1870–1942) was a social worker and founder of the American National Society for the Promotion of Occupational Therapy (NSPOT) in 1917; Doris Sym (1913–98) was the founder and first principal of the Glasgow School of Occupational Therapy and Fellow of Glasgow Caledonian University.

anniversaries have also afforded opportunities to look back.[101] A number of themes emerge from this literature, including the use of arts and crafts in the early days of the profession and how that impacted on the subsequent reputation of occupational therapy; the "identity crisis" that seems to have plagued the profession in the second half of the twentieth century; and occupational therapy's struggle to be seen as scientific. It is argued in this study that the rejection of occupational therapy by French psychiatrists during the interwar period was linked to a perceived need to demonstrate their scientific credentials (addressed in Chap. 6). They sought to do this by focusing on biological remedies for the curable and dismissed occupational therapy as unscientific.

Clare Hocking explores the philosophical foundations of occupational therapy in rationalism and Romanticism. In a series of three articles, she argues that an individual's sense of themselves is expressed through the items they own and use, an idea that strongly influenced the interwar pioneers of occupational therapy in England.[102] Hocking also examines how early British occupational therapists perceived themselves and their practice, focusing in particular on their use of craft activities and how to demonstrate the efficacy of their methods.[103] Ann Wilcock and Beryl Steeden examine the influence of the Arts and Crafts Movement on the profession of occupational therapy. They emphasise the importance of creativity to health and wellbeing, and to fulfilling people's potential for "doing, being

[101] Such as: G Kielhofner and J P Burke, "Occupational therapy after 60 years: An account of changing identity and knowledge," *American Journal of Occupational Therapy* 31 no. 10 (1977): 675–89; Catherine F. Paterson, "75th Anniversary of the Founding of the Scottish Association of Occupational Therapies," *British Journal of Occupational Therapy* 70 no. 12 (2007): 537–40; Paterson, "Occupational Therapy and the National Health Service, 1948–98," *British Journal of Occupational Therapy* 61 no. 7 (1998): 311–15; Peggy Jay and Hester Monteath, "The British Association of Occupational Therapists Celebrates its Silver Jubilee," *British Journal of Occupational Therapy* 61 no. 5 (1999): 189–91; Julia Scott, "British Journal of Occupational Therapy 70 years on," *British Journal of Occupational Therapy* 70 no. 7 (2007): 275; Clare Hocking, "Early Perspectives of Patients, Practice and the Profession," *British Journal of Occupational Therapy* 70 no. 7 (2007): 284–90.

[102] Clare Hocking, "The Way We Were: Romantic Assumptions of Pioneering Occupational Therapists in the United Kingdom," *British Journal of Occupational Therapy* 71 no. 4 (2008): 146–154; Hocking, "The Way We Were: Thinking Rationally," *British Journal of Occupational Therapy* 71 no. 5 (2008): 185–95; Hocking, "The Way We Were: The Ascendance of Rationalism," *British Journal of Occupational Therapy* 71 no. 6 (2008): 226–32.

[103] Hocking, "Early Perspectives of Patients, Practice and the Profession," *British Journal of Occupational Therapy* 70 no. 7 (2007): 284–90.

and becoming".[104] Exploring the same topic, Ruth Ellen Levine considers the similarities between the beliefs of the founders of occupational therapy in the USA and advocates of the Arts and Crafts Movement. Both, she argues, expressed contempt for mass-produced goods, and lauded the almost spiritual sense of satisfaction and the calming effects of making items by hand. This preoccupation with the spiritual made the occupational therapists vulnerable to criticism when required to explain the value of their activities to medical practitioners.[105] Kathlyn Reed also delves into the history of the profession and regards the Arts and Crafts Movement as a "means of revitalising the ideas of moral treatment in a new rationale".[106] The close links between moral treatment and occupational therapy are highlighted in this study.

The history of the profession is frequently used in attempts to identify occupational therapy's core beliefs or philosophy. Hélène Polatajko looks back to 1933, when the first issue of the *Canadian Journal of Occupational Therapy* was published, during the Great Depression.[107] The lead article lamented the "disease of unemployment" and the need to "remedy human dissatisfaction and mental unrest" by providing tasks to keep minds occupied and bodies healthy. Providing occupation has been at the core of occupational therapy since its beginnings, although the rationale for its provision has changed over time. Polatajko claims that occupation has been variously conceived as diversion, therapy, rehabilitation and re-education, all of which are apparent in this present study. Polatajko does not mention the economic benefits of patient occupation for institutions.

Robert K. Bing used his Eleanor Clarke Slagle lectureship in 1981 to draw lessons from the past, such as treating the individual holistically.[108] He traces the roots of occupational therapy back to the moral therapy of

[104] Ann A. Wilcock and Beryl Steeden, "Reflecting the Creative Roots of the Profession," *British Journal of Occupational Therapy* 61 no. 1 (1999): 1.

[105] Ruth Ellen Levine, "The Influence of the Arts-and-Crafts Movement on the Professional Status of Occupational Therapy," *American Journal of Occupational Therapy* 41 no. 4 (1987): 248–54.

[106] Kathlyn L. Reed, "Tools of Practice: Heritage or Baggage? 1986 Eleanor Clarke Slagle Lecture," *American Journal of Occupational Therapy* 40 no. 9 (1986): 597–605.

[107] Hélène Polatajko, "The evolution of our occupational perspective: the journey from diversion through therapeutic use to enablement," *Canadian Journal of Occupational Therapy* 68 no.4 (2001): 203–207.

[108] Robert K. Bing, "Eleanor Clarke Slagle Lectureship 1981—Occupational Therapy Revisited: A Paraphrastic Journey," *American Journal of Occupational Therapy* 35 no. 8 (1981): 499–518.

Tuke and Pinel. He highlights the "disappearance" of moral treatment in the late nineteenth century, and its re-emergence and development in the early twentieth as occupational therapy, led by the American pioneers, including Adolf Meyer and Eleanor Clarke Slagle (1870–1942). Bing then discusses the contribution of "second generation" occupational therapists in elaborating, codifying and applying the initial theory in the late 1920s and 1930s.[109] Kathlyn Reed also highlights the philosophy of occupational therapy and finds that the philosophical beliefs of one of the profession's founders, Eleanor Clarke Slagle, are still relevant to the profession today (in 2019). These included the belief that, through individualised activity programmes, occupational therapy could help people with mental health issues to change their habits and routines to improve their ability to integrate socially and function in society.[110] Reed's analysis of the core beliefs underpinning occupational therapy could equally apply to those of the moral therapists, as this study demonstrates.

French histories of occupational therapy, or *ergothérapie*, are more recent, which is unsurprising given that the profession in France is some 30 years younger than it is in the UK. M.C. Morel-Bracq *et al.* discuss the conceptual paradigms that underpinned the French profession of occupational therapy at various stages of its development, observing the parallels between the holism of the early French occupational therapists and that of the moral therapists.[111] This view supports the thesis presented in this study, which links holism within psychiatry as a pre-requisite to the adoption of occupational therapy. Lisbeth Charret and Sarah Thiébaut Samson also emphasise the holism of French occupational therapy in the late 1940s.[112] However, they also highlight the fact that by the time that French schools of occupational therapy were established in the mid-1950s, the prevailing conceptual paradigm had become biomedical (as it had

[109] Ibid., 512.

[110] Kathlyn L. Reed, "The Beliefs of Eleanor Clarke Slagle: Are they Current or History?" *Occupational Therapy in Healthcare* 33 no. 3 (2019): 265–85.

[111] M.-C. Morel-Bracq, A.-C. Delaisse, J.-F. Bodin, L. Charret and H. Hernandez, "Une approche historique du développement de l'identité professionnelle des ergothérapeutes en France: l'évolution des valeurs et intérêts à travers le temps," *La Revue Ergothérapies* 81 April (2021). https://revue.anfe.fr/2021/04/22/une-approche-historique-du-developpement-de-lidentite-professionnelle-des-ergotherapeutes-en-france-levolution-des-valeurs-et-interets-des-ergotherapeutes-a-travers-le-temps/ (accessed 20/3/2022).

[112] L. Charret and S. Thiébaut Samson, "Histoire, fondements et enjeux actuels de l'ergothérapie," *Contraste*, 45 (2017): 17–36.

been in the 1920s and 1930s), supporting the belief that human beings were an agglomeration of disparate structures and functions.[113] Only in the 1980s, Charret and Samson argue, did occupational therapy begin to reunite biological, psychological and social factors, and to consider all the needs of an individual trying to live in and interact with their environment.[114]

Jean-Pierre Goubert and Rémi Remondière situate the origins of occupational therapy in the recommendation of work for patients by Pinel, Esquirol and Jean-Baptiste-Maximien Parchappe (who was impressed by the work undertaken by patients in British asylums), and in the French asylum legislation of 1839 and 1857.[115] The authors also emphasise the influence of gymnastics, introduced into French schools in the 1850s, citing the benefits to the brain of making muscles move. Goubert and Remondière note that the vocational training offered by American occupational therapists in World War I acted as a stimulus to the profession of physiotherapy, while occupational therapy did not develop in France until after World War II.[116] Gabriel Gable examines the history of occupational therapy from 1900, when it was first conceived in the USA, citing the influence of psychiatrists John Hall, Adolph Meyer and William Rush Dunton and others including Eleanor Clarke Slagle, Susan Tracy and George Edward Barton. [117] Like Goubert and Remondière, Gable emphasises the role of American occupational therapists in returning physical function to injured soldiers in Paris during World War I. The interwar period did not see the development of the profession in France; this occurred after the re-assessment of psychiatry following World War II and the exposure of abuses in French asylums during the conflict.[118]

Most of the French histories of occupational therapy address the American influence on the origins of the profession, while few mention the contribution of the German psychiatrist, Hermann Simon (1867–1947).

[113] Ibid., 25.

[114] Ibid., 26.

[115] J.-P. Goubert and R. Remondière, "Les origines historiques de l'ergothérapie en France," *Sociologie Santé*, 20 June (2004): 247–268.

[116] Ibid., 262–3.

[117] Gabriel Gable, "Naissance de l'ergothérapie," 85–89.

[118] Ibid., 88.

Isabelle Pibarot is an exception.[119] In her account, she links moral therapist Philippe Pinel's theories to the methods advocated in works by chief medical officer Charles Ladame (1908 and 1926) and German psychiatrist Hermann Simon (1929).[120] Pibarot quotes the Spanish psychiatrist François Tosquelles (1912–1994), who sought refuge in France during the Spanish Civil War (1936–1939). Tosquelles, a keen advocate of French psychiatric reform after World War II, praised Hermann Simon's methods in his influential book of 1967.[121] Both Pibarot and Tosquelles see Simon as a key figure in the development of the therapeutic use of occupation and of the profession of occupational therapy.

Studies of occupational therapy by historians include those by Mark Jackson and John Hall. Jackson focuses on the links between prevailing attitudes towards work, leisure, unemployment and poverty on occupational therapy, with particular reference to an institution for children and adults with learning difficulties (the Sandlebridge Colony). He argues that it is the strength of the ideological and pragmatic links between therapy, health and work, rather than a rigid biomedical explanation of disease, that has constituted the basis for the professional expertise of occupational therapists.[122] This view resonates with the argument presented in this book that support for occupational therapy by English psychiatrists depended upon their adoption of a holistic approach to mental disorder after World War I. In France (outside Paris), a biomedical conception of mental disorder remained firmly in place and psychiatrists were unwilling to adopt a therapy based on a broader conception of mental disorder.

John Hall, a historian and clinical psychologist, focuses on the administrative frameworks that underpinned development of the profession of occupational therapy in the UK from the end of World War I.[123] Hall highlights the foundation of the first school of occupational therapy in

[119] Isabelle Pibarot, "Une histoire du soin par activité," in *Ergothérapie en psychiatrie: De la souffrance psychique à la réadaptation*, Second ed., ed. H. Hernandez, (Paris: De Boeck Supérieure, 2016), 28–36.

[120] Ibid., 29–30.

[121] François Tosquelles, *Le Travail Thérapeutique en psychiatrie* [1967] (Paris: Éditions érès, 2009).

[122] Mark Jackson, "From Work to Therapy: The changing politics of occupation in the twentieth century," *British Journal of Occupational Therapy* 56 no. 10 (1993): 361–4.

[123] John Hall, "From work and occupational therapy: the policies of professionalisation in English mental hospitals from 1919 to 1959," in *Work, Psychiatry and Society,* ed. Ernst, 314–333.

England in 1930 and the formation of the Association of Occupational Therapists in 1936. His account, which extends to 1959, identifies the shift from the early hospital-based provision of occupational therapy, with its focus on arts and crafts, to the incorporation of more rehabilitative activities after World War II that were designed to prepare patients for life in the community. Hall's account emphasises the influence of the American style of occupational therapy on British practice that led to the focus on arts and crafts, as opposed to the work practices that characterised European asylum regimes. Hall's observation corroborates the findings of this study in its comparison of occupation in England and France.

Class, Gender and Patient Work

Another important theme in this study is how class and gender influenced the type of work allocated to patients. As Joan Busfield has emphasised, class and gender are "embedded" in psychiatry, impacting on both aetiology and treatment, and thus on patient occupation.[124] She maintains that psychiatric practice cannot avoid "reflecting, incorporating, reproducing and sustaining class and gender divisions", since the aim of psychiatry was to return individuals to their place in society.[125] Busfield's observations are corroborated by studies focusing on the nineteenth century. In the early Victorian period, as Levine-Clark maintains, poor women were expected to work to stave off pauperism, and the work ascribed to male and female patients at Yorkshire's West Riding Asylum reflected the "expectation of society that work itself was gender- as well as class-specific". [126] In the late Victorian and early Edwardian era, normative concepts of masculinity and femininity remained very rigid, as Louise Hide observes. Patients at the Bexley and Claybury asylums were allocated work suitable for their sex and class, skills, previous occupation and physical and mental abilities.[127]

[124] Joan Busfield, "Class and Gender in Twentieth-Century British Psychiatry: Shell-Shock and Psychopathic Disorder," in *Sex and Seclusion, Class and Custody: Perspectives on Gender and Class in the History of British and Irish Psychiatry*, eds Anne Digby and Jonathan Andrews (Amsterdam, New York: Rodopi, 2004), 239.

[125] Ibid.

[126] Marjorie Levine-Clark, "Embarrassed Circumstances: Gender, Poverty and Insanity in the West Riding of England in the Early Victorian Years," in *Sex and Seclusion, Class and Custody*, eds Digby and Andrews, 123.

[127] Louise Hide, *Class and Gender in English Asylums, 1890–1914* (Basingstoke: Palgrave Macmillan, 2014).

Women were assigned work that reflected their domestic duties, while men were employed in more varied productive roles in the grounds or workshops, or on the farm. At the private Holloway Asylum, entertainment and "genteel occupations" were considered more in keeping with middle-class life than physical exertion through work.[128] Hide notes that, although they were encouraged to do so, few private patients at Bexley or Claybury worked.[129] These observations of the period before World War I raise the question of whether such rigid gender and class divides were still apparent in interwar mental hospitals, after so many Victorian "social norms" were allegedly swept away by the war.

Diana Gittins' study of Severalls Hospital includes the interwar period and suggests that narrowly defined concepts of class and gender still governed the assignation of patient work. Long-stay female patients with "no, or few, obvious disorders" were employed in the laundry and sewing room, while those more prone to violence or disruption worked in the ward or scrubbed corridor floors.[130] Understaffing meant that many female patients requiring supervision had nothing to do.[131] There was a much wider variety of occupations available to male patients, with work in the workshops, on the farm, in the gardens, or in the smithy.[132] Gittins pays scant attention to the introduction of occupational therapy during the interwar period, although she notes that a small department was established at Severalls in 1922 and nurses were encouraged to take up a handicraft.[133] More attention is paid to occupational therapy in the 1960s, by which time the department had expanded considerably but still reflected "a rigid and Victorian stereotype of gender".[134] The present study reveals that the traditional asylums tended to be more conservative regarding what was considered appropriate work for women and men than the more recently established hospitals, such as the Maudsley and Henri Rousselle hospitals.

[128] Anne Shepherd, "The Female Patient Experience in Two Late Nineteenth-Century Surrey Asylums," in *Sex and Seclusion, Class and* Custody, eds Digby and Andrews, 241.

[129] Hide, *Class and* Gender, 111.

[130] Ibid., 145.

[131] Ibid.

[132] Ibid., 159.

[133] Diana Gittins, *Madness in its Place: Narratives of Severalls Hospital, 1913–1997* (London and New York: Routledge, 1998), 149.

[134] Ibid., 107.

The Patient's View

Recent research in the history of psychiatry has been able to reveal the views and actions of patients, using letters, notes, poems, art, journals and other written items produced by patients and included in their medical records as evidence of their condition.[135] Allan Beveridge's study (1998) of over a thousand letters written by patients admitted to the Royal Edinburgh Asylum between 1873 and 1908 reveal the patients' frustration and dissatisfaction with institutional life.[136] Frequent complaints included the monotony of the daily routine, the brutality of the attendants, the behaviour of other inmates, and the prison-like characteristics.[137] Beveridge notes that the majority of the letters were written by upper-class patients who were not obliged to work, but in one letter a pauper patient complained that, "I've knitted stockings for the High persons and done seawing to [*sic*] … and got no thanks for it", providing a rare glimpse of attitudes towards work.[138] In a volume dedicated to the voices of the mad and their carers,[139] Rory du Plessis uses correspondence and casebooks to create a portrait of patient life in a South African mental institution between 1890 and 1910.[140] His account reveals evidence of racial discrimination, poor diet, a regimen of physical labour disguised as therapy and such personal indignities as a denial of clothing.[141] Violence and threats of violence by asylum staff are revealed in Tomas Vaiseta's account of the lives of mental patients in twentieth-century Lithuania.

[135] Alexandra Bacopoulos-Viau and Aude Fauvel, "The Patient's Turn. Roy Porter and Psychiatry's Tales, Thirty Years On," *Medical History* 60, no. 1 (2016): 1; Isabelle von Bueltzingsloewen, "Vers un Désenclavement de l'Histoire de la Psychiatrie," *Le Mouvement Social* 253 (2015): 6.

[136] Allan Beveridge, "Life in the Asylum: patients' letters from Morningside, 1873–1908," *History of Psychiatry* 36 no. 9 (1998): 431–70.

[137] Ibid., 462–3.

[138] Ibid., 453.

[139] Robert Ellis, Sarah Kendal and Steven J. Taylor, eds, *Voices in the History of Madness: Personal and Professional Perspectives on Mental Health and Illness* (Cham: Palgrave Macmillan, 2021). See: Part II Reconstructing Patient Perspectives, 115–216.

[140] Rory du Plessis, "'Tells his story quite rationally and collectedly': Examining the casebooks of Grahamstown Lunatic Asylum, 1890–1910, for cases of delusion where patients voiced their life stories," in *Voices in the History of Madness*, eds Ellis, Kendal and Taylor, 137–154.

[141] Ibid., 138, 150.

Patients' letters described not only brutal treatment, but also their struggle to be seen as individuals.[142]

Using similar sources, Monika Ankele highlights the attitudes of patients towards work therapy at the Hamburg-Langenhorn asylum, Germany, during the Weimar Republic.[143] Patients complained that they were not paid for their work in the institution, and felt that they should not have to work because they were ill.[144] They asked why they had not been released, since, if they were expected to work inside the asylum, they could also work outside.[145] Whilst patients were not paid, their food rations were linked to the amount and type of work they performed, with certain jobs that were considered more useful to the asylum attracting larger quantities of food. Hard-working patients could also be awarded privileges, such as tobacco or the opportunity to walk in the gardens.[146] Studies by Lee-Ann Monk and Geoffrey Reaume also address the issue of payment for work carried out by patients. Reaume highlights the unpaid labour of patients who constructed the boundary wall round the Toronto Asylum in the late nineteenth century. The architect of the wall, Kivas Tully, was commemorated in the 1970s, but the unpaid patients who constructed it were not.[147] At Kew Cottages, an institution for people with learning disabilities in Australia, patients worked in the workshops, grounds and kitchens and some helped nurse other patients. They were not paid and working conditions were often unsafe, leading to accusations of exploitation.[148] The present study reveals different patient reimbursement policies in French and English institutions. As Reaume has emphasised, the non-payment of patients, or the very low wages, could be explained or excused by referring to the work as therapy.[149]

Although patients at the Hamburg-Langenhorn asylum had little choice over whether they worked, they could control how much effort to

[142] Tomas Vaiseta, "Dehumanising Experience, Rehumanising Self-Awareness: Perceptions of Violence in Psychiatric Hospitals of Soviet Lithuania," in *Voices in the History of Madness*, eds Ellis, Kendal and Taylor, 155–172.

[143] Ankele, "The Patient's View of Work Therapy," 238–61.

[144] Ibid., 246–7.

[145] Ibid., 247.

[146] Ibid.

[147] Geoffrey Reaume, "A Wall's Heritage: Making Made People's History Public," *Public Disability History* 1 (2016): 20.

[148] Lee-Ann Monk, "Exploiting patient labour at Kew Cottages, Australia, 1887–1950," *British Journal of Learning Disabilities* 38 (2010): 86–94.

[149] Reaume, "A Wall's Heritage," 20.

put into the work, and in some cases, they could influence the type of work they did. Ankele cites the example of a shoemaker who asked to be transferred to a ward where he could work in the cobbler's workshop, thus enabling him to maintain his skills and his professional identity at the same time as repairing numerous shoes for the institution.[150] At Kew Cottages, the medical superintendent responded to criticism by maintaining that the work carried out by patients was voluntary and never imposed. But, as Monk highlights, patients were committed to the institution by law and had little power to negotiate, compared to the institution's paid staff or workers outside. Some patients refused to work, although the consequences of refusal are unclear.[151]

The patient's role as an actor—rather than as a passive recipient of care—is the focus of several recent studies. Scull (2006), Forsythe (1996), Wright (1998), Walsh (1999) and Michael (2003) dismiss the notion that patients and their families should be regarded as "submissive pawns" pushed about by officials, regarding them instead as active users and manipulators of the asylum system for their own ends.[152] In Australia and New Zealand during the late nineteenth and early twentieth centuries, families actively negotiated asylum care for their insane relatives, and regularly came to the asylum to check on patients' progress, treatment and welfare, as Catharine Coleborne observes.[153] In France, Patricia Prestwich notes that, following revision of the admissions procedure to mental asylums in the Seine department in 1876, some families began to take their

[150] Ankele, "The Patient's View of Work Therapy," 250.

[151] Monk, "Exploiting patient labour," 90.

[152] Andrew Scull, "Madfolk and Their Keepers. Roy Porter and the History of Psychiatry," in *The Insanity of Place, the Place of Insanity: Essays on the History of Psychiatry*, ed. Andrew Scull (London and New York: Routledge, 2006), 51; Bill Forsythe, Joseph Melling, and Richard Adair, "The New Poor Law and the County Pauper Lunatic Asylum—the Devon Experience, 1834–1884," *Social History of Medicine* 9 (1996); David Wright, "Family Strategies and the Institutional Commitment of Idiots in Victorian England," *Journal of Family History* 23 no. 2 (1998): 190–208; Oonagh Walsh, "'The Designs of Providence': Race, Religion and Irish Insanity," in *Insanity, Institutions and Society, 1800–1914*, eds Melling and Forsythe, 223–242; Pamela Michael, *Care and Treatment of the Mentally Ill in North Wales, 1800–2000* (Cardiff: University of Wales Press, 2003).

[153] Catharine Coleborne, "Challenging Institutional Hegemony: family Visitors to Hospitals for the Insane in Australia and New Zealand, 1880s–1910s," in *Permeable Walls: Historical Perspectives on Hospital and Asylum Visiting*, eds Graham Mooney and Jonathan Reinarz (Amsterdam and New York: Rodopi, 2009), 289.

insane relatives directly to the asylum to avoid the complicated, degrading and often traumatic committal system.[154] She claims that the new procedure shifted the balance of power between the family of these patients (although not the patients themselves) and the asylum.[155]

During the Great Depression of the 1930s, as this study reveals, some Parisian families chose to leave their relatives in the asylum despite their having been pronounced ready for discharge, to save on the costs of looking after them in the uncertain economic climate. Some Parisian psychiatrists actively sought to retain certain patients, knowing that they were unlikely to find work outside the asylum and that this could cause a return of their symptoms. One of the roles of the psychiatric social worker, employed by the metropolitan hospitals during the latter part of the interwar period, was to help patients find work after discharge. The high unemployment rate during the Depression was a recurring topic of communication between doctors, patients and family members at the Hamburg-Langenhorn asylum. Numerous patient letters contained requests to remain at the asylum. As Ankele highlights, for many patients and their families the asylum was not just a place of "control, expropriation, authoritarianism and confinement"; it was also perceived as a place of refuge in difficult circumstances, offering security, food, shelter and a sense of belonging.[156]

Two patients, one British and one American, whose experiences of asylum life were very negative used their voices to effect change. John Perceval (1806–1876), the son of a British prime minister, and Clifford W. Beers (1876–1943), a Harvard law graduate, both wrote autobiographical accounts of their mental breakdown and treatment. As Richard Hunter observes, published autobiographies by mental patients are rare, in part because few individuals wish to relive their experiences, and in part because they lack the ability to describe their ordeal in an "organised, readable form".[157] Perceval wrote of his experiences at two private asylums in Bristol and Sussex between 1830 and 1832, and Beers completed a book about his own mental breakdown and subsequent hospitalisation at three

[154] Prestwich, "Family Strategies," 802.

[155] Ibid., 809.

[156] Ibid., 252.

[157] Richard Hunter, Review of *Perceval's Narrative*, ed. Gregory Bateson, *British Medical Journal* 13/10/1962, 983.

different institutions in the USA between 1900 and 1905.[158] Both men claimed to have suffered abuses and harsh treatment, both resisted such abuses, and both successfully campaigned to improve conditions for other sufferers of mental disorder. Perceval's period of institutional confinement occurred just before the system of non-restraint was introduced; he endured weeks of being restrained in a strait-waistcoat fixed to a chair by day and secured to his bed at night.[159] Perceval helped found the Alleged Lunatics' Friend Society and was active in forcing the appointment of a Select Committee tasked with enquiring into the care and treatment of "lunatics".[160]

Clifford Beers' account is particularly relevant to this study as it led Beers to collaborate with psychiatrist Adolf Meyer and psychologist William James (1842–1910) to found the National Committee for Mental Hygiene in 1909 in the USA. Beers' book also captured the attention of French psychiatrist Édouard Toulouse, leading to an influential and close personal friendship between the two men.[161] Toulouse established the French League of Mental Hygiene in 1921. While Beers' original plan was to use his book to campaign for the improvement of hospital conditions for the mentally ill, Meyer persuaded him to broaden his remit to include promoting awareness of how to maintain mental health and to prevent the onset of mental disorder, in addition to improving hospital conditions. Meyer coined the phrase "mental hygiene".[162] The book itself caused a great stir in the contemporary press and whilst Beers' account was sensationalised by some newspapers, it served to draw attention to the "broad clinical picture" of the pre-war era.[163] Beers spoke of the crude, often cruel, behaviour of the attendants who showed no sympathy for his condition; the indignity of mechanical restraint; the frustration of the enforced rest-cure; the brutal attempts at force-feeding; the violence exhibited by other patients, attendants and even the medical staff; and the misery of seclusion. Beers does

[158] Gregory Bateson, ed., *Perceval's Narrative: A Patient's Account of his Psychosis, 1830–1832* [1838–40] (Stanford, CA: Stanford University Press, 1961); Clifford Whittingham Beers, *A Mind That Found Itself: An Autobiography* (New York: Longmans, Green, 1908).

[159] Hunter, Review of *Perceval's Narrative*, 983.

[160] Ibid.

[161] Michel Huteau, *Psychologie, Psychiatrie et Société sous la Troisième République: La biocratie d'Édouard Toulouse (1865–1947)* (Paris: L'Harmattan, 2002), 186–7.

[162] Eunice Winters, "Adolf Meyer and Clifford Beers, 1907–1910," *Bulletin of the History of Medicine* 43 no.5 (1969): 417.

[163] Ibid, 419.

not mention any form of work or therapeutic occupation in his book, although at one of the institutions where he stayed, the Hartford Retreat, there were amusements such as games, reading material (nonfiction only), a piano and the occasional dance.[164] As Norman Dain observes, during the early 1900s there were "no individualised programs to engage patients' interests and energies" and no other successful treatments had been identified.[165] Beers' claims regarding his treatment (nearly all of which have been verified by other sources) underscore the deterioration of the early nineteenth-century practices associated with moral treatment, as well as emphasising the difference in approach by psychiatrists such as Meyer.

Aims and Scope

This book compares approaches to patient occupation in France and England, approaches that remained similar during the nineteenth and early twentieth centuries, and then diverged after World War I. As such, it adds to the genre of comparative and transnational histories of psychiatry, such as those featured in edited works by Marijke Gijswijt-Hofstra and Harry Oosterhuis, Olé Peter Grell and Andrew Cunningham; Waltraud Ernst and Thomas Mueller; and Louise Westwood, Volker Roelcke and Paul Weindling.[166] These works demonstrate how modern psychiatry developed as a result of the constant transfer of ideas, perceptions and personnel across national borders. The present study shows how new ideas concerning occupational therapy were transferred from the USA and Germany to England and France but were received very differently in these countries. The book analyses the medical, socio-economic, cultural and legislative factors that caused this divergence. The comparison of patient occupation in England and France highlights the different

[164] Norman Dain, *Clifford W. Beers: Advocate for the Insane* (Pittsburgh: University of Pittsburgh Press, 1980), 26.

[165] Ibid.

[166] Marijke Gijswijt-Hofstra and Harry Oosterhuis, eds, *Psychiatric Cultures Compared: Psychiatry and Mental Health Care in the Twentieth Century: Comparisons and Approaches* (Amsterdam: Amsterdam University Press, 2005); Olé Peter Grell and Andrew Cunningham, eds, *Medicine and Religion in Enlightenment Europe* (Aldershot: Ashgate, 2007); Waltraud Ernst and Thomas Mueller, eds, *Transnational Psychiatries: Social and Cultural Histories of Psychiatry in Comparative Perspective, c.1800–2000* (Newcastle: Cambridge Scholars, 2010); Volker Roelcke, Paul J. Weindling and Louise Westwood, eds, *International Relations in Psychiatry: Britain, Germany, and the United States to World War II* (Cambridge: Cambridge University Press, 2010).

professional trajectories of English and French psychiatry that led to contrasting understandings of mental disorder during the interwar period.

What is lacking from this study is the views of the patients themselves. Clearly, their opinions on work and occupational therapy would offer valuable insights, since as Roy Porter reminded us nearly forty years ago, histories of medicine should not be written solely from the physicians' perspective.[167] However, without access to individual patient's medical records, to which letters, notes and journals are sometimes appended, the voices of patients have remained silent in this study.[168] But as Isabelle von Bueltzingsloewen maintains, it is possible to create a "history from below" using asylum records alongside official sources. [169] The present study provides a sense of what life was like for patients—such as the daily routines, the asylum environment and living conditions, the type of treatments prescribed, and the attitudes of staff—but does not reveal what patients actually felt about the occupations that were (or were not) allocated to them.

The origins of therapeutic work in the moral treatment of the early nineteenth century, and the waning of its popularity and perceived effectiveness as a cure by the end of the century are analysed in Chap. 2. Patient work was used very similarly by French and English psychiatrists between the early 1800s and the outbreak of World War I. In both countries, what began as a psychological treatment, carefully selected to suit individual patients, evolved into a bureaucratic system that commodified patient labour to suit institutional requirements. The changing perceptions of mental illness following World War I are the focus of analysis in Chap. 3. The reasons for the shift of psychiatric opinion from the organicist or physiological interpretation of mental disorder, that characterised late nineteenth-century psychiatric thinking in both England and France, to a psychosocial understanding of mental disorder that occurred in England and Paris, but not in provincial France, are explored. This ideological shift influenced the nature of occupation offered to patients, encouraging the adoption of occupational therapy. In institutions where a custodial model

[167] Roy Porter, "The Patient's View: Doing Medical History from Below," *Theory and Society* 14, no. 2 (1985): 175.

[168] Examples of recent studies drawing on patient letters and medical records include (in addition to those already mentioned): Hazel Morrison, "Constructing Patient Stories: 'Dynamic' Case Notes and Clinical Encounters at Glasgow's Gartnavel Mental Hospital, 1921–32," *Medical History* 60 no. 1 (2016): 67–86; Benoît Majerus, *Parmi Les Fous: Une histoire sociale de la psychiatrie au XXe siècle* (Rennes: Presses Universitaires de Rennes, 2013).

[169] von Bueltzingsloewen, "Vers un Désenclavement," 6.

of care, and an organicist interpretation of mental disorder prevailed, patient occupation continued much as it had done before World War I. The differences and similarities between the new methods of therapeutic occupation that emerged in Germany and the USA before World War I; patient work at the end of the nineteenth century; and occupation in the context of moral treatment, are assessed in Chap. 4.

Chapter 5 addresses the deleterious effects of World War I on the respective economies of England and France, and thus on asylum budgets. Tensions between the important economic role played by patient work, particularly in such straitened times, and the prescription of work or occupation as therapy are analysed. These tensions could be exacerbated by the different management structures in the French and English asylum systems. The issue of medical or administrative authority became key in decisions regarding how patients were occupied. The place of entertainments, which represented a cost to the asylum, could also be contested area. The role of the medical superintendent and chief medical officer, as the doctors who prescribed patient work and occupational therapy, is discussed in Chap. 6. Their therapeutic preferences, and the ideology that underpinned these preferences, were key to the type of occupations they prescribed, or indeed whether patients were occupied at all. The doctors' preferences were framed by the availability of effective remedies. In the 1920s these were minimal, but the late 1930s saw the introduction of shock therapies and psycho-surgery. As this chapter demonstrates, the French and English psychiatrists in charge of the hospitals in this study evaluated these various treatments, and their relative importance compared to occupation, differently. The psychiatrists who prescribed occupation spent very little time with patients, particularly in the provincial institutions where one psychiatrist was responsible for several hundred patients. The staff spending the most time with patients were the nurses, occupational therapists and workshop managers, whose supervision of the occupation prescribed by the doctors is the subject of Chap. 7. Analysis of staff competence, training, their commitment to the role, and the staff-to-patient ratios reveal marked differences between rural and metropolitan institutions as well as between French and English establishments.

The influence of a patient's mental condition, physical health, age, and class and gender on the occupation allocated to them in French and English institutions is investigated in Chap. 8. A patient's assumed curability or incurability, and the stage of their illness, affected whether they were prescribed occupation, and of what type. While the opinions of patients on work and occupation cannot be gleaned, this chapter gives an indication of their experiences of life in French and English institutions. The question of whether patients were being "moulded into productive citizens" is addressed in Chap. 9. Was this facilitated by the occupation prescribed to patients in the institution? Did the occupations inside reflect the types of work available outside it? The preparedness of patients in the metropolitan and rural institutions studied, for the new industries and the modern methods of production introduced in the interwar period, is discussed. Did occupational therapy, with its focus on arts and crafts, provide appropriate rehabilitation for a patient hoping to find work in an automobile factory? The issue of support for discharged patients, both charitable and state-funded, is also addressed in Chap. 9, which highlights differences in the availability of assistance both between England and France, and between rural and metropolitan areas.

The overall thesis of the book, drawn to a conclusion in Chap. 10, is that the nature of occupation prescribed to patients tells us so much more than simply how patients were occupied. The deployment of patient work highlights the financial situation in different institutions, revealing how some were more reliant on its economic contribution than others. It informs about the management structures inherent in different national institutions and how these could impact on the types of therapy prescribed, and about the multiplicity of factors involved in decisions to accept or reject ideas coming from overseas. The numbers of patients involved in work in different institutions suggested variations in the class, gender, age, physical fitness and severity of the mental disorder of patients. The type of occupation allocated to patients was also a matter of the personal preference of psychiatrists, reflecting their training and professional backgrounds. Whether occupation was prescribed at all is indicative of the level of competence of nursing staff and the existence of professional occupational therapists. The extent to which patients were prepared for the labour market in their locality raised questions about the about the rehabilitative value of patient work, and whether this was regarded as important by psychiatrists. But most crucially, patient occupation is indicative of levels of professionalisation within psychiatry, of attitudes towards the curability of mental disorder and its interpretation as an organic disease or a problem of adaptation to an individual's social environment. In the early days of moral treatment, the topic of Chap. 2, patients were encouraged to adapt their behaviour to enable them to integrate with society and earn their living. Such aims would re-emerge a century later.

Bibliography

Secondary Sources

Adams, Mark B., ed. *The Wellborn Science: Eugenics in Germany, France, Brazil and Russia*. Oxford and New York: Oxford University Press, 1990.

Andrews, Jonathan, Asa Briggs, Roy Porter, Penny Tucker, and Keir Waddington. *The History of Bethlem*. Abingdon and New York: Routledge, 1997.

Ankele, Monika. "The Patient's View of Work Therapy: The Mental Hospital Hamburg-Langenhorn during the Weimar Republic." In *Work, Psychiatry and Society, c.1750–2015*, edited by Waltraud Ernst, 238–61. Manchester: Manchester University Press, 2016.

Arveiller, J.-P. and Clément Bonnet. *Au Travail, Les Activités Productives dans le Traitement et la Vie du Malade Mental*. Toulouse: ERÈS, 1991.

Bacopoulos-Viau, Alexandra and Aude Fauvel. "The Patient's Turn. Roy Porter and Psychiatry's Tales, Thirty Years On." *Medical History* 60 no. 1 (2016): 1–18.

Bateson, Gregory, ed. *Perceval's Narrative: A Patient's Account of his Psychosis, 1830–1832* [1838–40]. Stanford, CA: Stanford University Press, 1961.

Beckman, Emily, Elizabeth Nelson and Modupe Labode. "'I like my job because it will get me out quicker': Work, Independence, and Disability at Indiana Central State Hospital (1986–1993)." In *Voices in the History of Madness: Personal and Professional Perspectives on Mental Health and Illness*, edited by Robert Ellis, Sarah Kendal and Steven J. Taylor, 173–90. Cham: Palgrave Macmillan, 2021.

Berg, Maxine and Pat Hudson. "Rehabilitating the Industrial Revolution." *Economic History Review* 45 no.1 (1992): 24–50.

Berrios, G. E. "Early Electroconvulsive Therapy in Britain, France and Germany: A Conceptual History." In *150 Years of British Psychiatry, Volume II: The Aftermath*, edited by Hugh Freeman and German E. Berrios, 3–15. London and New Jersey: Athlone, 1996.

Beveridge, Allan. "Life in the Asylum: patients' letters from Morningside, 1873–1908." *History of Psychiatry* 36 no. 9 (1998): 431–70.

Bewley, Thomas. *Madness to Mental Illness.* London: Royal College of Psychiatrists, 2008.

Bing, Robert K. "Eleanor Clarke Slagle Lectureship 1981—Occupational Therapy Revisited: A Paraphrastic Journey." *American Journal of Occupational Therapy* 35 no. 8 (1981): 499–518.

Blayney, Stephan. "Industrial Fatigue and the Productive Body: the science of work in Britain, 1900–1918." *Social History of Medicine* 32 no. 2 (2019): 310–28.

Blythe Doroshow, Deborah. "Performing a Cure for Schizophrenia: Insulin Coma Therapy on the Wards." *Journal of the History of Medicine and Allied Sciences* 62 no. 2 (2007): 213–43.

Borch-Jacobsen, Mikkel Sonu Shamdasani. *The Freud Files: an inquiry into the history of psychoanalysis.* Cambridge: Cambridge University Press, 2012.

Borsay, Anne and Pamela Dale, eds. *Mental Health Nursing: The working lives of paid carers in the nineteenth and twentieth centuries.* Manchester: Manchester University Press, 2013.

Boschma, Geertje. *The Rise of Mental Health Nursing: A history of psychiatric care in Dutch asylums, 1890–1930.* Amsterdam: Amsterdam University Press, 2003.

Broadberry, Stephen and Kevin O'Rourke, eds. *The Cambridge Economic History of Modern Europe, Volume II: 1870 to the Present.* Cambridge: Cambridge University Press, 2010.

Burnett, John. *Idle Hands: The Experience of Unemployment, 1790–1990.* London and New York: Routledge, 1994.

Busfield, Joan. "Class and Gender in Twentieth-Century British Psychiatry: Shell-Shock and Psychopathic Disorder." In *Sex and Seclusion, Class and Custody: Perspectives on Gender and Class in the History of British and Irish Psychiatry*, edited by Anne Digby and Jonathan Andrews, 295–322. Amsterdam and New York: Rodopi, 2004.

Buyst, Erik and Piotr Franaszek. "Sectoral Developments, 1914–1945." In *The Cambridge Economic History of Modern Europe, Volume 2: 1870 to the Present*, edited by Stephen Broadberry and Kevin O'Rourke, 208–231. Cambridge: Cambridge University Press, 2010.

Carlson, Eric and Norman Dain. "The Psychotherapy that was Moral Treatment." *The American Journal of Psychiatry* no. 117 (1960): 519–24.

Castel, Robert. *L'Ordre psychiatrique: L'âge d'or de l'aliénisme.* Paris: Minuit, 1976.

Paterson, Catherine F. "75th Anniversary of the Founding of the Scottish Association of Occupational Therapies." *British Journal of Occupational Therapy* 70 no.12 (2007): 537–40.

Chaney, Sarah. "Useful members of society, or motiveless malingerers? Occupation and malingering in British asylum psychiatry, 1870–1914." In *Work, Psychiatry and Society, c.1750–2015* edited by Waltraud Ernst, 277–97. Manchester: Manchester University Press, 2016.

Charland, Louis. "Benevolent Theory: Moral Therapy at the York Retreat." *History of Psychiatry* 18 no. 1 (2007): 61–80.

Charret, L. and S. Thiébaut Samson. "Histoire, fondements et enjeux actuels de l'ergothérapie." *Contraste* 45 (2017): 17–36.

Cherry, Steven. *Mental Health Care in Modern England: The Norfolk Lunatic Asylum/St. Andrew's Hospital, 1910–1998.* Woodbridge: Boydell, 2003.

Coffin, Jean-Christophe. *La Transmission de la Folie, 1850–1914.* Paris: L'Harmattan, 2003.

Coleborne, Catharine. “Challenging Institutional Hegemony: Family visitors to hospitals for the insane in Australia and New Zealand, 1880s–1910s.” In *Permeable Walls: Historical Perspectives on Hospital and Asylum Visiting*, edited by Graham Mooney and Jonathan Reinarz, 289-308. Amsterdam and New York: Rodopi, 2009.

Crammer, John. “Training and Education in British Psychiatry, 1770–1970.” In *150 Years of British Psychiatry, Volume II: The Aftermath*, edited by Hugh Freeman and German E. Berrios, 209–42. London and New Jersey: Athlone, 1996.

Crammer, John. *Asylum History: Buckinghamshire County Pauper Lunatic Asylum—St John’s*. London: Gaskell, 1990.

Creton, Anne. *Psychiatrie Française et Première Guerre Mondiale: Évolution des Idées et Situation au Sein des Établissements de Soins du Nord-Pas-De-Calais*. Thèse. Lille: Université de Lille, 2015.

Dain, Norman. *Clifford W. Beers: Advocate for the Insane*. Pittsburgh: University of Pittsburgh Press, 1980.

Derrien, Marie. *‘La Tête En Capilotade’: Les Soldats De La Grande Guerre Internés Dans Les Hôpitaux Psychiatriques Français (1914–1980)*. Thèse. Lyon: Université de Lyon 2, 2015.

Digby, Anne. “Moral Treatment at the Retreat, 1796–1846.” In *The Anatomy of Madness, Volume II: Institutions and Society*, edited by W.F. Bynum, Roy Porter and Michael Shepherd, 52–72. London and New York: Tavistock, 1985a.

Digby, Anne. *Madness, Morality and Medicine: A Study of the York Retreat, 1796–1914*. Cambridge: Cambridge University Press, 1985b.

Doerner, Klaus. *Madmen and the Bourgeoisie: A Social History of Insanity and Psychiatry* [1969], translated by Joachim Neugroschel and Jean Steinberg. Oxford: Blackwell, 1981.

Dowbiggin, Ian. “French Psychiatry and the search for a professional identity: the Société Médico-Psychologique 1840–1870.” *Bulletin of the History of Medicine* 63 no.3 (1989): 331–55.

Drake, Margaret. *Reconstructing Soldiers: An Occupational Therapist in World War I* (iUniverse, 2013).

Du Plessis, Rory. “‘Tells his story quite rationally and collectedly’: Examining the casebooks of Grahamstown Lunatic Asylum, 1890–1910, for cases of delusion where patients voiced their life stories.” In *Voices in the History of Madness: Personal and Professional Perspectives on Mental Health and Illness*, edited by Robert Ellis, Sarah Kendal and Steven J. Taylor, 137–54. Cham: Palgrave Macmillan, 2021.

Dutton, Paul V. *Origins of the French Welfare State: The Struggle for Social Reform in France, 1914–1947*. Cambridge: Cambridge University Press, 2002.

Eastoe, Stef. *Idiocy, Imbecility and Insanity in Victorian Society: Caterham Asylum, 1867–1911*. Cham: Palgrave Macmillan, 2020.

Ellis, Robert. *London and its Asylums, 1888–1914.* Cham: Palgrave Macmillan, 2020.

Eraso, Yolanda. "'A burden to the state': The reception of the German 'Active Therapy' in an Argentinian 'Colony-Asylum' in the 1920s and 1930s." In *Transnational Psychiatries: Social and Cultural Histories of Psychiatry in Comparative Perspective, c.1800–2000*, edited by Waltraud Ernst and Thomas Mueller, 51–79. Cambridge: Cambridge Scholars Publishing, 2010.

Ernst, Waltraud, ed. *Work, Psychiatry and Society, c.1750–2015.* Manchester: Manchester University Press, 2016.

Ernst, Waltraud. "Useful Both to the Patients as well as to the State": Patient Work in Colonial Mental Hospitals in South Asia, c.1818–1948." In *Work, Psychiatry and Society, c.1750–2015*, edited by Waltraud Ernst, 117–41. Manchester: Manchester University Press, 2016.

Ernst, Waltraud and Thomas Mueller, eds. *Transnational Psychiatries: Social and Cultural Histories of Psychiatry in Comparative Perspective, c.1800–2000.* Newcastle: Cambridge Scholars, 2010.

Fauvel, Aude. "Aliénistes contre psychiatres. La médecine mentale en crise (1890–1914)." *Psychologie Clinique* 17 (2004): 61–76.

Fau-Vincenti, Véronique. "Valeur du Travail à la Troisième Section de l'Hôpital de Villejuif: Entre thérapie et instrument de discipline." *Criminocorpus* [online] *Savoirs, Politiques et Pratiques de l'exécution des peines en France au XXe siècle* (12 September 2014). Accessed 03/03/2018. http://criminocorpus.revues.org/2788.

Feudtner, Chris. "'Minds the Dead Have Ravished': Shell-shock, History and the Ecology of Disease Systems." *History of Science* 31 no. 4 (1993): 337–420.

Fine, Reuben. *A History of Psychoanalysis.* New York: Continuum Publishing, 1990.

Forsythe, Bill, Joseph Melling and Richard Adair. "The New Poor Law and the County Pauper Lunatic Asylum—the Devon Experience, 1834–1884." *Social History of Medicine* 9 no. 3 (1996): 335–55.

Foucault, Michel. *Madness and Civilization: A History of Insanity in the Age of Reason* [1961], translated by Richard Howard. New York: Random House, 1965.

Freebody, Jane. "'The Root of All Evil is Inactivity': The Response of French Psychiatrists to New Approaches to Patient Work and Occupation, 1918–1939." In *Voices in the History of Madness: Personal and Professional Perspectives on Mental Health and Illness*, edited by Robert Ellis, Sarah Kendal and Steven J. Taylor, 71–94. Cham: Palgrave Macmillan, 2021.

Freebody, Jane. "The role of work in late eighteenth- and early nineteenth-century treatises on moral therapy in France, Tuscany and Britain." In *Work, Psychiatry and Society, c.1750–2015*, edited by Waltraud Ernst, 31–54. Manchester: Manchester University Press, 2016.

Freeman, Hugh. "Psychiatry in Britain, *c*.1900." *History of Psychiatry* 21 no. 3 (2010): 12–24.

Freidson, Eliot. *Profession of Medicine: a study of the sociology of applied knowledge.* New York: Dodd Mead, 1970.

Freis, David. *Psycho-Politics between the Wars: Psychiatry and Society in Germany, Austria and Switzerland.* Cham: Palgrave Macmillan, 2019.

Gable, Gabriel. "Naissance de l'ergothérapie." In *Nouveau Guide de Pratique en Ergothérapie: entre concepts et réalité,* edited by J.-M. Caire, 85–89. Marseille, Solal: 2008.

Gijswijt-Hofstra, Marijke and Harry Oosterhuis, eds. *Psychiatric Cultures Compared: Psychiatry and Mental Health Care in the Twentieth Century: Comparisons and Approaches.* Amsterdam: Amsterdam University Press, 2005.

Gillette, Nicole P. "Occupational Therapy in Mental Health: A Historical Perspective." In *Occupational Therapy,* Fourth edition, edited by Helen S. Willard and Clare S. Spackman, 51–60. Philadelphia, J.B. Lippincott & Co, 1963.

Gittins, Diana. *Madness in Its Place: Narratives of Severalls Hospital, 1913–1997.* London and New York: Routledge, 1998.

Goffman, Erving. *Asylums: Essays on the Social Situation of Mental Patients and Other Inmates.* Garden City, NY: Anchor, 1961.

Goldstein, Jan. *Console and Classify: The French Psychiatric Profession in the Nineteenth Century.* Chicago and London: Chicago University Press, 2001.

Goubert, J.-P. and R. Remondière. "Les origins historiques de l'ergothérapie en France." *Sociologie Santé* 20 June (2004): 247–68.

Greaves, Julian. "The First World War and its Aftermath." In *Twentieth-Century Britain: Economic, Cultural and Social Change,* Second edition, edited by Francesca Carnevali and Julie-Marie Strange, 127–144. Harlow and New York: Pearson Education, 2007.

Grell, Olé Peter and Andrew Cunningham, eds. *Medicine and Religion in Enlightenment Europe.* Aldershot: Ashgate, 2007.

Guihard, Jean-Philippe. "Quelle évolution pour l'ergothérapie en psychiatrie?" Conference paper delivered at the Institute of Occupational Therapy Training, Rennes, 2000. Accessed 03/02/2022. http://jp.guihard.pagesperso-orange.fr/articles/evolution/evolution.html.

Guillemain, Hervé and Stéphanie Tison. *Du Front à l'Asile, 1914–1918.* Paris: Alma éditeur, 2013.

Guyotat, J. "La formation des psychiatres en France." *Raison Présente* 83 (1987): 69–81.

Hall, John. "From work and occupational therapy: the policies of professionalisation in English mental hospitals from 1919 to 1959." In *Work, Psychiatry and Society, c.1750–2015,* edited by Waltraud Ernst, 314–333. Manchester: Manchester University Press, 2016.

Harris, Bernard. *The Origins of the British Welfare State: Social Welfare in England and Wales, 1800–1945.* Basingstoke: Palgrave Macmillan, 2004.

Healy, David. "The History of British Psychopharmacology." In *150 Years of British Psychiatry, Volume II: The Aftermath*, edited by Hugh Freeman and German E. Berrios, 61–88. London and New Jersey: Athlone, 1996.

Henckes, Nicholas. "Réformer et Soigner. L'émergence de la psychothérapie institutionelle en France, 1944–55." In *Psychiatries dans l'histoire*, edited by Jacques Arveiller, 277–288. Caen: PUC, 2008.

Hide, Louise. *Class and Gender in English Asylums, 1890–1914*. Basingstoke: Palgrave Macmillan, 2014.

Hocking, Clare. "The Way We Were: Romantic Assumptions of Pioneering Occupational Therapists in the United Kingdom." *British Journal of Occupational Therapy* 71 no. 4 (2008a): 146–154.

Hocking, Clare. "The Way We Were: The Ascendance of Rationalism." *British Journal of Occupational Therapy* 71 no. 6 (2008b): 226–32.

Hocking, Clare. "The Way We Were: Thinking Rationally." *British Journal of Occupational Therapy* 71 no. 5 (2008c): 185–95.

Hocking, Clare. "Early Perspectives of Patients, Practice and the Profession," *British Journal of Occupational Therapy* 70 no. 7 (2007): 284–90.

Hollen Lees, Lynn. *The Solidarities of Strangers: The English Poor Laws and the People, 1700–1948*. Cambridge: Cambridge University Press, 1998.

Hunter, Richard. Review of *Perceval's Narrative*, edited by Gregory Bateson. *British Medical Journal* 13 October (1962): 983.

Huteau, Michel. *Psychologie, Psychiatrie et Société sous la Troisième République: La biocratie d'Édouard Toulouse (1865–1947)*. Paris: L'Harmattan, 2002.

Jackson, Mark. "From Work to Therapy: The changing politics of occupation in the twentieth century." *British Journal of Occupational Therapy* 56 no. 10 (1993): 360–364.

Jay, Peggy and Hester Monteath. "The British Association of Occupational Therapists Celebrates its Silver Jubilee." *British Journal of Occupational Therapy* 61 no. 5 (1999): 189–91.

Kielhofner, G. and J.P. Burke. "Occupational therapy after 60 years: An account of changing identity and knowledge." *American Journal of Occupational Therapy* 31 no. 10 (1977): 675–89.

King, Lucy. "The Best Possible Means of Benefiting the Incurable: Walter Bruetsch and the Malaria Treatment of Paresis." *Annals of Clinical Psychiatry* 12 no. 4 (2004): 197–203.

Klein, Alexandre. "Théodore Simon (1873–1961). Itinéraire d'un psychiatre engage pour la professionnalisation des infirmiers et infirmières d'asile." *Association de recherche en soins infirmiers* 135 (2018): 91–106.

Krzyzaniak, Patrice. *Georges Daumézon (1912–1979): un camisard psychiatre et pedagogue: une contribution singulière aux sciences de l'éducation*. Thèse. Lille: Université Charles de Gaulle, 2017.

Laing, R.D. *The Divided Self: A Study of Sanity and Madness.* London: Tavistock, 1960.

Larkin, Gerald. "The Emergence of Para-Medical Professions." In *Companion Encyclopaedia of the History of Medicine,* edited by W.F. Bynum and R. Porter, 1329–40. London and New York: Routledge, 1994.

Laws, Jennifer. "Crackpots and Basket-Cases: A history of therapeutic work and occupation." *History of the Human Sciences* 24 no. 2 (2011): 65–81.

Leese, Peter. *Shell Shock: Traumatic Neurosis and the British Soldiers of the First World War.* Basingstoke: Palgrave Macmillan, 2002.

Lerner, Paul. *Hysterical Men: War, Psychiatry and the Politics of Trauma in Germany, 1890–1930.* Ithaca, NY: Cornell University Press, 2003.

Levine, Ruth Ellen. "The Influence of the Arts-and-Crafts Movement on the Professional Status of Occupational Therapy." *American Journal of Occupational Therapy* 41 no. 4 (1987): 248–54.

Levine-Clark, Marjorie. "Embarrassed Circumstances: Gender, Poverty and Insanity in the West Riding of England in the Early Victorian Years." In *Sex and Seclusion, Class and Custody: Perspectives on Gender and Class in the History of British and Irish Psychiatry,* edited by Anne Digby and Jonathan Andrews, 123–48. Amsterdam and New York: Rodopi, 2004.

Long, Vicky. "Rethinking Post-War Mental Health Care: Industrial Therapy and the Chronic Mental Patient in Britain." *Social History of Medicine* 26 no. 4 (2013): 738–58.

Long, Vicky. "'A Satisfactory Job is the Best Psychotherapist': Employment and Mental Health 1939–1960." In *Mental Illness and Learning Disability since 1850: Finding a Place for Mental Disorder in the United Kingdom,* edited by J. Melling and P. Dale, 179–99. Abingdon: Routledge, 2006.

Loughran, Tracey. "Shell Shock, Trauma, and the First World War: The Making of a Diagnosis and Its Histories." *Journal of the History of Medicine & Allied Sciences* 67 no. 1 (2012): 94–119.

Macdonald, E.M. *Occupational Therapy in Rehabilitation.* London: Baillière, Tindall and Cox, 1961.

MacKenzie, Charlotte. *Psychiatry for the Rich: A History of Ticehurst Private Asylum, 1792–1917.* London: Routledge, 1992.

Majerus, Benoît. "Surveiller, Punir et Soigner? Pratiques psychiatriques en Europe de l'ouest du 19e siècle aux années 1950." *Histoire, Médecine et Santé* 7 (2015): 51–62.

Mayr, Ernst. "When is historiography Whiggish?" *Journal of the History of Ideas* 5 no. 2 (1990): 301–09.

McCrae, Niall. "A Violent Thunderstorm: Cardiazol Treatment in British Mental Hospitals." *History of Psychiatry* 17 no. 1 (2006): 67–90.

McIvor, Arthur J. *A History of Work in Britain, 1880–1950.* Basingstoke and New York: Palgrave, 2001.

McKay, Kathryn. "From Blasting Powder to Tomato Pickles: Patient work at the provincial mental hospitals in British Columbia, Canada, 1855–1920." In *Work, Psychiatry and Society, c.1750–2015*, edited by Waltraud Ernst, 99–116. Manchester: Manchester University Press, 2016.

McMillan, James F. *Twentieth-Century France: Politics and Society 1898–1991*. London: Arnold, 1992.

Melling, Joseph. "Accommodating Madness: New Research in the Social History of Insanity and Institutions." In *Insanity, Institutions and Society, 1800–1914: A social history of madness in comparative perspective*, edited by Joseph Melling and Bill Forsythe, 1–30. London and New York: Routledge, 1999.

Melling, Joseph and Bill Forsythe, eds. *Insanity, Institutions and Society, 1800–1914: A social history of madness in comparative perspective*. London and New York: Routledge, 1999.

Micale, Mark. "The Psychiatric Body." In *Companion to Medicine in the Twentieth Century*, edited by Roger Cooter and John Pickstone, 323–46. London and New York: Routledge, 2000.

Michael, Pamela and David Hirst. "Establishing the 'Rule of Kindness': The Foundation of the North Wales Lunatic Asylum, Denbigh." In *Insanity, Institutions and Society, 1800–1914: A social history of madness in comparative perspective*, edited by Joseph Melling and Bill Forsythe, 159–79. London and New York: Routledge, 1999.

Michael, Pamela. *Care and Treatment of the Mentally Ill in North Wales, 1800–2000*. Cardiff: University of Wales Press, 2003.

Moncrieff, Joanna. "An Investigation into the Precedents of Modern Drug Treatment in Psychiatry." *History of Psychiatry* 10 no. 4 (1999): 475–90.

Monk, Lee-Ann. "Exploiting patient labour at Kew Cottages, Australia, 1887–1950." *British Journal of Learning Disabilities* 38 (2010): 86–94.

Moran, Frances M. *The Paradoxical Legacy of Sigmund Freud*. Abingdon and New York: Routledge, 2018.

Morel-Bracq, M.-C., A.-C. Delaisse, J.-F. Bodin, L. Charret and H. Hernandez. "Une approche historique du développement de l'identité professionnelle des ergothérapeutes en France: l'évolution des valeurs et intérêts à travers le temps." *La Revue Ergothérapies* 81 April (2021). Accessed 20/03/2022. https://revue.anfe.fr/2021/04/22/une-approche-historique-du-developpement-de-lidentite-professionnelle-des-ergotherapeutes-en-france-levolution-des-valeurs-et-interets-des-ergotherapeutes-a-travers-le-temps/.

Morrison, Hazel. "Constructing Patient Stories: 'Dynamic' Case Notes and Clinical Encounters at Glasgow's Gartnavel Mental Hospital, 1921–32." *Medical History* 60 no.1 (2016): 67–86.

Nolan, Peter. *A History of Mental Nursing*. London: Chapman and Hall, 1993.

Nord, Philip. "The Welfare State in France, 1870–1914." *French Historical Studies* 18 no. 3 (1994): 821–38.

Paterson, C.F. "A Short History of Occupational Therapy in Psychiatry." In *Occupational Therapy and Mental Health*, Third Edition, edited by Jennifer Creek, 3–15. Edinburgh and London: Churchill Livingstone, 2002.

Paterson, Catherine F. *Opportunities not Prescriptions: The Development of Occupational Therapy in Scotland 1900–1960.* Aberdeen: Aberdeen History of Medicine Publications, 2010.

Paterson, Catherine F. "Occupational Therapy and the National Health Service, 1948–98." *British Journal of Occupational Therapy* 61 no.7 (1998): 311–15.

Pederson, Susan. *Family, Dependence, and the Origins of the Welfare State: Britain and France, 1914–1945.* Cambridge: Cambridge University Press, 1993.

Perkin, Harold. *The Rise of Professional Society: England since 1880.* Abingdon: Routledge, 2002.

Pibarot, Isabelle. "Une Histoire de soin par activité." In *Ergothérapie en psychiatrie: De la souffrance à la réadaptation*, Second edition, edited by H. Hernandez, 28–36. Paris: De Boeck Supérieure, 2016.

Polatajko, Hélène. "The evolution of our occupational perspective: the journey from diversion through therapeutic use to enablement." *Canadian Journal of Occupational Therapy* 68 no. 4 (2001): 203–207.

Porter, Roy and Mark Micale. "Introduction: Reflections on Psychiatry and Its Histories." In *Discovering the History of Psychiatry*, edited by Mark Micale and Roy Porter, 3–38. Oxford and New York: Oxford University Press, 1994.

Porter, Roy. "The Patient's View: Doing Medical History from Below." *Theory and Society* 14 no. 2 (1985): 175–98.

Porter, Roy. "Was there a moral therapy in eighteenth-century psychiatry?" *Lychnos* (1981): 12–26.

Porter, Roy. *Mind Forg'd Manacles—A History of Madness in England from the Restoration to the Regency.* London: Penguin Books, 1987.

Pressman, Jack. *Last Resort: Psychosurgery and the Limits of Medicine.* New York: Cambridge University Press, 1998.

Prestwich, Patricia E. "Family Strategies and Medical Power: 'voluntary' committal in a Parisian asylum, 1876–1914." *Journal of Social History* 27 no. 4 (1994): 799–818.

Quevillon, É. "Construire la science de l'activité humaine en France." In *L'activité humaine: un potentiel pour la santé?* edited by M.-C. Morel-Bracq and E. Trouvé, 323–330. Paris: De Boeck-Solal, 2015.

Quiroga, Virginia. *Occupational Therapy: The First Thirty Years, 1900–1930.* Bethesda: American Occupational Therapy Association, 1995.

Rabinbach, Anson. *The Human Motor: Energy, Fatigue and the Origins of Modernity.* Berkeley: University of California Press, 1992.

Reaume, Geoffrey. "A Wall's Heritage: Making Made People's History Public." *Public Disability History* 1 (2016): 20. Accessed 12/02/2022. https://www.public-disabilityhistory.org/2016/11/a-walls-heritage-making-mad-peoples.html

Reaume, Geoffrey. "Patients at Work: Insane Asylum Inmates' Labour in Ontario, 1841–1900." In *Mental Health in Canadian Society: Historical Perspectives*, edited by James Moran and David Wright, 69–116. Montreal: McGill Queen's University Press, 2006.

Reed, Kathlyn L. "The Beliefs of Eleanor Clarke Slagle: Are they Current or History?" *Occupational Therapy in Healthcare* 33 no. 3 (2019): 265–85.

Reed, Kathlyn L. "Tools of Practice: Heritage or Baggage? 1986 Eleanor Clarke Slagle Lecture." *American Journal of Occupational Therapy* 40 no. 9 (1986): 597–605.

Reid, Fiona. *Broken Men: Shell Shock, Treatment and Recovery in Britain, 1914–1930*. London: Continuum, 2010.

Robert, Ellis, Sarah Kendall and Steven J. Taylor, eds. *Voices in the History of Madness: Personal and Professional Perspectives on Mental Health and Illness*. Cham: Palgrave Macmillan, 2021. See Part II: Reconstructing Patient Perspectives, 115–216.

Roelcke, Volker, Paul J. Weindling and Louise Westwood, eds. *International Relations in Psychiatry: Britain, Germany and the Unite States to World War II*. Rochester: University of Rochester Press, 2010.

Rothman, David. *The Discovery of the Asylum: Social Order and Disorder in the New Republic*. Boston: Little Brown, 1971.

Roudebush, Marc. "A Battle of Nerves: Hysteria and Its Treatments in France During World War I." In *Traumatic Pasts: History, Psychiatry and Trauma in the Modern Age, 1870–1930* edited by Paul Lerner and Mark S. Micale, 253–79. Cambridge: Cambridge University Press, 2001.

Royle, Edward. *Modern Britain: A Social History 1750–2011*. Third edition. London and New York: Bloomsbury, 2012.

Scott, Julia. "British Journal of Occupational Therapy 70 years on." *British Journal of Occupational Therapy* 70 no. 7 (2007): 275.

Scull, Andrew T. *Museums of Madness: The Social Organisation of Insanity in Nineteenth-Century England*. London: Allen Lane, 1979.

Scull, Andrew. "Focal Sepsis and Psychosis: The Career of Thomas Chivers Graves (1883–1964)." In *150 Years of British Psychiatry, Volume II: The Aftermath*, edited by Hugh Freeman and German E. Berrios, 517–36. London and New Jersey: Athlone, 1996.

Scull, Andrew. "Madfolk and Their Keepers. Roy Porter and the History of Psychiatry." In *The Insanity of Place, the Place of Insanity: Essays on the History of Psychiatry*, edited by Andrew Scull, 38–53. London and New York: Routledge, 2006.

Scull, Andrew. *The Most Solitary of Afflictions: Madness and Society in Britain, 1700–1900*. New Haven and London: Yale University Press, 1993.

Searle, G.R. "The Politics of National Efficiency and War, 1900–1918." In *A Companion to Early Twentieth-Century Britain*, edited by Chris Wrigley, 56–71. Oxford: Blackwells, 2003.

Shepherd, Anne. "The Female Patient Experience in Two Late Nineteenth-Century Surrey Asylums." In *Sex and Seclusion, Class and Custody: Perspectives on Gender and Class in the History of British and Irish Psychiatry*, edited by Anne Digby and Jonathan Andrews, 223–48. Amsterdam and New York: Rodopi, 2004.

Simonnot, Anne-Laure. *Hygiénisme et Eugénisme au XXe Siècle: à travers la psychiatrie française*. Paris: Éditions Seli Arslan, 1999.

Smith, Leonard. "Experiences of the Madhouse in England 1650–1810." In *Voices in the History of Madness: Personal and Professional Perspectives on Mental Health and Illness*, edited by Robert Ellis, Sarah Kendal and Steven J. Taylor, 117–36. Cham: Palgrave Macmillan, 2021.

Smith, Leonard. *Lunatic Asylums in Georgian England, 1750–1830*. London and New York: Routledge, 2007.

Sturdy, Steve and Roger Cooter. "Science, Scientific Management and the Transformation of Medicine in Britain *c.*1870–1950." *History of Science* 36 no. 1 (1998): 421–46.

Szasz, Thomas. *The Myth of Mental Illness: Foundations of a Theory of Personal Conduct*. New York: Hoeber-Harper, 1961.

Tatu, Laurent and Julien Bogousslavsky. *La Folie Au Front: La Grande Batailledes Névroses De Guerre (1914–1918)*. Paris: Éditions Imago, 2012

Thomson, Mathew. "Disability, Psychiatry and Eugenics." In *The Oxford Handbook of the History of Eugenics*, edited by Alison Bashford and Philippa Levine, 116–133. Oxford and New York: Oxford University Press, 2010a.

Thomson, Mathew. "Mental Hygiene in Britain during the First Half of the Twentieth Century." In *International Relations in Psychiatry: Britain, Germany and the United States to World War II*, edited by Volker Roelcke, Paul Weindling and Louise Westwood, 134–155. Rochester: University of Rochester Press, 2010b.

Toms, Jonathan. *Mental Hygiene and Psychiatry in Modern Britain*. Basingstoke: Palgrave Macmillan, 2013.

Tosquelles, François. *Le Travail Thérapeutique en psychiatrie* [1967]. Paris: Éditions érès, 2009.

Vaiseta, Tomas. "Dehumanising Experience, Rehumanising Self-Awareness: Perceptions of Violence in Psychiatric Hospitals of Soviet Lithuania." In *Voices in the History of Madness: Personal and Professional Perspectives on Mental Health and Illness*, edited by Robert Ellis, Sarah Kendal and Steven J. Taylor, 155–72. Cham: Palgrave Macmillan, 2021.

Von Bueltzingsloewen, Isabelle. "Réalité et Perspectives de la Médicalisation de la Folie dans la France de L'Entre-Deux-Guerres." *Genèses* 1 no. 82 (2011): 52–74.

Von Bueltzingsloewen, Isabelle. "Vers un Désenclavement de l'Histoire de la Psychiatrie." *Le Mouvement Social* 253 (2015): 3–11.

Von Bueltzingsloewen, Isabelle. *L'Hécatombe des Fous: La Famine dans les hôpitaux psychiatriques français sous l'occupation*. Paris: Éditions Flammarion, 2007.

Waddington, Ivan. "The Movement towards the Professionalisation of Medicine." *British Medical Journal* 301 (1990): 688–90.

Walk, Alexander. "Some Aspects of the 'Moral Treatment' of the Insane up to 1854." *Journal of Mental Science* no. 421 (1954): 807–37.

Walsh, Oonagh. "'The Designs of Providence': Race, Religion and Irish Insanity." In *Insanity, Institutions and Society, 1800–1914: A social history of madness in comparative perspective*, edited by Joseph Melling and Bill Forsythe, 223–242. London and New York: Routledge, 1999.

Warner, Richard. *Recovery from Schizophrenia: Psychiatry and Political Economy*. London and New York: Routledge and Kegan Paul, 1985.

Welshman, John and Jan Walmsley, eds. *Community Care in Perspective: Care, Control and Citizenship*. Basingstoke and New York: Palgrave Macmillan, 2006.

Wilcock, Ann A. *Occupation for Health: A Journey from Prescription to Self Health, Volumes I and II*. London: British College of Occupational Therapists, 2002.

Winters, Eunice. "Adolf Meyer and Clifford Beers, 1907–1910." *Bulletin of the History of Medicine* 43 no.5 (1969): 414–43.

Wojciechowski, Jean-Bernard. *Hygiène Mentale et Hygiène Sociale: contribution à l'histoire de l'hygiénisme, tomes I et II*. Paris and Montreal: Éditions L'Harmattan, 1997.

Wright, David. "Family Strategies and the Institutional Confinement of 'Idiot' Children in Victorian England." *Journal of Family History* 23 no. 2 (1998): 190–208.

Zaretsky, Eli. *Political Freud: A History*. New York: Columbia University Press, 2015.

Archival Sources

Beers, Clifford Whittingham. *A Mind That Found Itself: An Autobiography*. New York: Longmans, Green, 1908.

Dausset, Louis. Rapport Général au nom de la 3e Commission sur les propositions budgétaires pour le Service des Aliénés (budget de 1919). Conseil Général de la Seine, 1918. ADP/D.10K3/27/20.

Lomax, Montagu. *Experiences of an Asylum Doctor: With suggestions for asylum and lunacy law reform*. London: G. Allen and Unwin Ltd, 1921.

Meyer, Adolf. "The Philosophy of Occupational Therapy." In *The Collected Papers of Adolf Meyer*, edited by Eunice Winters, 86–92. Baltimore: Johns Hopkins Press, 1952.

CHAPTER 2

Patient Work before World War I

The health benefits of activity and exercise have been understood since Graeco-Roman times, but it is only relatively recently (in the last two hundred years) that work has been medically prescribed for mentally ill patients. During the early modern period, physicians emphasised the importance of balancing the six "non-naturals" to remain healthy in mind and body. These non-naturals included air (as in the freshness of the air), food and drink, exercise and rest, sleep and waking, repletion and evacuation, and the passions and emotions, over which an individual had some control. Exercise might involve walking, dancing, sport or manual labour.[1] Such recommendations for exercise and labour also informed the views of non-medical writers. Robert Burton (1577–1640), Oxford scholar and cleric, recommended bodily exercise, labour, industry and keeping active as antidotes to melancholy.[2] Burton referred to idleness as "the bane of body and mind".[3] Philosopher John Locke (1632–1704) also extolled the benefits of labour, which he believed protected people from the "ills of idleness or the diseases that attend constant study in a sedentary life".[4]

[1] Louise Hill Curth, "Lessons from the Past: Preventative Medicine in Early Modern England," *Medical Humanities* 29, 1 (2003): 16–20.

[2] Robert Burton, *The Anatomy of Melancholy* [1621] (Penguin Classics, 2021), 857.

[3] Ibid., 240.

[4] John Locke, "Labour," in *The Oxford Book of Work,* ed. Keith Thomas (Oxford: Oxford University Press, 1999), 125.

J. Freebody, *Work and Occupation in French and English Mental Hospitals, c.1918–1939*, Mental Health in Historical Perspective,
https://doi.org/10.1007/978-3-031-13105-9_2

Pangloss in Voltaire's *Candide* (published in 1759) maintained that, "man was not born to repose".[5] Denis Diderot (1713–1784), editor of the French *Encyclopédie*, believed that man owed "his health, his subsistence, his peace of mind, his good sense, and perhaps his virtue" to daily occupation.[6] But it was not until the late-eighteenth century, when new ways of treating the mentally ill emerged from the humanitarian teaching of the Enlightenment, that work was incorporated as a therapy for the mentally ill under the umbrella of moral treatment.[7]

The introduction of work in asylums may have been linked to the beginnings of important and far-reaching changes to the economy associated with the transition from pre-industrial to industrial economic organisation. These changes began in the mid-eighteenth century in England, and the mid-nineteenth century in France, prompting the emergence of new attitudes towards work and workers. The independence and autonomy enjoyed by artisans were gradually sacrificed as workers were brought together in a factory or workshop and subordinated to the control of their employers. Technological advances led to the division of labour as work became more specialised, leading to the de-skilling and fragmentation of the workforce. Industrial workers had to adapt from a task-oriented approach to work, involving bursts of intense labour followed by periods of rest, to one which was governed by the clock. Employers expected long hours, six days per week of hard, disciplined labour, uninterrupted by opportunities to converse, run an errand or take a break. Irregular working rhythms were replaced by the discipline of clock time, rigorously enforced by employers.[8] Whole families, including women and children from the age of six, began to spend more time working. In England between 1750 and 1800, annual working hours increased by at least one fifth.[9] A new class of urban labourer was emerging whose labour was their only source of income. This made workers vulnerable to the vagaries of the marketplace and led to poverty on an unprecedented scale.

Poverty was putting increasing pressure on the English Poor Law by the early nineteenth century, leading to calls for a radical reform of the old

[5] Voltaire, "Candide," in *The Oxford Book of Work*, ed. Thomas, 160.

[6] Denis Diderot, "Encyclopédie," in *The Oxford Book of Work*, ed. Thomas, 122.

[7] Jennifer Laws, "Crackpots and Basket-Cases: A History of Therapeutic Work and Occupation," *History of the Human Sciences* 24 no. 2 (2011): 67.

[8] E.P. Thompson, "Time, Work-Discipline and Industrial Capitalism," *Past and Present* 38 no. 1 (1967): 61–73.

[9] Hans-Joachim Voth, "Time and Work in Eighteenth-Century London," *Journal of Economic History* 58, no. 1 (1998): 4.

system. Thomas Malthus' *Essay on the Principle of Population* (1798) criticised the existing system, claiming that the provision of generous allowances subsidised the work-shy and would result in even larger families unable to feed themselves. Debates raged over whether poverty was the result of "indolence" or "improvidence" and whether paupers were deserving of relief. Reformers agonised over how to combine compassion with "incentive-compatibility".[10] In 1832 a Royal Commission concluded that the system was too expensive and morally flawed in allowing men capable of work to claim relief. The resulting Poor Law Amendment Act of 1834 (the New Poor Law) abolished outdoor relief for able-bodied indigents and instituted a system of workhouses where only the most destitute would be fed and housed. The system was undoubtedly harsh; conditions inside the workhouse for the able-bodied pauper were designed to be worse than those of the poorest-paid labourer living outside. The aim of the 1834 Poor Law, according to Peter Bartlett, was to "root out and dismantle a culture of poverty, perceived in terms of immorality, intemperance and promiscuity, and replace it with a culture of self-help, respectability, sobriety and hard work".[11] The prospect of entering the workhouse was to be a humiliating last resort, designed to discourage those capable of working. For Foucault, the existence of English workhouses provided "a certain ethical consciousness of labour" and a moral symbol affirming the value of work.[12] Whilst training in the habits of work was considered important, especially for children entering the workhouse, the real 'value' of the institution was its symbolic presence as a place of degradation and brutality that awaited those who failed to work.[13] The message was clear; the able-bodied were expected to work for their living and idleness was not to be tolerated. Such attitudes provided the backdrop to the establishment of the English asylum system.

New theories about work and the role of labour were developed by political economists who studied the changes taking place in the way wealth was created and distributed in capitalist economies. They sought to "understand, control and predict both the economics and the politics of the

[10] Joel Mokyr, *The Enlightened Economy: Britain and the Industrial Revolution, 1700–1850* (London: Penguin Books, 2009), 441–3.

[11] Peter Bartlett, "The Asylum and the Poor Law: The Productive Alliance," in *Insanity, Institutions and Society, 1800–1914: A Social History of Madness in Comparative Perspective*, eds Joseph Melling and Bill Forsythe (London and New York: Routledge, 1999), 53.

[12] Michel Foucault, *Madness and Civilization: A History of Insanity in the Age of Reason*, trans. Richard Howard (New York: Random House, 1965), 55.

[13] Joan Busfield, *Managing Madness: Changing Ideas and Practice* (London and Melbourne: Hutchinson, 1986), 231.

market".[14] One of the earliest political economists, Adam Smith (1723–1790) claimed, in his famous work the *Wealth of Nations* (1776), that work was "the real price of everything" since it was the "toil and trouble" of labour that enabled an individual to acquire goods or services.[15] James Mill (1773–1836) agreed. As he expressed in his *Essay on Government* (1820), labour—although "painful" to perform—was the key to obtaining happiness, since it enabled the individual to make or procure what they desired.[16] Smith maintained that "the annual labour of every nation is the fund which originally supplies it with all the necessaries of life which it annually consumes".[17] It was not "gold and silver" that was responsible for generating the original "wealth of the world", but labour.[18] The notion that labour was central to national and individual prosperity provided the backdrop to the development of the asylum system. It made sense to ensure that work formed an integral part of institutional life, equipping the recovering mental patient or prisoner with the attitudes and skills necessary to make their economic contribution to society and to earn their living.

Industrialisation began later in France, in part due to the political, social and economic upheaval generated by the French Revolution (1789) and the Napoleonic Wars (1803–1815). Following the redistribution of land after the Revolution of 1789, France became a nation of small farmers. In 1862, 85% of French farms were under 10 hectares.[19] This had enduring social and economic consequences, including labour shortages throughout the economy, the persistence of craft industries and a "household economy" whereby family members held other jobs, such as handweaving, as well as working on the smallholding.[20] Coal mines and factories were often viewed as means of earning money during bad weather or agricultural slack periods, rather than as a labourers' only source of income.[21]

[14] For a history of economic thought in France, see Gilbert Faccarello, ed., *Studies in the History of French Political Economy* (London and New York: Routledge, 1998); Cynthia Koepp, "Learning to Calculate: A Review Article," *Eighteenth-Century Studies* 37, no. 1 (2003): 145.

[15] Adam Smith, *An Inquiry into the Nature and Causes of the Wealth of Nations* (Ware, Herts.: Wordsworth Editions, 2012), 34.

[16] Victor Bianchini, "Production and Education according to James Mill: The Precious Middle Point," *Journal of the History of Economic Thought* 38 no.2 (2016): 163.

[17] Smith, *Wealth of Nations*, 3.

[18] Ibid., 34.

[19] Robert Tombs, *France 1814–1914* (Harlow: Pearson, 1996), 269.

[20] Ibid.

[21] Ibid.

Significant industrial development did not occur until the onset of a period of relative political stability following the establishment of the Third Republic in 1871.[22] Industrialisation, therefore, was not the context from which moral treatment and work therapy emerged. However, in post-revolutionary France, good citizenship involved working hard, supporting oneself and contributing to the new Republic. Idleness, which was associated with the aristocracy, was shunned.

It was in this context of economic and political change that the asylum system developed. For the pioneers of moral treatment, William Tuke (1732–1822) in England and Philippe Pinel (1745–1826) in France, work was a fundamental aspect of their therapeutic regimes, helping patients to master their symptoms and enabling them to maintain their professional skills and thus their means of subsistence. As asylums proliferated during the early nineteenth century, the principles of moral treatment became the ideal to which most asylum superintendents and chief medical officers aspired. Work programmes for patients were introduced in the context of moral treatment at a time when work was expected of the inmates of most institutions (including prisons, workhouses and orphanages) and when there were no other successful remedies for mental disorder.[23]

Moral Treatment

Moral treatment represented a significant departure from previous methods of treating the mentally disordered. Before the mid-eighteenth century, it was believed that mentally disordered individuals were possessed by the Devil or, having lost their reason, reduced to wild beasts, rather than as people who were ill and in need of compassion and support. Earlier methods of treatment infamously included the use of mechanical restraint; intimidation, coercion and fear; regular bloodletting, the administration of emetics and laxatives; cupping and blistering; cold baths or showers; and a restricted diet. "Lunatics" were regularly beaten and were thought to be insensitive to the cold. In France, a nationwide survey of institutions for the mentally ill was undertaken in 1817 by the physician Jean-Étienne Dominique Esquirol (1772–1840), former student of Pinel. His report revealed the grim conditions endured by mental patients: "I have seen them naked, clad in rags … coarsely fed, lacking air to breathe, water to

[22] Roger Magraw, *France 1800–1914: A Social History* (London and New York: Longman, 2002), 8.

[23] Ibid., 6.

quench their thirst, wanting the basic necessities of life. I have seen them at the mercy of veritable jailers, victims of their brutal supervision. I have seen them in narrow, dirty, infested dungeons without air or light, chained in caverns where one would fear to lock up wild beasts…".[24] Esquirol sought to establish an asylum system in France where Pinel's teaching would be implemented.

According to the principles of moral treatment, the mentally disordered deserved compassion and should be treated as human beings who were still capable of rational behaviour, albeit intermittently. Corporal punishments, harsh treatment and indiscriminate physical remedies were not to be used. The creation of a clean, comfortable, homely environment, the provision of nutritious food and decent bedding, and the establishment of a regular daily routine were key aspects of moral treatment. Removal from the patient's home was considered essential since it separated the patient from the difficulties that may have caused their mental illness, such as family problems, business or financial worries, intemperance or religious fanaticism.[25] The orderly, structured environment of the asylum was believed to be curative in itself. The asylum grounds and setting were considered important aids to recovery, the landscaped gardens and surrounding countryside affording plentiful opportunities for exercise, fresh air and exposure to nature.[26] Patients were encouraged to develop self-control through a system of rewards and the withdrawal of certain privileges. Work was considered a particularly effective means of instilling self-control; it distracted the patient from their troubles, focused their attention on the fulfilment of a task and provided opportunities for social interaction and cooperation with staff and other patients. A patient's day was scheduled to include regular hours for work, balanced with periods for recreation and amusements.[27]

[24] Cited in Dora B. Weiner, "'Le geste de Pinel': The History of a Psychiatric Myth," in *Discovering the History of Psychiatry*, eds Mark Micale and Roy Porter (New York and Oxford: Oxford University Press, 1994), 234.

[25] Busfield, *Managing Madness*, 214.

[26] See: Clare Hickman, "The Picturesque at Brislington House: The Role of Landscape in Relation to the Treatment of Mental Illness in the Early Nineteenth-Century Asylum," *Garden History* 33, 1 (2005): 47–60.

[27] Recreation and amusements in the context of moral therapy are the topic of research by Ute Oswald. See: Ute Oswald, "'Distraction from Hurtful Thoughts': Recreational Activities as Agents of Healing in Nineteenth-Century British Asylums," *Medizinhistorisches Journal* 56 no. 1–2 (2021): 30–57.

While there is some debate amongst historians of medicine over the precise origins of moral treatment,[28] its teaching was most famously and comprehensively outlined in England by Samuel Tuke's account of his grandfather's methods, published in 1813, and in France by Philippe Pinel's *Traité* of 1800.[29] Tuke described the moral treatment practised at the Retreat at York, which opened in 1796. Here, he claimed, "it has [been] demonstrated, beyond all contradiction, the superior efficacy, both in respect of cure and security, of a mild system of treatment in all cases of mental disorder".[30] Tuke rejected the "drugs and medicaments" that were trialled initially because a "moral regimen" that helped a patient control themselves, proved more effective. Pinel warned against all forms of violence towards or physical punishment of patients, not simply because they were inhumane, but because they exacerbated the patient's condition.[31] Both Tuke and Pinel regarded treating the insane rather like rearing children. Tuke spoke of restraining patients from certain activities and encouraging them in others, while Pinel observed that when children were at play, they ceased to be lazy and disobedient and became active, focused and keen to obey the rules. Patients who were engaged in some form of activity were distracted from their morbid thoughts.[32]

Moral treatment acted on the mind, encouraging patients to behave in a way that enabled them to fit in with the rest of society. As such it was a psychological method of treatment, although the term "psychological" was not in use at that time. However, as Roy Porter reminds us, moral

[28] See: William F. Bynum, "Rationales for therapy in British Psychiatry, 1780–1835," *Medical History* 18 no.4 (1974): 317–34; Roy Porter, "Was there a moral therapy in eighteenth-century psychiatry?" *Lychnos* (1981): 12–26; Dora B. Weiner, "The madman in the light of reason. Enlightenment psychiatry, Part I," in *History of Psychiatry and Medical Psychology*, eds Edwin R. Wallace and John Gach (Boston: Springer, 2008), 281–303; Jan Goldstein, *Console and Classify: The French Psychiatric Profession in the Nineteenth Century* (Chicago: University of Chicago Press, 2001).

[29] Samuel Tuke, *Description of the Retreat: An Institution near York for Insane Persons of the Society of Friends* [1813] (Bibliolife: LCC); Philippe Pinel, *Traité medico-philosophique sur l'aliéntation mentale, ou la manie* (Paris: Chez Richard, Caille et Ravier, 1800). For a discussion of other works on moral treatment, see: Jane Freebody, "The role of work in later eighteenth- and early nineteenth-century treatises on moral therapy in France, Tuscany and Britain," in *Work, Psychiatry and Society, c.1750–2015*, ed. Waltraud Ernst (Manchester: Manchester University Press, 2016, 31–54.

[30] Tuke, *Description of the Retreat*, vi.

[31] Pinel, *Traité*, 195.

[32] Freebody, "Role of Work," 40–4.

treatment should not be confused with modern psychotherapy since neither Tuke nor Pinel were interested in "talking cures" or in "working through" problems.[33] They sort to influence behaviour by making patients "want to be good".[34] Tuke and Pinel's theories were similar, although interpretation of the word "moral" in Tuke's "moral treatment" and Pinel's "*traitement morale*" differed in emphasis, as Louis Charland explains.[35] At the Retreat, the term "moral" had an ethical dimension linked to the Quaker faith of the institution's founder, William Tuke. The non-medical Tuke believed that all individuals were imbued with a "moral sense", an innate sense of right and wrong, and that the mentally disordered could be persuaded to control themselves by appealing to this moral sense. Pinel's *traitement morale* was designed to act on the passions, sentiments, and emotions.[36] As a qualified physician, whose beliefs were firmly rooted in the sectarianism of revolutionary France, Pinel was not influenced by religious concerns. That said, in early nineteenth-century France, the emotions were considered closely linked to ethics and morality.[37] The aim of both versions of moral treatment was to influence behaviour without the use of drugs, physical treatments or mechanical restraint, but at the Retreat, ethics and religion played a particularly important role.[38]

For non-conformists, including Quakers, work was a social duty, a source of dignity and moral worth, as Max Weber (1864–1920) observed in his essay *The Protestant Ethic and the Spirit of Capitalism* (1904–1905). The accumulation of wealth was morally acceptable provided it was combined with a sober, industrious career and not squandered on dissolute living.[39]

Weber observed that thrift, industriousness, self-discipline and sobriety characterised the growing class of non-conformist entrepreneurs. This observation led him to believe that the non-conformist or puritan outlook "stood at the cradle of modern economic man."[40] The philanthropist

[33] Roy Porter, *Mind Forg'd Manacles—A History of Madness in England from the Restoration to the Regency* (London: Penguin Books, 1987), 225–6.

[34] Ibid., 226.

[35] Louis Charland, "Benevolent theory: moral treatment at the York Retreat," *History of Psychiatry* 18, 1 (2007): 75–6.

[36] Ibid., 76.

[37] Goldstein, *Console and Classify*, 72.

[38] Admission to the Retreat was restricted to Quakers until 1820.

[39] Anthony Giddens, "Introduction," in *The Protestant Work Ethic and the Spirit of Capitalism (1930) by Max Weber*, ed. Anthony Giddens (London: Taylor & Francis, 2005), xii-xiv.

[40] P. D. Anthony, *The Ideology of Work* (London and New York Tavistock Publications, 1977), 42–4.

William Tuke was a successful wholesale trader of tea, coffee and cocoa, remaining in business until the age of eighty-six. No stranger to hard work, Tuke was also a patron of the Bible Society, treasurer of the York Society of Friends, a campaigner for the abolition of the slave trade and the founder of three schools.[41] Hard work, as opposed to aristocratic idleness, and the duty of citizens to be useful members of society, were post-revolutionary principles that influenced Pinel's theories too. Work, for both Tuke and Pinel, was key in encouraging the mentally disordered to restrain themselves. Pinel regarded manual labour as one of the "most effective and reliable means of restoring reason". He added that aristocrats who rejected this form of therapy were merely perpetuating their condition.[42] Pinel noted that when his patients at the Bicêtre were provided with appropriate work, they immediately became calm, responsive and lucid. He insisted that all asylums for the insane should provide employment for their patients and that even the "furious" should be given some sort of physical work. Idleness exacerbated symptoms, while physical work fixed the attention and helped maintain self-control. Farming, or working a plot of land, was a particularly effective form of labour, given "the pleasure man derives from growing his own food and providing for his own needs".[43]

In a similar vein, Tuke maintained that "indolence has a natural tendency to weaken the mind and to induce ennui and discontent". Activities, such as walking, reading, conversation and physical exercise were all effective ways of diverting a patient's attention away from "their illusions". Tuke's frequently cited remark that "of all the modes by which patients may be induced to restrain themselves, regular employment is perhaps the most generally efficacious", particularly employment that involved "considerable bodily action", shows him to be in accord with Pinel.[44] Tuke emphasised the role of work in promoting self-esteem. The latter, Tuke insisted, was a far more powerful tool than fear in encouraging patients to control their behaviour.[45] It was up to the attendant to ascertain the type of work or amusements most appropriate for the patient, such as "active and exciting" activities for the melancholic and more sedentary

[41] Anne Digby, "Tuke, William (1732–1822), Philanthropists and founder of the York Retreat," *Oxford Dictionary of National Biography*, https://www.oxforddnb.com (accessed 10/04/2021).

[42] Pinel, *Traité*, 226.

[43] Ibid., 225–6.

[44] Tuke, *Description of the Retreat*, 156.

[45] Ibid., 157.

occupations for the "maniacal class". There were no hard and fast rules, but the "inclination of the patient" should guide the choice of employment unless it appeared to exacerbate his or her condition.[46] Tuke described how the condition of one patient, a gardener by profession, was greatly improved by involving him in the management of the asylum grounds, and worsened when this form of employment was no longer available.[47] For Tuke, work or amusements were to be tailored to suit the patient, based on the nature of their condition, their preferences and previous occupation, and on whether the choice of occupation proved beneficial. The successful allocation and supervision of occupation required that the attendant know the patient well and had the time to oversee the activities.

The Adoption of Moral Treatment in French and English Asylums

The basic principles of moral treatment, as set out by Tuke and Pinel, influenced how the mentally disordered were treated in asylums for the next *c.*150 years. The teaching of Pinel and Tuke informed subsequent generations of psychiatrists, including Jean-Étienne Dominique Esquirol and Guillaume Ferrus (1784–1861) in France, and Sir William Charles Ellis (1780–1839) and John Conolly (1794–1866) in England. In England, the Select Committees of 1815 and 1827, established to investigate care of the mentally disordered in institutions, recommended the adoption of moral treatment methods. As Kathleen Jones highlights, Tuke's system "was set as the ideal" and was regarded by members of the 1827 committee as key to "creating an environment in which patients could live with some personal satisfaction and dignity".[48] The report produced by the 1827 Select Committee included an Appendix that set out what asylums should provide, including manual labour, intellectual pursuits, and hobbies. The report stated that, "In the moral treatment of the patients, it is considered an object of importance to encourage their own

[46] Ibid., 181.

[47] Ibid., 152–6. Tuke's gardener-patient is discussed by Jennifer Laws, "The hollow gardener and other stories: Reason and relation in the work cure," in *Work, psychiatry and society, c.1750–*2015, ed. Ernst, 351–68.

[48] Kathleen Jones, *Asylums and After. A Revised History of the Mental Health Services: From the Early 18th Century to the 1990s* (London and Atlantic Highlands, NJ: Athlone Press, 1993), 64.

efforts of self-restraint...", thereby underlining the psychological nature of the treatment.[49] The campaign to establish the French asylum system and the drafting of the 1838 legislation regarding how asylums should be managed was led by Esquirol. He ensured that the legislation (refined but not substantially changed in 1857) was based on the principles of moral treatment and included the provision of work for patients.[50] The law stipulated that work and other occupations should be prescribed for patients by a doctor as a means of therapy.[51]

In England, the "non-restraint" movement of the 1830s and 1840s, owed much to Tuke and Pinel. The campaign to abolish mechanical restraint, pioneered by Robert Gardiner Hill (1811–78) and John Conolly, was based on the aim of replacing strait jackets and other methods of physical restraint with the introduction of elaborate patient work programmes to encourage self-restraint.[52] Gardiner Hill, who abolished restraint at the Lincoln Asylum, declared, "I wish to complete what Pinel began".[53] As a result, many English asylums became "hives of activity" in the mid-nineteenth century. Echoing Tuke, Sir William Charles Ellis, who took over stewardship of the Hanwell Asylum in 1831, shortly before the New Poor Law was passed, maintained that "nothing is found so efficacious as employment". He advocated the provision of workshops in all asylums for the poor, where "patients may perform different branches of mechanical labour to which they have previously been accustomed".[54] Failing that, patients could be taught a new skill, such as shoemaking or twine spinning.

Esquirol supported Pinel's view that every asylum should have a farm where patients could work the land and emphasised the moderating effects of manual labour on the passions. His female patients at the Salpêtrière Hospital benefited from tending the garden or engaging in domestic

[49] Appendix 1., *Report of the Select Committee on Pauper Lunatics in Middlesex and on Lunatic Asylums* (London, 1827), 11. https://archive.org/details/b30459291/page/10/mode/2up?view=theater (accessed on 17/02/2022).

[50] Goldstein, *Console and Classify*, 132, 277.

[51] *Bulletin Officiel du Ministère de l'Intérieur, Circulaire No. 7* (Paris, 1857), 42–79. See in particular: Section XXI: "Travail", 57–9, and Section XXII "Occupations intellectuelles et distractions", 59. https://gallica.bnf.fr/ark:/12148/bpt6k5539701n?rk=150215;2 (accessed on 17/02/2022).

[52] See: Kathleen Jones, "Robert Gardiner Hill and the non-restraint movement," *Canadian Journal of Psychiatry* 29, 2 (1984): 121–4; Akihito Suzuki, "The Politics and Ideology of Non-restraint: the case of the Hanwell Asylum," *Medical History* 39 (1995): 1–17.

[53] Leonard D. Smith, '*Cure, Comfort and Safe Custody*': *Public Lunatic Asylums in Early Nineteenth-Century England* (London and New York: Leicester University Press, 1999), 264.

[54] Cited in Freebody, *Role of Work*, 46–7.

chores.[55] Esquirol's protegés, whom he helped to become chief medical officers in the new provincial asylums, adopted his methods and became "missionaries" for his version of moral treatment.[56] Camille Bouchet, for example, who became chief medical officer of the asylum in Nantes, maintained that work alone could "provide a 'sustained distraction from delirious impressions and thoughts'".[57] Another physician influenced by Esquirol's methods, Gustave-François Étoc-Demazy (1806–1893), became chief medical officer of the Asile de la Sarthe in 1834, remaining there until his retirement in 1872. Étoc-Demazy had trained under Guillaume Ferrus at the Bicêtre hospital. Ferrus' commitment to patient work was demonstrated by his foundation of La Ferme Ste Anne, where patients from the Bicêtre could engage in agricultural work.[58] Throughout his long career, Étoc-Demazy maintained his belief in the psychological origins of mental disorder and remained committed to moral treatment, including the provision of work for his patients. He was not swayed by the growing prevalence of physiological interpretations of mental disorder based on heredity towards the end of his tenure.[59]

In England, the first annual report of the Littlemore (produced in 1847) suggested that great importance was attached to the occupations made available to patients. The medical superintendent, Dr William Ley noted that around half of the patients worked. In suitable weather, men worked "in the garden or in some other outdoor occupation" while patients of both sexes were employed "in the domestic work of the House".[60] Three tailors and a carpenter had been admitted for whom work in their respective fields had been found. Ley noted that it was not always possible to find work for patients that matched their previous professions. He could not, for example, find work ideally suited to the soldier, butcher, hawker, fellmonger, schoolmaster, clerk, bookbinder, wheelwright, linen draper and "more than one medical man" who had been admitted, although many patients were happy to work in the garden. For women, needlework, working in the laundry and housework seemed to

[55] Freebody, "Role of Work," 48.

[56] Goldstein, *Console and Classify*, 143.

[57] Ibid.

[58] La Ferme Ste Anne became the site upon which the Ste Anne's asylum complex was constructed in 1867.

[59] Hervé Guillemain, *Chronique de la Psychiatrie Ordinaire: Patients, Soignants et Institutions en Sarthe du 19e au 21e Siècle* (Tours: Éditions de la Reinette, 2010), 34.

[60] Medical Superintendent's Report, Littlemore Annual Reports (Oxford, 1847), 9.

bring satisfaction, particularly from a social perspective.[61] Ley observed that patients welcomed the opportunity to have "their hands and eyes occupied in what they retain a certain knowledge of". Patients took "much interest" in the yield of the crops they cultivated and "also contemplate with pleasure the participation in the fruits of their labour." Ley also highlighted his patients' enjoyment of the music and dancing that formed part of the asylum's entertainment programme.[62] Patients of both sexes attended reading classes with the Chaplain.[63]

Treatment practices at Bethlem changed dramatically after the damning evidence gathered by the 1815 Select Committee. This revealed that cramped, unsanitary conditions devoid of any comfort, widespread use of mechanical restraint, inadequate staff numbers, and a catalogue of abuses were the norm at Bethlem.[64] Committee members noted how unfavourably Bethlem compared with the Retreat at York. Bethlem's move to new premises shortly after the enquiry, and a change of personnel, resulted in the introduction of moral treatment, although it took several years before therapeutic work was provided for patients.[65] The 1830s saw the addition of eleven workshops and patients were encouraged to help with household chores. The 1843 hospital report maintained that work had proved beneficial, but that its use was limited because Bethlem did not possess extensive grounds.[66] Sport and recreation were added to Bethlem's regime following the 1815 Select Committee report; patients were able to play football, battledore, trap-ball and cricket outside while cards and dominoes were provided indoors.[67] By the 1820s, Bethlem's reputation had improved so much that physicians were being sent to Bethlem and the Retreat at York to observe best practice.[68] Not everyone agreed that this reputation was deserved. The physician Alexander Halliday observed in 1827 that, despite the adoption of some aspects of moral treatment, "there is too little space for exercise and employment for it [Bethlem] ever to prove an efficient hospital" and that there existed "too rigid a system of

[61] Ibid.
[62] Ibid., 10.
[63] Ibid.
[64] Andrews *et al.*, *History of Bethlem*, 427–9.
[65] Ibid., 447.
[66] Ibid., 448.
[67] Ibid., 449.
[68] Ibid.

quackery".[69] It was not until the arrival of superintendent William Charles Hood (1824–70) in 1852 that Bethlem's practices were brought fully in line with "modern sentiments and requirements".[70] Hood was a firm believer in moral treatment and in the benefits of work for his patients' mental and physical health. He found it "lamentable to see strong and healthy men, in the prime of life, idling away their time from morning till night, lounging listlessly about the wards, doing nothing."[71] Hood improved the recreational facilities, food and general living conditions at Bethlem and introduced excursions to places of interest.

Unlike the long-standing Bethlem, Ste Anne's was only established in 1867. Ste Anne's was the flagship institution, which incorporated the Asile Clinique, in Baron Haussmann's plans for reorganising provision for the mentally disordered in the Paris region. Despite being created some thirty years after its heyday, the asylum was designed according to the principles of moral treatment set out by psychiatrist François Leuret (1797–1851). Leuret, a former pupil of Esquirol, has been described as France's "last moral therapist".[72] He did not adhere to the new ideas that attributed mental illness to a somatic disorder and remained committed to psychological treatment. As a result, Ste Anne's had plenty of green spaces where patients could take exercise, enjoy recreation in the fresh air and work in the orchards and gardens.[73] Patients also worked in the various workshops. Intellectual pursuits were encouraged; there was a library; and musical and theatrical performances were organised.[74]

Initially, moral treatment was deemed highly effective. Its initial success led to a general sense of therapeutic optimism regarding the curability of mental disorder, regardless of the social status, age or sex of the affected individual, as long as treatment began soon after the onset of symptoms.[75] Some historians have attributed this optimism to the inaccurate early

[69] Ibid., 451.

[70] Ibid., 484.

[71] Ibid., 494.

[72] Edward M. Brown, "François Leuret: the last moral therapist," *History of Psychiatry* 29, 1 (2018): 38–48.

[73] Stéphane Henry and Pierre-Louis Laget, "L'Asile Sainte-Anne: une architecture, une organisation, une existence," in *L'Hôpital Sainte-Anne: Pionnier de la psychiatrie et des neurosciences au coeur de Paris,* eds Stéphane Henry, Catherine Lavielle and Florence Patenotte (Paris: Somogy, 2016), 23–4.

[74] Michel Caire, "Les traitements du temps de l'asile," in *L'Hôpital Sainte-Anne*, eds Henry, Lavielle and Patenotte, 62–63.

[75] Busfield, *Managing Madness*, 246.

reporting methods of asylum superintendents, keen to demonstrate their success, and others to the higher proportion of patients with favourable prognoses admitted to asylums in the first half of the nineteenth century, before institutions became overcrowded with chronic and incurable cases.[76] The efficacy of moral treatment, and thus of patient work, was barely questioned until the middle of the century.[77] By this time, a *modus operandi* had been established and asylums in both France and England were managed according to the framework provided by moral treatment. Patient work formed part of the daily asylum routine that included regular hours for waking, meals, exercise, recreation and social activities, and sleeping (Figs. 2.1 and 2.2).

Changing Views Regarding the Curability of Mental Disorder

Doubts about the efficacy of moral treatment began to creep in as the numbers of individuals detained in the asylum system, notably those with incurable or chronic conditions, kept increasing each year. Moral treatment did not seem to be as successful as it first appeared. Such doubts were compounded by changes in the way mental disorder was perceived. During the early nineteenth century it was generally accepted that individuals who had been mentally disordered since birth; had developed mental disease in old age (such as senile dementia); or had suffered a mentally debilitating physical injury or illness (such as syphilis), were unlikely to improve significantly. Other cases were believed to stand a good chance of recovery *if treated early*. This therapeutic optimism was widespread, as Joan Busfield emphasises. Asylum reform in the early nineteenth century was based on a "cult of curability", the belief that asylums could achieve a high rate of cure.[78] William Browne, medical superintendent of the Montrose Asylum, remarked in 1837 that certain physicians were claiming an ability to cure 90 in every 100 cases, "proving" that mental disorder was "the most curable of all diseases".[79] This impressive claim only applied to recent cases, Browne added, those whose conditions had existed for three months or less before they began medical treatment. For these cases,

[76] Ibid., 246, 256.
[77] Ibid., 246.
[78] Ibid.
[79] Ibid.

ASILE CLINIQUE (SAINTE-ANNE). — COUR DES ATELIERS.
Edité par le Patronage des Asiles de la Seine.

Fig. 2.1 "Les Services Techniques" or Technical Services Workshops, Asile Clinique, Paris, 1900. (© *Collection Bibliothèque Henri Ey (don Gérard Proust), GHU Paris, photographie Direction de la communication du GHU*)

the chances of recovery were high, regardless of the age, social rank or sex of the patient.[80]

Early institutional treatment was considered economically prudent since an increased likelihood of cure lessened the chance of long-term dependency. Len Smith reports that throughout the 1820s and 1830s, medical superintendents complained that parish officials tried to save money by delaying the transfer of the mentally disordered from workhouse to asylum. But waiting until cases had descended into chronicity was far more costly in the longer term since chronic patients would end up staying far longer in the asylum (possibly for life) and their families would soon became pauperised.[81] As W.J. Gilbert, an Assistant Poor Law Commissioner in Devon observed in 1839, it was ill-advised to allow the

[80] Ibid., 247.
[81] Smith, '*Cure, Comfort and Safe Custody*', 114–5.

Fig. 2.2 Attendants and patients working in the grounds, Littlemore Hospital, Oxford, 1910s. (© *Oxfordshire County Council, Oxfordshire History Centre, POX0571824*)

insane patient's condition to linger without proper treatment since the disease would soon become "inveterate and recovery hopeless".[82]

As asylums became overcrowded with incurable patients in the late 1840s, they became places of detention rather than cure. The sheer numbers of patients made it hard for doctors to treat those whose conditions might be curable, leading to doubts about the curability of mental disorder.[83] Theories claiming that mental disease was a condition of the mind lost ground. Psychiatrists began to regard mental disorder solely as an organic disease, a physiological condition of the brain and central nervous system, rather than as a problem of the psyche. This physiological interpretation of mental disorder was reinforced by the circulation in the late 1850s of theories of heredity, developed by the French psychiatrist Bénédict-Augustin Morel (1809–1873). He believed that the majority of cases of mental disorder were caused by a hereditary "defect" to the brain

[82] Busfield, *Managing Madness*, 247–8.

[83] Smith, '*Cure, Comfort and Safe Custody*', 117.

or central nervous system. Mental disorder was inherited either directly as "mental deficiency" (or intellectual impairment in today's parlance), in which case individuals were impaired from birth, or indirectly as a "predisposition" towards developing some form of mental illness in the future. If the latter, this would worsen or "degenerate" with each generation.[84]

The widespread influence of Morel's theory of heredity, both in England and France, encouraged psychiatrists to turn away from psychological modes of treatment, such as moral treatment, and towards trying to identify a biological cure. From the mid-nineteenth century, articles in the *Journal of Mental Science* and *L'Aliénation Mentale* increasingly focused on the search for physical causes of mental disorder and on the quest for successful biological remedies. Moral treatment and patient work were considered outmoded as methods likely to bring about a cure. In 1887, an author writing in *The British Medical Journal* observed that "the physical treatment of insanity is put prominently forward". He added, "Formerly we heard much about the moral treatment of insanity; nowadays it is rarely mentioned".[85] In 1900, in an article in the *Journal of Mental Science* referring to the provision of farm-work for patients, the author remarked that "It was late in the day to advocate that primitive measure".[86] In France, Pinel's son, the physician Scipion Pinel (1795–1859), attempted to divert attention away from his father's reputation as the founder of moral treatment by focusing on Pinel's alleged breaking of the chains from patients at the Bicêtre.[87] This supposedly heroic gesture, for which there is little evidence, became legendary. According to Dora Weiner, perpetuation of the myth was motivated by Scipion's embarrassment at his father's emphasis on the psychological causes and treatment of mental illness. Scipion himself was convinced that mental illness was caused by physiological factors.[88]

[84] See: Ian Dowbiggin, *Inheriting Madness: Professionalisation and Psychiatric Knowledge in Nineteenth-Century France* (Berkeley and Oxford: University of California Press, 1991).

[85] AU, "The Address in Psychological Medicine," *The British Medical Journal*, 13 August (1887): 367.

[86] AU, "The Employment of the Insane," Notes and News Section, *Journal of Mental Science* Jan (1900): 208.

[87] Weiner, "'Le geste de Pinel'," 233.

[88] Ibid.

The Decline of Moral Treatment

In line with this change from a psychological to a physiological orientation, the practices associated with early nineteenth-century moral treatment evolved.[89] Anne Digby observes that at the York Retreat, the increasing size of the institution resulted in the introduction of more formal rules and codes of behaviour. Patients began to be treated less as individuals and more as "inmates in a bureaucratic regime".[90] The framework provided by moral treatment remained, while faith in its ability to cure mental disorder had diminished. The more bureaucratic, disciplinarian methods, associated with a custodial model of care, became an effective means of managing large numbers of patients in an efficient and humane way. Patient numbers, particularly in the large public asylums, prevented the allocation of work according to individual needs, preferences, or previous professions. The provision of work, exercise and amusements for the more tractable patients did not equate to an adherence to the principles of moral treatment, as Busfield has highlighted.[91] This required individual attention from staff and a response to the needs of each patient, rather than the imposition of a uniform, regimented routine.[92] The work performed by patients represented a considerable cost-saving for the asylum, but the nature of work in the late nineteenth century was, as Dr David Henderson remarked, often "mere drudgery".[93] Over half of working patients in both French and English establishments were engaged in ward work (essentially cleaning), the potential of which to stimulate a patient's creativity, intelligence or self-esteem was limited. Far from being a "cornerstone" of moral therapy prescribed according to patient need, work had simply become part of the daily asylum routine, organised to supplement the smooth-running of the institution, to distract calm, chronic and incurable patients and to prepare convalescent patients for life outside the institution.[94] Unlike during the early

[89] Anne Digby, "The changing profile of a nineteenth-century asylum: the York Retreat," *Psychological Medicine* 14 (1984): 739–748.

[90] Ibid., 747.

[91] Busfield, *Managing Madness*, 261.

[92] Ibid.

[93] D. K. Henderson and R. D. Gillespie, *A Text-book of Psychiatry for Students and Practitioners*, Fifth edition, (London: Oxford University Press, 1940), 593.

[94] Andrew Scull, *The Most Solitary of Afflictions: Madness and Society in Britain 1700–1900* (New Haven: Yale University Press, 1993), 102.

nineteenth century, psychiatrists' "interest in the *therapeutic* value of … employment was minimal".[95] Textbooks, such as Daniel Hack Tuke's two-volume *Dictionary of Psychological Medicine* (1892), continued to recommend work and other occupations but devoted far less attention to them than to physical or biological remedies.[96] By the last quarter of the nineteenth century, the benefits of patient work to the institution appeared at least as important as its benefit to the patient (as discussed in Chap. 5).

Changes to the perceived curative value of moral treatment and patient work took place at different times in different locations, depending on the views of the medical superintendent or chief medical officer, patient numbers, the extent of overcrowding and the proportion of patients with poor prognoses. At the Littlemore, it was clear from the dearth of comments about moral treatment or the therapeutic value of occupation from the 1870s onwards that these were no longer priorities. This can be linked to Dr Ley's retirement in 1868 and to the fact that patient numbers at the Littlemore had increased from 286 in 1850 to 527 in 1870 without a proportionate increase in staffing levels. Work was only mentioned in the Committee of Visitors' report where the standard phrase, "Employment for the Patients capable of being engaged, is found the House, Garden, and Grounds of the Asylum; besides work in certain Trades at which they are from time to time enabled to work", was repeated each year.[97] Although Littlemore patients continued to be employed during the last quarter of the nineteenth century, as Table 2.1 indicates, the work did not appear to be allocated on the basis of patient preference or need. The type of work (such as cleaning and hair-picking) suggested that it was geared to the needs of the institution. As the Commissioners in Lunacy noted in 1910, the amount and type of work provided for patients depended on the "the extent to which the medical superintendent … takes a real and lively interest in these matters so essential to the care and treatment of the insane".[98] Entertainments, an essential aspect of the therapeutic arsenal in the eyes of the moral therapists, also seemed to be neglected at the Littlemore by the end of the nineteenth century. The Commissioners in Lunacy noted in 1901 that only one recreational event had been held that year, apart from

[95] Laws, "Crackpots and Basket-Cases," 69.

[96] Daniel Hack Tuke, *A Dictionary of Psychological Medicine, Vol. II* (Philadelphia: P. Blakiston, 1892), 1315.

[97] Visitors' Report, Annual Reports of the Littlemore Asylum, Oxford, 1873, 9.

[98] Sixty-fourth Annual Report of the Commissioners in Lunacy (London: HMSO, 1910), 31.

Table 2.1 Table to show the number and percentage of male and female patients who worked at the Littlemore Asylum, Oxford, 1870–1895 and 1900–1913

	1870	*1873*	*1876*	*1878*	*1881*	*1883*	*1887*	*1890*	*1892*	*1895*
Total no. female patients	291	264	281	296	258	283	279	300	296	322
No. female workers	199	183	183	184	110	113	125	121	131	149
% of women who worked	68%	69%	65%	62%	43%	40%	45%	40%	44%	46%
Total no. male patients	236	210	205	215	207	226	219	223	215	213
No. male workers	147	164	121	118	117	117	139	127	139	139
% of men who worked	62%	78%	59%	55%	57%	52%	63%	57%	65%	65%
Total Population	527	474	486	511	465	509	498	523	511	535
Total no. workers	346	347	304	302	227	230	264	248	270	288
% Working Population	66%	73%	63%	59%	49%	45%	53%	47%	53%	54%
	1900	*1901*	*1902*	*1903*		*1909*	*1910*	*1911*	*1912*	*1913*
Total no. female patients	315	312	325	320		366	380	404	407	414
No. female workers	153	164	150	160		211	251	278	284	241
% of women who worked	49%	53%	46%	50%		58%	66%	69%	70%	58%
Total no. male patients	229	241	239	244		284	257	301	295	309
No. male workers	154	165	152	150		184	192	194	199	195
% of men who worked	67%	68%	64%	61%		65%	75%	64%	68%	63%
Total Population	544	553	564	564		650	637	705	702	723
Total no. workers	307	329	302	310		395	443	472	483	436
% Working Population	56%	59%	54%	55%		61%	70%	67%	69%	60%

Source: Littlemore Hospital Annual Reports, 1870–1895 and 1900–1913. OHA LA1/1/15-35 and OHA LA1/1/45–53

the regular dances.[99] Concern was expressed in 1908 that almost half the patients were confined to the airing courts for fresh air and exercise.[100] Requests to provide more reading material to Littlemore patients "whereby their lives may be greatly brightened at a comparatively trifling cost" cropped up in 1901, 1910 and 1912, suggesting that this form of amusement was not a priority for the superintendent.[101]

At the Asile de la Sarthe, belief in the effectiveness of moral treatment *appeared* to have lasted considerably longer. The rules for staff set out in the *Règlement* of 1893 demonstrate a routine closely aligned with the principles of moral treatment, with set times for meals, work, prayer, recreation and the doctor's daily visit. Echoing the legislation regarding patient work compiled in 1857, the rules stated that, "Work is provided in the asylum as a means of treatment and distraction for patients".[102] Which patients were given work, the nature of that work and the length of time each patient was to spend on it were to be decided by the chief medical officer. The types of work, both indoor and outdoor, that patients were permitted to perform were indicated in the *Règlement*; for example, patients should not be given work that relied solely on muscular force, such as operating the pumps or carousel.[103] The working day was limited to eight hours in winter and nine hours in the summer.[104] These regulations suggest that patient work was conceived as a therapeutic activity, but it was also made clear that the product of the patients' labour belonged to the asylum.[105] "Intellectual occupations", games and physical exercises were also to be provided for patients, as directed by the chief medical officer and supervised by the nurses and attendants (Fig. 2.3).[106]

In his 1899 and 1901 annual reports, Dr Petit (chief medical officer of the Asile de la Sarthe from 1898 to 1904) highlighted patient work and entertainments as the most effective therapies. These methods were considered more important than pharmaceuticals, described as simply an "accessory" to moral treatment.[107] However, while Dr Petit may have considered

[99] Inspection Report of the Commissioners in Lunacy, Annual Reports of the Littlemore Asylum (Oxford, 1901).

[100] Ibid., 1908.

[101] Ibid., 1901, 1910 and 1912.

[102] *Règlement*, Asile de la Sarthe (Le Mans, 1893), 38.

[103] Ibid.

[104] Ibid., 39.

[105] Ibid.

[106] Ibid., 42.

[107] Rapport Médical, Asile de la Sarthe, 1899 (Le Mans, 1900).

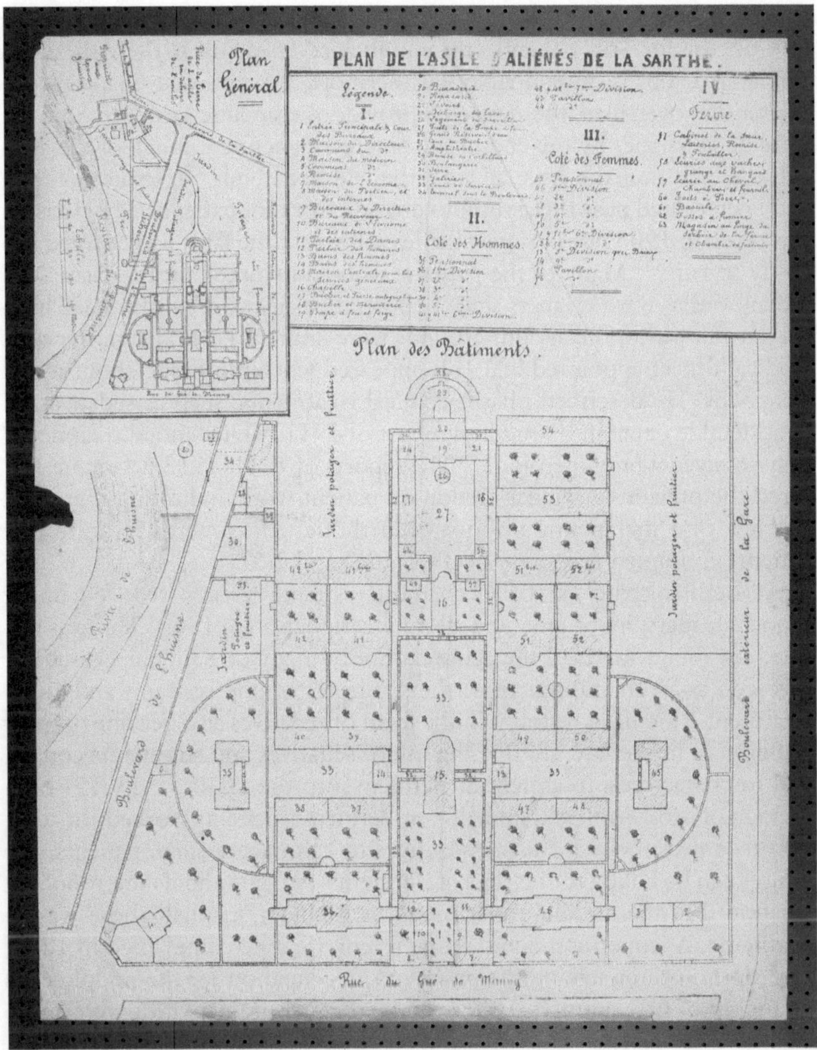

Fig. 2.3 Plan of the Asile de la Sarthe, Le Mans, 1891. The plan shows the areas allocated to market gardening ("jardin potager et frutier") and the workshops ("Maison centrale pour les services généraux"). (© *Arch. Dép. Sarthe, 4 N 158/8*)

moral treatment his preferred methodology, his ability to deliver it, as the sole doctor for 865 patients (the number of residents at the Asile de la Sarthe in 1903) is questionable. Furthermore, it seems that financial considerations overruled the desire to provide therapeutic farm work at the Asile de la Sarthe. By the end of the nineteenth century, much of the agricultural land near the Asile de la Sarthe had been built on, so a small farm had been rented near the asylum to provide work for patients. The decision was taken in 1903 not to renew the lease because the farm was deemed too costly to run.[108] Many of the patients who had worked on the farm were redeployed as ward cleaners since opportunities to work on the small market garden within the asylum grounds were limited. Dr Bourdin, the chief medical officer appointed in 1911, appeared less enthusiastic about moral treatment. He described pharmaceutical treatments, such as sedatives, in considerable detail in his medical report of 1911, while "moral treatment" only received a brief mention.[109] The reports of 1911–1913 no longer featured the previously regular section on patient work, indicating that work was less of a therapeutic priority, even though the records indicate that patients continued to work (See Tables 2.2 and 2.3).

At Bethlem, the use of mechanical restraint, greatly reduced under Hood's leadership, re-emerged after his resignation in 1862. This was justified on the grounds of the acute nature of many of Bethlem's cases but was seen as an abandonment of the principles of non-restraint and moral treatment. Chemical restraint in the form of sedatives also became popular at Bethlem in the late 1870s. The use of sedatives and stimulants continued to characterise treatment at Bethlem into the 1930s.[110] By 1914, an average of 40%—or less than half—of Bethlem's patients worked. In addition to a preference for using sedatives to calm acute-stage patients, the figures can be linked to the class of Bethlem's patients and its metropolitan location. For the middle-class patients at Bethlem, manual labour was an anathema. As superintendent George Henry Savage complained in 1882, "we are no nearer solving the problem of occupation for the middle-class insane".[111] Entertainments, sport and other leisure activities were more

[108] Rapport du Directeur sur la Ferme d'Angevinière, Compte Administratif, Asile de la Sarthe, 1902 (Le Mans, 1903).

[109] Rapport Médical, Asile de la Sarthe, 1911 (Le Mans, 1912).

[110] Andrews *et al.*, *History of Bethlem*, 522–3.

[111] Cited in Sarah Chaney, "Useful members of society or motiveless malingerers? Occupation and malingering in British asylum psychiatry, 1870–1914," in *Work, Psychiatry and Society, c.1750–2015*, ed. Waltraud Ernst (Manchester: Manchester University Press, 2016), 278.

Table 2.2 Table to show the number and percentage of male and female pauper patients who worked at the Asile de la Sarthe, Le Mans, 1900–1913

	1900	*1901*	*1902*	*1903*	*1904*	*1905*	*1913*
Total no. female patients	310	319	331	325	325	327	352
No. female workers	220	217	247	249	223	221	218
% of women who worked	71%	68%	75%	77%	69%	68%	62%
Total no. male patients	262	254	257	262	267	269	277
No. male workers	136	139	130	136	143	138	155
% of men who worked	52%	55%	51%	52%	53%	51%	56%
Total Population	572	573	588	587	592	596	629
Total no. workers	356	356	377	385	366	359	373
% Working Population	62%	62%	64%	66%	62%	60%	59%

Source: Reports of the Chief Medical Officer, Asile de la Sarthe, 1900–1913. ADS-1X961

Table 2.3 Table to show the number and percentage of male and female paying patients who worked at the Asile de la Sarthe, Le Mans, 1900–1913

	1900	*1901*	*1902*	*1903*	*1904*	*1905*	*1913*
Total no. female patients	121	118	124	125	118	127	141
No. female workers	15	16	19	12	15	12	14
% of women who worked	12%	14%	15%	10%	13%	9%	10%
Total no. male patients	48	48	48	53	64	69	63
No. male workers	17	17	16	16	18	16	21
% of men who worked	35%	35%	33%	30%	28%	23%	33%
Total Population	169	166	172	178	182	196	204
Total no. workers	32	33	35	28	33	28	35
% Working Population	19%	20%	20%	16%	18%	14%	17%

Source: Reports of the Chief Medical Officer, Asile de la Sarthe, 1900–1913. ADS-1X961

appropriate occupations than work for this class of patient. As Oswald has argued, psychiatrists recognised that, to be effective, occupations had to be suited to the class of patient, as well as their condition.[112] In the private section of the Asile de la Sarthe, just 17% of paying patients worked during the first decade of the twentieth century (Table 2.4).

Like Bethlem, the Asile Clinique was located in a city centre where the availability of land for farm work was limited. Other forms of work were available but were only being prescribed to a relatively small proportion of patients. In 1907, General Councillor Henri Rousselle felt compelled to

[112] Oswald, "Distraction from Hurtful Thoughts," 36.

Table 2.4 Table to show the number and percentage of patients who worked at the Bethlem Royal Hospital, London, 1900–1913 (no gender breakdown available)

	1900	*1901*	*1902*	*1906*	*1907*	*1908*	*1909*	*1910*	*1911*	*1912*	*1913*
Total Population	198	219	224	240	229	232	233	191	244	185	202
Total No. workers	120	96	87	132	92	98	117	99	80	70	75
% Working Population	61%	44%	39%	55%	40%	42%	50%	52%	33%	38%	37%

Source: Bethlem Royal Hospital Annual Reports, 1900–1913. BMM-BAR-53

remind the medical staff that it was "perfectly legitimate" to expect patients to work "to lessen the enormous cost of their care".[113] He argued that a balance needed to be reached enabling doctors to reconcile "the needs of humanity with those of an efficient administration".[114] The Asile Clinique doctors' apparent unwillingness to prescribe work for their patients may have been linked to a preference for alternative methods of treatment, such as continuous bedrest, the method introduced in 1896 by Valentin Magnan, the revered psychiatrist in charge of the Admissions department at Ste Anne's from its opening in 1867 until 1916.[115] In the decade before the outbreak of World War I in 1914, an average of just 29% of Asile Clinique patients worked. This contrasts with over 60% of pauper patients at the Littlemore and the public section of the Asile de la Sarthe, where there was a high proportion of chronic and incurable cases (Table 2.5).[116]

Psychiatrists at the Seine's Villejuif Asylum, who included the reformers Édouard Toulouse (1865–1947) and Auguste Marie (1865–1934), were greatly in favour of patient work as a means of therapy. They disagreed with colleagues who claimed that there was "no scientific evidence" for its efficacy and who maintained that "the best exercise was rest". [117] Marie emphasised in his report of 1905 that "the main purpose of work in asylums is the well-being of the patients; its usefulness as a means of

[113] Henri Rousselle, Rapport au nom de la 3e Commission sur les budgets et comptes de l'Asile Clinique (Conseil Général de la Seine: Paris, 1911), 5. ADP/D.10K3/21/42.

[114] Ibid.

[115] Caire, "Les traitements," 63.

[116] Figures for the numbers of working patients for all four asylums were gleaned from their annual reports 1900–13.

[117] Auguste Marie, Rapport de la division des hommes, 2e section, l'Asile Villejuif (Paris: Préfecture de la Seine, 1905), 218–9. ADP/PER-566-4.

Table 2.5 Table to show the number and percentage of male and female patients who worked at the Asile Clinique, Paris, 1900–1913

	1903	*1904*	*1905*	*1906*	*1907*	*1908*	*1909*	*1910*	*1911*	*1912*	*1913*
Total no. female patients	528	528	527	529	534	521	516	523	523	514	517
No. female workers	110	108	110	110	105	103	98	101	105	101	98
% of women who worked	21%	20%	21%	21%	20%	20%	19%	19%	20%	20%	19%
Total no. male patients	565	570	574	568	570	570	569	575	569	594	596
No. male workers	205	217	209	204	205	209	218	214	224	233	245
% of men who worked	36%	38%	36%	36%	36%	37%	38%	37%	39%	39%	41%
Total Population	1093	1097	1101	1097	1104	1091	1085	1098	1092	1108	1113
Total no. workers	315	325	319	314	310	312	316	315	329	334	343
% Working Population	29%	30%	29%	29%	28%	29%	29%	29%	30%	30%	31%

Source: Reports of the Chief Medical Officer to the Prefecture, 1900–1913. ADP/PER-566-4

production should be entirely subsidiary."[118] There needed to be a clear distinction between *therapeutic work* and *work-for-profit*. Unfortunately, this principle was all too often misunderstood. Marie warned that the "profit motive" could cloud the judgement of asylum physicians and administrators and push the goal of medical treatment into second place.[119] All staff, including the director, the bursar and their employees, needed to understand that the work carried out by asylum patients was therapeutic, not a means of making or saving money.[120] French asylums, Marie maintained, were often regarded by the public, the administrative authorities and by the patients themselves as "work colonies", rather than as hospitals where people were treated for mental illness.[121] In the colonies, patients were expected to work to contribute to the costs of their care whilst

[118] Ibid.
[119] Ibid., 217.
[120] Ibid.
[121] Ibid., 216.

Table 2.6 Table to show the average number of male and female patient workers per institution, 1900–1913

	Male Patients	*Female Patients*	*All Patients*
Littlemore	66%	58%	61%
Bethlem Royal	–	–	45%
Asile de la Sarthe—pauper	53%	70%	62%
Asile de la Sarthe—paying	31%	12%	18%
Asile Clinique	30%	20%	29%

Source: institutional reports, above

benefiting from a healthy activity in the fresh air.[122] In an asylum, on the other hand, doctors needed to consider the patients' preferences and aptitudes, and assign work that was relaxing and agreeable.[123] Toulouse agreed, insisting in 1905 and again in 1913, that the occupations and amusements provided for patients should be increased in frequency and variety, as these activities were essential to recovery (Table 2.6).[124]

Separation of Curable and Incurable Patients

Reformers argued that separating curable and incurable patients would lead to greater clarity in terms of establishing the rationale for prescribing patient work. It would also facilitate more effective treatment of curable cases, and more efficient allocation of limited financial resources. Calls for the separate treatment of curable and incurable cases had begun soon after the establishment of the asylum system. Psychiatrists recognised that treating curable cases separately would allow doctors to focus their attention on those most likely to make a recovery. An accumulation of chronic and incurable patients, who were likely to have to spend the rest of their lives in the asylum, would inevitably lead to overcrowding. In England, this situation had been foreseen by Thomas Bakewell who launched a twenty-year campaign for the establishment of an alternative system of state-run hospitals for curable patients.[125] Bakewell, proprietor of the Spring Vale private asylum in Staffordshire, maintained as early as 1814 that it was

[122] Ibid.

[123] Ibid., 222–4.

[124] Édouard Toulouse, Rapport de la division des femmes, 1ere section, l'Asile Villejuif (Paris: Préfecture de la Seine, 1913), 184. ADP/PER-566-4.

[125] Leonard D. Smith, "Close confinement in a mighty prison: Thomas Bakewell and his campaign against public asylums, 1810–1830," *History of Psychiatry* 5, 18 (1994): 191–214.

essential to separate the recent, curable cases from the chronic and incurable. A large public asylum, which was obliged to accept all types of patient, was "a great deal more calculated to prevent recovery than to promote it".[126] The campaign to treat curable and incurable patients separately was stimulated towards the end of the nineteenth century by Morel's theory of heredity and concerns regarding degeneration.

Anxiety about degeneration infiltrated all areas of society in both France and England. In England, many of the "social ills" associated with degeneration were attributed to a section of the "mentally deficient" population identified as the "feeble-minded". Segregation of this part of the population, as argued by Mary Dendy in her 1899 pamphlet, *The Importance of Permanence in the Care of the Feeble-Minded*, would be beneficial both for society and the feeble-minded individuals themselves.[127] Segregation would protect the public from the crimes supposedly committed by the feeble-minded; relieve overcrowding in asylums, workhouses and gaols; and prevent the feeble-minded from transmitting their condition to future generations. Ultimately, the measure would save money at the same time as protecting the feeble-minded from themselves and the rest of society.[128] A Royal Commission on the Care of the Feeble-Minded was appointed in 1904 and reported in 1908. "Mental defectives", defined as "idiots", "imbeciles", the "feeble-minded", and "moral imbeciles", were regarded as a "totally distinct and pathological group" whose conditions were "congenital" in character.[129] The Mental Deficiency Act of 1913, which embodied the Commission's recommendations, set in motion the establishment of "mental deficiency colonies" as a means of segregating the "mental defectives" from the rest of the population into custom-built, but inexpensive specialist institutions located in sparsely populated areas.[130] Eugenicists such as psychiatrist Alfred Tredgold, maintained that the segregation of these "mentally defective" (or intellectually impaired) individuals would prevent them from introducing "tainted strains" into the

[126] Smith, '*Cure, Comfort and Safe Custody*', 41.

[127] Mark Jackson, "Institutional Provision for the Feeble-Minded in Edwardian England," in *From Idiocy to Mental Deficiency: Historical Perspectives on People with Learning Difficulties*, eds David Wright and Anne Digby (London and New York: Routledge, 1996), 161.

[128] Ibid.

[129] The Sixty-eighth Report of the Commissioners in Lunacy to the Lord Chancellor (HMSO: London, 1914), 2.

[130] Jones, *Asylums and After*, 123.

population.[131] Segregation would also facilitate the separate care of the intellectually impaired (in "colonies") and the mentally ill (in asylums or mental hospitals).[132] Colonies for the intellectually impaired were to focus on care and self-sufficiency and were to provide a "simple and wholesome life" for inmates. [133] Adults who were capable of work would be provided with agricultural or simple industrial work, as they were at the David Lewis (established in 1904) and the Chalfont (1894) epileptic colonies, where "the labour of inmates produce[d] a considerable profit" and reduced the costs of their maintenance.[134] There was little time for the Mental Deficiency Act to take effect before the outbreak of the war in 1914, and little money after the conflict for new institutions to be created, but the principle of caring for curable and incurable patients separately had been established. At the same time, the principle of patient work as a means of reducing maintenance costs for incurable—as opposed to curable—patients was also established.

In France, calls for the separation of acute, curable cases and chronic, incurable cases into different institutions intensified in the 1890s amid a torrent of anti-alienist literature criticising asylum conditions and poor prognoses, as well as widespread concerns about national decline and degeneration.[135] Popular fears about the "quantity and quality" of the French population were expressed in Max Nordau's book, *Degeneration* (1892), while right-wing journalist Maurice Barrès warned in 1908 of the "moral feebleness" and "weakening of the will" that were early signs of

[131] Jackson, "Institutional Provision for the Feeble-Minded," 170. Dr Tredgold, incidentally, was part of the medical team at the Bethlem Outpatients Department (1918–1927).

[132] Mathew Thomson, "Disability, Psychiatry and Eugenics," in *The Oxford Handbook of the History of Eugenics*, eds Alison Bashford and Philippa Levine (Oxford: Oxford University Press, 2010), 118.

[133] John Welshman, "Ideology, Ideas and Care in the Community, 1948–1971," in *Community Care in Perspective: Care, Control and Citizenship*, eds John Welshman and Jan Walmsley (Basingstoke: Palgrave Macmillan, 2006), 19.

[134] Report of the Royal Commission on the Care and Control of the Feebleminded (London: HMSO, 1908), 345.

[135] Elizabeth Nelson, "Running in Circles: A Return to an Old Idea about Asylum Reform in Nineteenth-Century France," *Proceedings of the Western Society for French History* 42 (2014): 121–2.

mental illness and degeneration.[136] France's low birth-rate, the lowest in Europe, was also a source of anxiety for which degeneration was believed to be responsible. The population of France grew by just 11% between 1851 and 1901, while that of England and Wales increased by 81% in the same fifty-year period.[137] Between 1901 and 1911 the population of England and Wales increased by a further 11%, but that of France by only 1.8%.[138] The French, according to Robert Nye, became renowned for the "sheer obsessiveness" with which they pursued the problems of social deviance, degeneration and national decline.[139]

These fears were exacerbated by the rising number of cases of insanity in France, as noted by General Councillor Navarre in 1908. Asylums in the Seine were overcrowded, despite the existence of colonies for incurable patients just outside the capital. Based on similar principles as the epileptic colony in Britain, institutions for the Seine department's incurable patients had been established in the rural areas of Dun-sur-Auron in 1892 and at Ainay-le-Château in 1897 by Auguste Marie.[140] However, these institutions could not accommodate all the Seine's incurable patients. Increasing numbers of incurable and chronic patients were therefore remaining in the Seine asylums, where the costs of care were higher.[141] The policy of transferring some incurable patients from Paris to the provinces was being compromised by overcrowding in provincial asylums, which could no longer accept as many patients from the capital.[142] Maintaining incurable patients

[136] Robert Nye, *Crime, Madness and Politics in Modern France: The Medical Concept of National Decline* (Princeton, NJ: Princeton University Press, 1984), 330, 8. See also: William Schneider, "The Eugenics Movement in France, 1890–1940," in *The Wellborn Science: Eugenics in Germany, France, Brazil and Russia*, ed. Mark B. Adams (Oxford and New York: Oxford University Press, 1990), 69–109.

[137] 1911 Census of England and Wales, Vision of Britain, http://www.visionofbritain.org.uk/census/table/EW1911POP2_M5 (accessed 25/5/2018).

[138] Roger Price, *A Social History of Nineteenth-Century France* (London: Hutchinson, 1987), 45.

[139] Nye, *Crime, Madness and Politics*, 330.

[140] Michel Caire, "Armand Victor Auguste Marie," in *Histoire de la Psychiatrie en France* (www.psychiatrie.histoire.free/fr/pers/bio/htm, 2014). (Accessed 03/03/2019).

See also: Juliet Rigondet, *Un Village Pour Aliénés Tranquilles* (Paris: Fayard, 2019); E. Drouin and P. Hautecoeur, "Auguste Marie: un grand psychiatre, créateur des Colonies familiales," *Annales Médico-psychologiques* 180 no. 8 (2022): 832-4.

[141] M. Navarre, Rapport au nom de la 3e Commission sur les propositions budgétaires pour le service des Aliénés (Paris: Conseil Général de la Seine, 1908), 2. ADP/D.10K3/18/35.

[142] In 1899 44% of patients passing through the Ste Anne's Admissions Service in Paris were sent to the provincial asylums to relieve overcrowding in metropolitan asylums. By 1913, this percentage had decreased to 33%.

in asylums did not make economic sense, according to General Councillor Henri Rousselle. He referred to the incurable insane as *des non-valeurs sociales*; they were never going to be productive citizens, and yet the costs of their care (in the expensive asylums) were roughly equal to the amount earned by a labourer, who worked hard and contributed to society.[143] Rousselle sought to divide the Seine's asylums into two groups, one reserved for acute and curable cases, and the other exclusively for chronic and incurable patients.[144] The first group would justify a higher level of investment because patients being treated here would, once recovered, be able to re-join the labour market as productive citizens. Rousselle's plans would eventually (in 1927, a year after his death) be realised in the transformation of the Asile Clinique into an acute hospital. Édouard Toulouse, one of the Parisian psychiatrists committed to the separation of cases, was also convinced of the curability of mental disorder provided asylum doctors had the opportunity, training and facilities to actively treat those most likely to benefit. His ambitions to establish a hospital for incipient cases of mental disorder would also have to wait until after World War I. Both of these plans would have a significant impact on patient work.

Conclusion

Therapeutic work for mentally ill patients emerged in the context of moral treatment at the end of the eighteenth century. The early moral therapists perceived mental illness in psychological terms and used work and occupation as a means of helping patients to control their symptoms and to adopt behaviours that equated with contemporary social norms. The principles of moral treatment provided the administrative and medical framework for the newly established asylum systems of France and England, so that even when the efficacy of moral treatment was thrown into doubt, work and recreational activities continued to be organised for patients. They became accepted aspects of asylum regimes and in France, their provision was enshrined in law. That said, the nature of occupation underwent subtle changes in emphasis as asylums gradually evolved from places of treatment and cure to custodial institutions, and as the therapeutic optimism associated with moral treatment descended into the pessimism of hereditary

[143] Henri Rousselle, Rapport au nom de la 3e Commission sur les budgets et comptes de l'Asile Clinique (Paris: Conseil Général de la Seine, 1911), 4. ADP/D.10K3/21/42.

[144] Ibid.

degeneration. The benefit of work as a method of disciplining large numbers of patients and of offsetting institutional running costs was given equal or greater priority than its role as a means of therapy. While patient work featured prominently in the early asylum annual reports, it had become almost a footnote in reports at the end of the nineteenth century, with little evidence of work being found to suit patients' aptitudes or previous occupations. Most late-nineteenth and early twentieth-century psychiatrists supported theories of hereditary degeneration and did not believe that mental illness could be cured. There were some, however, who maintained faith in its curability and campaigned for institutional reform, such as the separation of curable and incurable cases. Only by separating cases could the curable receive the medical attention they required to make a recovery. The need for reform became more pressing after World War I and the war itself led to a reappraisal of the causes of and treatment for mental illness.

Bibliography

Secondary Sources

Andrews, Jonathan, Asa Briggs, Roy Porter, Penny Tucker, and Keir Waddington. *The History of Bethlem*. Abingdon and New York: Routledge, 1997.

Anthony, P. D. *The Ideology of Work*. London and New York: Tavistock Publications, 1977.

Bartlett, Peter. "The Asylum and the Poor Law: The Productive Alliance." In *Insanity, Institutions and Society, 1800–1914: A social history of madness in comparative perspective*, edited by Joseph Melling and Bill Forsythe, 48–67. London and New York: Routledge, 1999.

Bianchini, Victor. "Production and Education according to James Mill: The Precious Middle Point." *Journal of the History of Economic Thought*, 38 no.2 (2016): 153–173.

Brown, Edward M. "François Leuret: the last moral therapist." *History of Psychiatry*, 29 no. 1 (2018): 38–48.

Busfield, Joan. *Managing Madness: Changing Ideas and Practice*. London and Melbourne: Hutchinson, 1986.

Bynum, William F. "Rationales for therapy in British Psychiatry, 1780–1835." *Medical History*, 18 no. 4 (1974): 317–334.

Caire, Michel. "Armand Victor Auguste Marie." *Histoire de la Psychiatrie en France*, 2014. Accessed 03/03/2019. www.psychiatrie.histoire.free/fr/pers/bio/htm.

Caire, Michel. "Les traitements du temps de l'asile." In *L'Hôpital Sainte-Anne: Pionnier de la psychiatrie et des neurosciences au coeur de Paris* edited by Stéphane Henry, Catherine Lavielle and Florence Patenotte, 62–65. Paris: Somogy, 2016.

Chaney, Sarah. "Useful members of society, or motiveless malingerers? Occupation and malingering in British asylum psychiatry, 1870–1914." In *Work, Psychiatry and Society, c.1750–2015* edited by Waltraud Ernst, 277–297. Manchester: Manchester University Press, 2016.

Charland, Louis. "Benevolent theory: Moral treatment at the York Retreat." *History of Psychiatry*, 18 no. 1 (2007): 61–80.

Curth, Louise Hill. "Lessons from the Past: Preventative Medicine in Early Modern England." *Medical Humanities*, 29 no. 1 (2003): 16–20.

Digby, Anne. "The changing profile of a nineteenth-century asylum: the York Retreat." *Psychological Medicine*, 14 (1984): 739–748.

Digby, Anne. "Tuke, William (1732–1822), philanthropist and founder of the York Retreat." *Oxford Dictionary of National Biography*, 23 September 2004. Accessed 10/04/2021. https://www.oxforddnb.com/view/10.1093/ref:odnb/9780198614128.001.0001/odnb-9780198614128-e-27810.

Dowbiggin, Ian. *Inheriting Madness: Professionalisation and Psychiatric Knowledge in Nineteenth-Century France.* Berkeley and Oxford: University of California Press, 1991.

Faccarello, Gilbert, ed. *Studies in the History of French Political Economy.* London and New York: Routledge, 1998.

Flechard, Cécile. *Ainay le Château: de la colonie familiale au centre psychothérapique.* Thèse. Paris: Faculté de médecine, Cochin Port-Royal, 1977.

Foucault, Michel. *Madness and Civilisation: A History of Insanity in the Age of Reason.* Translated by Richard Howard. New York: Random House, 1965.

Freebody, Jane. "The role of work in late eighteenth- and early nineteenth-century treatises on moral therapy in France, Tuscany and Britain." In *Work, Psychiatry and Society, c.1750–2015*, edited by Waltraud Ernst, 31–54. Manchester: Manchester University Press, 2016.

Goldstein, Jan. *Console and Classify: The French Psychiatric Profession in the Nineteenth Century.* Chicago: University of Chicago Press, 2001.

Guillemain, Hervé. *Chronique de la Psychiatrie Ordinaire: Patients, Soignants et Institutions en Sarthe du 19e au 21e Siècle.* Tours: Éditions de la Reinette, 2010.

Henry, Stéphane and Pierre-Louis Laget. "L'Asile Sainte-Anne: une architecture, une organisation, une existence." In *L'Hôpital Sainte-Anne: Pionnier de la psychiatrie et des neurosciences au coeur de Paris* edited by Stéphane Henry, Catherine Lavielle and Florence Patenotte, 22–30. Paris: Somogy, 2016.

Hickman, Clare. "The Picturesque at Brislington House: The Role of Landscape in Relation to the Treatment of Mental Illness in the Early Nineteenth-Century Asylum." *Garden History*, 33 no. 1 (2005): 47–60.

Jackson, Mark. "Institutional Provision for the Feeble-Minded in Edwardian England." In *From Idiocy to Mental Deficiency: Historical Perspectives on People with Learning Difficulties*, edited by David Wright and Anne Digby, 161–183. London and New York: Routledge, 1996.

Jones, K. *Asylums and After: A Revised History of the Mental Health Services: From the Early 18th Century to the 1990s*. London and Atlantic Highlands, NJ: Athlone Press, 1993.

Jones, K. "Robert Gardiner Hill and the non-restraint movement." *Canadian Journal of Psychiatry*, 29 no. 2 (1984): 121–124.

Koepp, Cynthia. Review of *Learning to Calculate* by Carol Blum, Yves Citton, Gilbert Faccarello and Gérard Klotz. *Eighteenth-Century Studies*, 37 no. 1 (2003): 141–146.

Laws, Jennifer. "Crackpots and Basket-Cases: A History of Therapeutic Work and Occupation." *History of the Human Sciences*, 24 no. 2 (2011): 65–81.

Laws, Jennifer. "The hollow gardener and other stories: Reason and relation in the work cure." In *Work, Psychiatry and Society, c.1750–2015*, edited by Waltraud Ernst, 351–368. Manchester: Manchester University Press, 2016.

Magraw, Robert. *France 1800–1914: A Social History*. London and New York: Longman, 2002.

Mokyr, Joel. *The Enlightened Economy: Britain and the Industrial Revolution, 1700–1850*. London: Penguin Books, 2009.

Nelson, Elizabeth. "Running in Circles: A Return to an Old Idea about Asylum Reform in Nineteenth-Century France." *Proceedings of the Western Society for French History*, 42 (2014): 115–125.

Nye, Robert. *Crime, Madness and Politics in Modern France: The Medical Concept of National Decline*. Princeton, NJ: Princeton University Press, 1984.

Oswald, Ute. "'Distraction from Hurtful Thoughts': Recreational Activities as Agents of Healing in Nineteenth-Century British Asylums." *Medizinhistorisches Journal* 56 no. 1–2 (2021): 30–57.

Porter, Roy. "Was there a moral therapy in eighteenth-century psychiatry?" *Lychnos* (1981): 12–26.

Porter, Roy. *Mind Forg'd Manacles—A History of Madness in England from the Restoration to the Regency*. London: Penguin Books, 1987.

Price, Roger. *A Social History of Nineteenth-Century France*. London: Hutchinson, 1987.

Rigondet, Juliet. *Un Village Pour Aliénés Tranquilles*. Paris: Fayard, 2019.

Schneider, William. "The Eugenics Movement in France, 1890–1940." In *The Wellborn Science: Eugenics in Germany, France, Brazil and Russia*, edited by Mark B. Adams, 69–109. Oxford and New York: Oxford University Press, 1990.

Scull, Andrew. *The Most Solitary of Afflictions: Madness and Society in Britain 1700–1900*. New Haven: Yale University Press, 1993.

Smith, Adam. *An Inquiry into the Nature and Causes of the Wealth of Nations* [1776]. Ware, Herts: Wordsworth Editions, 2012.
Smith, Leonard D. *'Cure, Comfort and Safe Custody': Public Lunatic Asylums in Early Nineteenth-Century England*. London and New York: Leicester University Press, 1999.
Smith, Leonard D. "Close confinement in a mighty prison: Thomas Bakewell and his campaign against public asylums, 1810–1830." *History of Psychiatry*, 5 no. 18 (1994): 191–214.
Suzuki, Akihito. "The Politics and Ideology of Non-restraint: the case of the Hanwell Asylum." *Medical History*, 39 (1995): 1–17.
Thompson, E.P. "Time, Work-Discipline, and Industrial Capitalism." *Past and Present*, 38 no. 1 (1967): 56–97.
Thomson, Mathew. "Disability, Psychiatry and Eugenics." In *The Oxford Handbook of the History of Eugenics,* edited by Alison Bashford and Philippa Levine, 116–133. Oxford: Oxford University Press, 2010.
Tombs, Robert. *France 1814–1914*. Harlow: Pearson, 1996.
Voth, Hans-Joachim. "Time and Work in Eighteenth-Century London." *Journal of Economic History*, 58 no. 1 (1998): 29–58.
Weiner, Dora B. "'Le geste de Pinel': The History of a Psychiatric Myth." In *Discovering the History of Psychiatry*, edited by Mark Micale and Roy Porter, 232–247. New York and Oxford: Oxford University Press, 1994.
Weiner, Dora B. "The madman in the light of reason. Enlightenment psychiatry, Part I." In *History of Psychiatry and Medical Psychology* edited by Edwin R. Wallace and John Gach, 281–303. Boston: Springer, 2008.
Welshman, John. "Ideology, Ideas and Care in the Community, 1948–1971." In *Community Care in Perspective: Care, Control and Citizenship*, edited by John Welshman and Jan Walmsley, 17–37. Basingstoke: Palgrave Macmillan, 2006.

Archival Sources

1911 Census of England and Wales. Vision of Britain. Accessed 25/05/2018. http://www.visionofbritain.org.uk/census/table/EW1911POP2_M5.
Sixty-fourth Annual Report of the Commissioners in Lunacy. London: HMSO, 1910.
Sixty-eighth Annual Report of the Commissioners in Lunacy. London: HMSO, 1914.
Appendix 1. *Report from the Select Committee on Pauper Lunatics in the County of Middlesex and on Lunatic Asylums*, 9–12. HMSO: London, 1827. Accessed 17/02/2022. https://archive.org/details/b30459291/page/12/mode/2up?view=theater&q=Appendix+1.
Author Unknown. "The Address in Psychological Medicine." *The British Medical Journal*, 2 no. 1389 (1887): 367.

Bulletin Officiel du Ministère de l'Intérieur. Circulaire No. 7 du 20 mars 1857, 42–79. Paris, 1857. Accessed 12/02/2018. https://gallica.bnf.fr/ark:/12148/bpt6k5539701n/f2.item.zoom.

Burton, Robert. *The Anatomy of Melancholy* [1621]. London: Penguin Classics, 2021.

Diderot, Denis. "Encyclopédie." In *The Oxford Book of Work* edited by Keith Thomas, 122. Oxford: Oxford University Press, 1999.

Hack Tuke, Daniel. *A Dictionary of Psychological Medicine, Vol. II.* Philadelphia: P. Blakiston, 1892.

Henderson, D.K. and R.D. Gillespie. *A Text-book of Psychiatry for Students and Practitioners*, 5th edition. London: Oxford University Press, 1940.

Locke, John. "Labour," In *The Oxford Book of Work* edited by Keith Thomas, 125. Oxford: Oxford University Press, 1999.

Marie, Auguste. Rapport de la Division des Hommes, 2e Section, l'Asile Villejuif. Préfecture de la Seine: Paris, 1905. ADP/PER-566-4.

Navare, M. Rapport au nom de la 3e Commission sur les propositions budgétaires pour le service des Aliénés. Paris: Conseil Général de la Seine, 1908. ADP/D.10K3/18/35.

Pinel, Philippe. *Traité medico-philosophique sur l'aliéntation mentale, ou la manie.* Paris: Chez Richard, Caille et Ravier, 1800.

Rapport Médical. Comptes Administratives, 1899. Asile de la Sarthe, Le Mans, 1900. ADS-1X961.

Rapport Médical. Comptes Administratives, 1911. Asile de la Sarthe, Le Mans, 1912. ADS-1X963.

Règlement. Asile de la Sarthe, Le Mans, 1893. ADS-1X961.

Report of the Medical Superintendent. Annual Reports of the Littlemore Asylum, Oxford, 1847. OHA-L1/A1/1.

Report of the Royal Commission on the Care and Control of the Feebleminded. London: HMSO, 1908.

Rousselle, Henri. Rapport au nom de la 3e Commission sur les budgets et comptes de l'Asile Clinique. Conseil Général de la Seine: Paris, 1901. ADP/D.10K3/14/52.

Rousselle, Henri. Rapport au nom de la 3e Commission sur les budgets et comptes de l'Asile Clinique. Conseil Général de la Seine: Paris, 1911. ADP/D.10K3/21/42.

Tuke, Samuel. *Description of the Retreat: An Institution near York for Insane Persons of the Society of Friends* [1813]. Bibliolife: LCC.

Vistors' Report. Annual Reports of the Littlemore Asylum, 1873. OHA-L1/A1/27.

Voltaire. "Candide." In *The Oxford Book of Work* edited by Keith Thomas, 160. Oxford: Oxford University Press, 1999.

Weber, Max. *The Protestant Ethic and the Spirit of Capitalism.* New York: Scribner, 1930.

For the References

Section XXI: "Travail." Bulletin Officiel du Ministère de l'Intérieur, Circulaire No. 7, 57–59. Paris, 1857. Accessed 17/02/2022. https://gallica.bnf.fr/ark:/12148/bpt6k5539701n?rk=150215;2.

Section XXII: "Occupations intellectuelles et distractions." Bulletin Officiel du Ministère de l'Intérieur, Circulaire No. 7, 59. Paris, 1857. Accessed 17/02/2022. https://gallica.bnf.fr/ark:/12148/bpt6k5539701n?rk=150215;2.

Add In

Rapport du Directeur sur la Ferme d'Angevinière, Compte Administratif, Asile de la Sarthe, 1902 (Le Mans, 1903). ADS-1X961.

CHAPTER 3

From Alienism to Psychiatry

Throughout the nineteenth century, the origins of, and the treatment for, mental disorder were perceived similarly by psychiatrists in both France and England. In both countries, moral treatment had been embraced as the answer to curing the scourge of mental disease. Later, disenchantment with moral treatment and a move towards a physiological interpretation of mental disorder, based on theories of heredity, occurred in both countries at roughly the same time. Attitudes towards patient work and occupation followed similar paths, in line with these changing views. But after World War I, psychiatry in France and England appeared to diverge. This chapter explores the nature and causes of this divergence and its effect on how French and English patients were occupied during the interwar period. It is argued that different models of care evolved in each country after the war. Divergent pathways to professionalisation taken by French and English psychiatry and the different emphases placed on the psychological and physiological causes of mental disorder led to the persistence of an alienist, or custodial, model in one country and to the evolution of a psychiatric model in another.

J. Freebody, *Work and Occupation in French and English Mental Hospitals, c.1918–1939*, Mental Health in Historical Perspective,
https://doi.org/10.1007/978-3-031-13105-9_3

The Professionalisation of Psychiatry

In England, psychiatry and neurology were separate disciplines. Neurology had a centre of excellence in London at the National Hospital, Queen Square, which opened in 1860.[1] As psychiatrist David Henderson put it, neurologists dealt with "nerves" and knew "nothing about the disorders of the mind".[2] In his view, their rigid and objective training led them to think solely in terms of structure and pathology. Neurologists were able to rule out the existence of an organic lesion, but that was the limit of their usefulness to psychiatry.[3] British neurologists did not tend to become involved in asylum medicine and, unlike in France, neuropsychiatry did not develop as a combined major specialty.[4] While neither specialist training nor a qualification in mental medicine were essential requirements for medical superintendents in English mental hospitals (although they had to be qualified doctors), the Medico-Psychological Association (MPA) sought to introduce professional standards with its Certificate in Psychological Medicine, established in 1885. It was never very popular, and was based on vocational rather than academic knowledge, but by 1896 some 240 doctors held the certificate.[5] In 1892, the MPA, aware of the shortcomings of the Certificate, established an Educational Committee to explore options for expanding and improving training and qualification.

The Diploma in Psychological Medicine (DPM) was created in 1908–1910, and universities were invited to participate. The DPM involved written and practical tests in psychology, anatomy and physiology, together with questions on asylum administration and legislation.[6] The examination could only be taken two years after medical qualification and following at least three months' clinical experience in a mental hospital. The first DPM examinations were first taken at Leeds University in 1911 and at Cambridge and Edinburgh Universities in 1912, following

[1] Hugh Freeman, "Psychiatry in Britain, *c.*1900," *History of Psychiatry* 21 no. 3 (2010): 318.

[2] David K. Henderson, "Nineteenth Maudsley Lecture: A Revaluation of Psychiatry," *Journal of Mental Science* 85 no. 354 (1939): 16.

[3] Ibid.

[4] Freeman, "Psychiatry in Britain," 317.

[5] Thomas Bewley, *Madness to Mental Illness* (London: The Royal College of Psychiatrists, 2008), 126.

[6] Crammer, "Training and Education in British Psychiatry, 1770–1970," in *150 Years of British Psychiatry, Vol II: The Aftermath*, eds Hugh Freeman and German E. Berrios (London: Athlone, 1996), 221.

courses of instruction at those institutions.[7] Candidate numbers remained small before the outbreak of World War I, and it was not until the opening in 1923 of the Maudsley Hospital, which offered a six-month course of intensive lectures and demonstrations delivered by high-profile teachers, that the DPM began to attract candidates in double figures.[8] Nonetheless, the qualification, while it did not become a pre-requisite for a post in a mental hospital during the 1920s and 1930s, marked a significant step in the professionalisation of psychiatry in Britain and its establishment as an independent, academic discipline.

In France, there were no specialist academic courses of instruction, nor any qualifications in psychiatry until after World War II.[9] Asylums were staffed by full-time physicians appointed by the Ministry of the Interior and a system of internship ensured the transmission of psychiatric knowledge. Medical students entered the asylum as an "*interne*" in their final year of study.[10] The fact that psychiatrists were unable to identify the alleged physiological causes of mental disorder, or to effect successful cures, led to frequent attacks on the profession by the press from the 1860s.[11] Journalists cited therapeutic failure, inappropriate sequestration and the neglect of patients. Morale among alienists plummeted and asylums fell into disrepair. Forging links with the more prestigious, scientifically advanced specialism of neurology was a means of countering this criticism.[12] Neurologists, benefiting from advances in microscopy, improved laboratory techniques and experimental methods, had enhanced their understanding of neuroanatomy and neurological disease during the latter half of the nineteenth century. These developments led to neurology's reputation as the leading medical specialty in *fin-de-siècle* France, with alienism its poor relation.[13] Alienism and neurology overlapped in many ways. Both interpreted mental disorder from an organicist perspective,

[7] Ibid.

[8] Ibid., 222.

[9] A Certificate in Neuro-psychiatry was created in 1949, although the syllabus was usually taught by neurologists, and psychiatry only became a fully autonomous discipline in 1968.

[10] Jean Guyotat, "La Formation Des Psychiatres En France," *Raison Présente* no. 83 (1987): 70.

[11] Gregory M. Thomas, *Treating the Trauma of the Great War: Soldiers, Civilians and Psychiatrists in France, 1914–1940* (Baton Rouge: Louisiana State University Press, 2009), 30.

[12] Ibid., 31.

[13] See: Toby Gelfand, "Neurologist or Psychiatrist? The Public and Private Domains of Jean-Martin Charcot," *Journal of the History of the Behavioural Sciences* 46 no. 3 (2000): 215–29.

and it was possible for doctors to switch from one to the other with ease as professional qualifications were yet to be established in either specialism.

Many of the alienists who worked in the Seine asylums had studied neurology at the Salpêtrière before taking up their asylum posts. The post of Chair of Mental Disorder, based at the Faculty Clinic at Ste Anne's, was given to individuals with a predominantly neurological background, including Benjamin Ball (1833–1893), Alexis Joffroy (1844–1908), Gilbert Ballet (1853–1916), Ernest Dupré (1862–1921) and Henri Claude (1869–1945).[14] From 1894, neurologists were included in the annual alienists' conference, which became known as the *Congrès des aliénistes et neurologistes.*[15] Ballet, who became Chair of Mental Disorder in 1909, was instrumental in developing the alliance between neurology and asylum medicine. Ballet founded a new journal, *L'Encéphale* (meaning "brain"), in 1906, with the aim of overcoming psychiatry's isolation from the rest of medicine and highlighting the many commonalities between neurology and the treatment of mental illness.[16]

Neurology's high status in France was in part due to the work of Jean-Martin Charcot (1825–1893), the first Professor of Clinical Diseases of the Nervous System at the University of Paris (from 1882). Practising at the Salpêtrière hospital, Charcot pioneered a systematic approach to the clinical analysis of many neurological conditions, but was most famous for his research into hysteria, one of the so-called 'functional nervous disorders'. Charcot, who supported the theory of hereditary degeneration prevalent in the late nineteenth century, attributed hysteria to the psychical effects of traumatic events on a degenerate individual.[17] In other words, the underlying cause of hysteria was physiological, linked to a weakness of the central nervous system, while the symptoms were psychological, triggered by a traumatic experience.[18] No physical lesion could be found to indicate a physiological weakness, however, prompting consideration of alternative, non-organic conceptualisations of the condition by the next generation of neurologists.

[14] F. Clarac and F. Boller, "History of Neurology in France," in *Handbook of Clinical Neurology* eds S. Finger, F. Boller, and K. L. Tyler (Amsterdam: Elsevier B.V, 2010), 645.

[15] Jacques Postel and Claude Quétel, *Nouvelle Histoire de la Psychiatrie* (Paris: Dunod, 2012), 341.

[16] Ibid.

[17] Rhodri Hayward, *The Transformation of the Psyche in British Primary Care, 1880–1970* (London and New York: Bloomsbury, 2014), 17.

[18] Ibid.

The Emergence of More Psychological Approaches to Psychiatry

A new discipline—that of psychology—emerged in Europe and the USA in the last quarter of the nineteenth century. Initially, the study of psychology remained academic, a branch of philosophy, and its findings were not applied to the "conduct disorders" that constituted psychiatry.[19] For many years, alienists were indifferent to psychology.[20] This period of indifference was ended by the famous German psychiatrist Emil Kraepelin (1856–1926). Kraepelin adopted a multiple approach to psychiatry that incorporated not only neurology and brain anatomy, but also experimental psychology and a thorough investigation of a patient's life history.[21] The Swiss psychiatrist Eugen Bleuler (1857–1939), who developed new theories regarding schizophrenia, also sought to introduce greater psychological understanding in the treatment of mental patients between 1890 and 1900.[22] As medical director of the Rheinau, and later the Burghölzli Mental Hospitals, Bleuler developed psychological treatment regimes that included occupation, which he considered essential to a patient's health. Bleuler pioneered the approach that would later be developed by the Swiss-born, American psychiatrist, Adolf Meyer, one of the founders of the Mental Hygiene Movement.[23] Psychodynamic theories were most famously developed by the Viennese neurologist, Sigmund Freud (1856–1939) and the Swiss psychiatrist Carl Jung (1875–1961), who had worked as Bleuler's assistant. Freud's ideas were not greeted favourably by his contemporaries, however, with the notable exception of Bleuler.[24] Less well-known were the psychoanalytic theories of French psychiatrist Pierre Janet (1859–1947). In fact, Janet developed the concept of psychological analysis seven years before Freud, although it was Freud who coined the term "psychoanalysis" in 1896.[25]

[19] Henderson, "Nineteenth Maudsley Lecture," 9.

[20] Postel and Quétel, *Nouvelle Histoire*, 251.

[21] Henri F. Ellenberger, *The Discovery of the Unconscious: The History and Evolution of Dynamic Psychiatry* (London: Fontana, 1994), 284.

[22] Ibid., 288.

[23] Meyer, who conducted his medical training in his native Switzerland before emigrating to the USA, studied under August Forel (1848-1931), Bleuler's predecessor at the Burghölzli Hospital and professor of psychiatry at the University of Zurich, in the late 1880s. See S.D. Lamb, *Pathologist of the Mind: Adolf Meyer and the Origins of American Psychiatry* (Baltimore: Johns Hopkins University Press, 2014), 36.

[24] Postel and Quétel, *Nouvelle Histoire*, 351.

[25] Michael Fitzgerald, "Why Did Sigmund Freud Refuse to See Pierre Janet? Origins of Psychoanalysis: Janet, Freud or Both?," *History of Psychiatry* 28 no. 3 (2017): 358.

Whilst far from mainstream, psychological interpretations of mental disorder were therefore in circulation on the continent, and in France and England, before World War I. Experimental psychology laboratories were established in Paris (1889) and Cambridge (1897); national psychological societies were founded in both countries in 1901; and journals dedicated to psychology in 1904. While the evolution of psychology appeared to follow similar trajectories in France and Britain, significant differences lay in the pre-war familiarity with psychoanalysis and in the impact of psychology on asylum medicine. Freud's theories were being read in England well before 1914, whereas in France they were almost unknown until after World War I. Traditional French hostility to Germanic scholarship, coupled with linguistic barriers and a preference for physiological explanations, may explain this.[26] The English psychiatrist David Eder read a paper on Freud's methods to the British Medical Association in 1911; Freud's *Papers on Psycho-Analysis* were published in Britain in 1912 and the London Society of Psychoanalysis was founded in 1913.[27] Bernard Hart (1879–1966), who, before World War I worked at the Long Grove Asylum with Edward Mapother, published *The Psychology of Insanity* in 1912 that included references to the works of Freud, Jung and Janet.[28]

Despite Pierre Janet's development of a similar psychodynamic approach to that of Freud, psychoanalysis and other psychological approaches to treating mental disorder did not impinge upon French psychiatry or asylum medicine before World War I. In England, notably at the new Long Grove Asylum in Epsom, Surrey (founded in 1907), psychiatrists responded more favourably.[29] The English psychiatrist Charles Mercier (1851–1919) anticipated the psycho-biological approach that would characterise psychiatry in England after World War I in his work on conduct disorders published in 1911.[30] A psychodynamic approach was influential in the development of treatment for traumatised British, but not French,

[26] Postel and Quétel, *Nouvelle Histoire*, 350.

[27] Roy Porter, "Two Cheers for Psychiatry! The Social History of Mental Disorder in Twentieth-Century Britain," in *150 Years of British Psychiatry, Vol. II: The Aftermath*, eds Hugh Freeman and German E. Berrios (London: Athlone, 1996), 388–9.

[28] Bernard Hart, *The Psychology of Insanity* (Cambridge: Cambridge University Press, 1912). A fourth edition of the work was published in 1930.

[29] Philip Kuhn, *Psychoanalysis in Britain 1893–1913: Histories and Historiography* (New York: Lexington Books, 2016), 272.

[30] Henderson, "Nineteenth Maudsley Lecture," 11; Charles Mercier, *Conduct and Its Disorders: Biologically Considered* (London: Macmillan & Co., 1911).

soldiers during World War I. This would have implications not only for the introduction of occupational therapy and the practice of patient work, but for the development of psychiatry in each country between the wars.

The Influence of World War I

During World War I, approaches to treating soldiers suffering from war neuroses (the term "shell-shock" was only used in Britain) reflected the different psychiatric traditions of each of the combatant countries.[31] They also depended on the physicians' previous training and social networks.[32] The different experiences of French and English physicians in these areas led to different emphases on the use of occupation as a treatment for war neuroses in France and England. The treatment of war neuroses as a purely neurological problem by French psychiatrists, and the greater willingness by English psychiatrists to treat shell-shock with psychotherapy, appeared to affect the readiness of psychiatrists in each country to adopt new methods of occupying patients after the conflict.

Conventional historical accounts hold that before World War I, all British psychiatrists were "physicalists" who regarded war neurosis as a "functional nervous disease", the underlying cause of which was physiological and linked to notions of hereditary degeneration, and that by the end of the conflict, psychiatrists had become psychologists, incorporating the methodology of Freud and psychoanalysis.[33] This interpretation rightly emphasises the role of the war in boosting the subsequent spread of depth psychology, but as Chris Feudtner argues, it oversimplifies the situation.[34] Partly physical and partly psychological diagnoses of traumatic hysteria and neurasthenia occurred before 1914.[35] Physicians who were known to have a strong physical, neurological view of war neurosis, such as Frederick Mott, also recognised the psychological aspects of the condition, while some psychologists, such as William McDougall, linked mental disorders with underlying physiological issues. There was no "crisp dividing line"

[31] L. Crocq, "La Psychiatrie De La Première Guerre Mondiale. Tableaux Cliniques, Options Pathogéniques, Doctrine Thérapeutiques," *Annales Médico-psychologiques* 163 no. 3 (2005): 285.

[32] Chris Feudtner, "'Minds the Dead Have Ravished': Shell-shock, the History and Ecology of Disease Systems," *History of Science* 31 no. 4 (1993): 388.

[33] Ibid., 413 footnote 37; 386.

[34] Ibid., 386.

[35] Ibid., 387.

between the physicalists and the psychologists; the reality was more nuanced.[36] There were also variations in the types of treatment that constituted "psychological" therapy, which included techniques of persuasion and psychoanalysis.

The different training and networks of those involved in treating shell-shocked soldiers led to different approaches. Historian Eric Leed identified two different techniques in the British treatment of shell-shock, "disciplinary" and "analytic", which were "rooted in two different conceptual frameworks and visions of human nature".[37] The disciplinary method, or "quick cure" was the preferred method in the French military. It was based on a 'moral' view of war neurosis as a form of malingering. In England, this method was associated with the neurological National Hospital at Queen Square where a specialist unit had been opened for severe cases of shell-shock, such as mutism or paralysis. The treatment comprised a mix of high pressure techniques of persuasion; shouted commands were accompanied by the use of strong, painful electric shocks (faradisation), icy showers and isolation.[38] The analytic method, associated with psychologists based at the Red Cross Hospital, Maghull, including W.H.R. Rivers, T.H. Pear and William McDougall, won the "grudging support" of the military authorities towards the end of war, and the Maghull was given responsibility for training military psychiatrists.[39] This method, which included re-education and psychotherapy, involved paying close attention to the whole patient and their state of mind, while the disciplinary method focused solely on their neurological symptoms.[40]

The emphasis placed upon the re-education of the patient, a key aspect of which was the prescription of work and occupation, contributed to the greater interest shown in therapeutic occupation in England during the interwar period. The emphasis on occupation is revealed in a number of English publications produced during or just after the war. Elliot Smith and T.H. Pear's *Shell Shock and its Lessons* (1917), for example, advocated therapeutic work for the shell-shocked patient to prevent him from

[36] Ibid., 388.

[37] Peter Leese, *Shell Shock: Traumatic Neurosis and the British Soldiers of the First World War* (Basingstoke: Palgrave Macmillan, 2002), 73.

[38] Ibid., 74.

[39] Edgar Jones and Simon Wessely, *Shell Shock to PTSD: Military Psychiatry from 1900 to the Gulf War* (Hove and New York: Psychology Press, 2005), 33.

[40] Leese, *Shell Shock: Traumatic Neurosis*, 81–4.

"dwelling upon his subjective troubles".[41] A "suitable occupation" should be identified comprising "useful work" to stop the patient from feeling that he was a burden, reflecting contemporary notions of usefulness, efficiency and duty. The work should be interesting and occupy the patient's mind, not just his body, and combined with attempts to identify the root cause of his trouble through psychotherapy.[42]

A report by American psychiatrist, T.W. Salmon, who visited British treatment centres in France before the US entered the war in 1917, believed that "re-education by physical means is a valuable adjunct to treatment" and that this was best achieved by occupation.[43] While the British facilities for treatment were quite limited, the Americans introduced a range of clinical interventions, which included the provision of psychotherapy and occupational therapy workshops.[44] Activities could be conducted in bed (such as basket-making, net-making, polishing and sand-papering), indoors (such as carpentry, wood carving, metal work, printing, book-binding and cigarette making) or outdoors (including farming, gardening, animal care and building work).[45] Salmon emphasised that "shell-shock" was a disorder of "will" as well as function and that "progressive achievement" was the only means by which "manhood and self-respect" could be restored.[46] Patients were encouraged to undertake physical tasks such as the cultivation of farmland, wood cutting and road construction, and art therapy which helped them come to terms with their traumatic experiences.[47]

British physician Millais Culpin, who had treated British troops in France in treatment centres established by the British military, also emphasised the value of work for shell-shocked patients in *The Psychoneuroses of War and Peace* (1920). Work and hobbies formed part of a programme of psychotherapy and re-education in which "all methods converge[d] and

[41] G. Elliot Smith and T. H. Pear, *Shell Shock and its Lessons*, Second ed. (Manchester: Manchester University Press, 1917), 51.

[42] Ibid., 51–2.

[43] Thomas W. Salmon, *The Care and Treatment of Mental Diseases and War Neuroses ("shell-shock") in the British Army* (New York: War Work Committee of the National Committee for Mental Hygiene, 1917), 39.

[44] Jones and Wessely, *Shell Shock to PTSD*, 32. [Occupational therapy had evolved as a profession in the USA just before the Americans' entry into the war in 1917.]

[45] Salmon, *Care and Treatment*, 40.

[46] Ibid.

[47] Jones and Wessely, *Shell Shock to PTSD*, 32.

overlap[ped] in order to make the patient efficient again and to enable him to cope with himself and his environment". Doctors sought to re-establish patients' self-confidence, which was usually "painfully lacking", by assigning chores and small projects that fostered a sense of achievement.[48] Work helped the patient feel that he was capable of "taking part in the world again" and provided a "useful gauge of his progress towards active citizenship".[49] The War Office Committee agreed, maintaining that at the re-education stage of treatment, "the patient should be occupied consistently and not allowed to slip back into unprofitable habits by neglect or lack of mental diversion".[50] At the Maudsley Hospital, under the direction of Frederick Mott, "an atmosphere of cure" was emphasised through "purposeful activity".[51] Occupational therapy and social activities were encouraged; soldiers grew vegetables in the hospital grounds and constructed a poultry house to provide a supply of fresh produce. Patients were taught carpentry and woodwork in a large, fully equipped workshop (see Fig. 3.1). Mott donated a piano and advocated choral singing as an "uplifting mental diversion" which he believed would promote "that sense of wellbeing so essential for mental and bodily recuperation".[52] These re-educative methods involving occupation were rare in French military psychiatry.

The French Army's *Service de Santé* equated war neuroses with hysteria, the condition made famous by the flamboyant French neurologist Jean-Martin Charcot (1825–1893). The authorities regarded the incidence of such functional disorders as highly contagious, posing a threat to the morale of the army that needed to be contained.[53] The "disciplinary" approach to treatment, associated with neurology (the dominant medical specialty in France and with which psychiatry had sought to align itself)

[48] Feudtner, "'Minds the Dead Have Ravished'," 401.

[49] Ibid.

[50] Report of the War Office Committee Enquiry into "Shell-shock" (London: HMSO, 1922), 130.

[51] The hospital was loaned by the Royal Army Medical Corps for the duration of the war, and from 1919–20, by the Ministry of Pensions, for the treatment of servicemen and ex-servicemen suffering from war neuroses.

[52] Frederick Mott, *War Neuroses and Shell Shock* (London, Henry Frowde and Hodder & Stoughton, 1919), 297.

[53] Marc Roudebush, "A Battle of Nerves: Hysteria and its Treatments in France during World War I," in *Traumatic Pasts: History, Psychiatry and Trauma in the Modern Age, 1870–1930*, eds Paul Lerner and Mark Micale (Cambridge: Cambridge University Press, 2001), 254.

Fig. 3.1 Carpentry Workshop at the Maudsley Hospital, London, 1918. During World War I the hospital was used by the Royal Army Medical Corps to treat soldiers suffering from "shell shock". (© *By permission of Bethlem Museum of the Mind, HPC-19*)

was preferred by the French military authorities. Military neuropsychiatric treatment centres were established in the main cities and run by neurologists. In Paris, for example, neurologists Jules Dejerine and Pierre Marie assumed responsibility for the military neurological department at the Salpêtrière Hospital; Joseph Babinski and Jules Froment took over the military unit at the Pitié; Gilbert Ballet at the Maison Blanche and Achille Souques at the Paul-Brousse Hospital.[54] Neuropsychiatric centres were also established near the Front, overseen by neurologist Gustave Roussy.

The disciplinary approach, similar to that adopted at the English neurological hospital at Queen Square, was championed by neurologist Joseph

[54] Julien Bogousslavsky and Laurent Tatu, "French Neuropsychiatry in the Great War: Between moral support and electricity," *Journal of the History of the Neurosciences* 22 no. 2 (2013): 147.

Babinski, Charcot's former pupil. It was this approach that came to dominate French treatment of war neuroses.[55] Babinski's methods were "virile and correctional", sometimes involving the use of electricity to produce intense pain. This technique, which became known as *torpillage* (from the French word *torpille* meaning electric eel) was controversial and some soldiers refused treatment.[56] These disciplinary methods did not treat the underlying psychological causes of the trauma, but were effective in removing, in the short-term at least, physical symptoms such as mutism or an inability to walk. Recidivism was common, however. It has been suggested that in countries where the fighting was on home territory, such as France, firm physical methods of treatment predominated, while elsewhere (such as Britain) patient management was focused on longer-term psychological methods.[57] French physical methods of persuasion, whether or not they involved the more extreme measures of *torpillage*, were based on a medical-military attitude that sought the rapid return of soldiers to the Front to defend French territory.[58] Physical therapies were preferred in France, although some doctors, such as Dejerine, did opt for more psychological methods.

A French military treatment centre for soldiers who had sustained physical injuries, such as the loss or paralysis of a limb, rather than for those suffering from war neuroses, used occupation therapeutically. At the neurological centre in Montpellier, Dr Villaret developed a programme of "professional re-education" involving a form of occupational therapy (*ergothérapie*).[59] Agricultural work, artisanal activities such as metalwork, leatherwork, woodwork and upholstery formed part of a programme that included training in typing, accounting and languages. Recovering patients assisted the nurses with training new patients.[60] A film made at the centre in 1919 shows soldiers with paralysed hands looking after the pigs and learning agricultural skills.[61] Agricultural work was also used to rehabilitate

[55] See: Joseph Babinski, *Hystérie-pithiatisme et troubles nerveux d'ordre réflexe* (Paris: Masson, 1917).

[56] Bogousslavsky and Tatu, "French Neuropsychiatry," 149.

[57] Ibid., 150.

[58] Ibid.

[59] Hervé Guillemain and Stéphanie Tison, *Du Front à L'Asile, 1914–1918* (Paris: Alma éditeur, 2013), 322.

[60] Ibid.

[61] Le Centre Neurologique de Montpellier; Dr Villaret; Référence B 189. https://images-defense.gouv.fr/fr/le-centre-neurologique-de-montpellier-docteur-villaret.html# (accessed 24/04/2022).

those who had suffered severe facial injuries during the war. These individuals, known as *les gueules cassées* (or broken faces) found their disfigurement a barrier to returning to their former employment, and in some cases to their families. In 1921, a group of veterans established *L'Union des blessés de la face et de la tête* that organised work on farms for such individuals, enabling them to support themselves whilst coming to terms with their transformed appearance and avoiding public stigma.[62]

By the end of the war, British servicemen suffering from war neurosis were more likely to receive holistic treatment that focused on their psychological condition as well as their physical symptoms than their French counterparts. The psychodynamic treatment offered at the Maghull was perceived as the most effective, long-term method of treatment.[63] The British, influenced by the Americans, who established sophisticated occupational therapy workshops within their treatment centres near the Front, recognised the value of in-depth psychotherapy coupled with re-educative techniques involving work and occupational therapy. In France, the disciplinary or quick cure, based on physical treatment methods, was preferred. These did not incorporate work therapy or occupational therapy and did not deal with the underlying psychological causes of the traumatic response. The different approaches to dealing with war neuroses by French and English psychiatrists drew on the different psychiatric traditions of each country; these traditions led to divergent responses to new theories of therapeutic occupation emerging after World War I.

The Mental Hygiene Movement

The war highlighted the need for services where mildly troubled former soldiers and traumatised civilians could be treated without the threat of internment in an asylum.[64] The psychological effect of the war on civilians was particularly noticeable in France, where so many homes and livelihoods had been destroyed as a result of the fighting. The cure of these individuals was essential to revitalise the French nation. Reformers, such as Édouard Toulouse, called for the establishment of new, medicalised

[62] See: M. D. Colas, F. Benslama, and M. Daudin, "Les « Gueules Cassées », D'une Génération À Une Autre : Approche Sociologique Et Psychopathologique Des Blessures De Guerre," *Annales Médico-psychologiques* 170, no. 4 (2012); S. Delaporte, *Les Gueules Cassées* (Paris: Noesis, 1996).

[63] The "quick cure" resulted in many cases of recidivism.

[64] Thomas, *Treating the Trauma of the Great War*, 153.

psychiatric facilities.[65] The post-war period in both France and England saw the introduction of social hygiene measures to improve the physical health of their populations, including strategies to tackle tuberculosis, alcoholism and syphilis, and thereby boost the productivity of their workforces.[66] Efforts to improve the populations' mental health (or "mental hygiene") were stimulated by the Mental Hygiene Movement. The movement originated in the USA, where Clifford Beers, aided by the professional expertise of psychiatrist Adolf Meyer, co-founded the National Committee for Mental Hygiene in 1909. Beers' experiences in three different mental hospitals in the early 1900s had persuaded him to devote his energies to campaigning for improved services for the mentally ill. The movement spawned national organisations in France and England in 1920 and 1922 respectively. The chief purposes of the movement, as reported by the *Journal of Mental Science* in 1923, were to conserve mental health; promote the study of mental disorders and intellectual impairment; to obtain and disseminate information regarding mental health; raise standards of care and treatment; and co-ordinate the activities of national organisations.[67] The movement was associated with the growing interest in psychology and an awareness of the importance of conserving the mental health of a population to the national economy.[68] It emphasised the important role that psychiatry had come to play in the social life of the community.[69]

Adolf Meyer, co-founder of the Mental Hygiene Movement in the USA, was an early adopter of occupational therapy at the Henry Phipps Clinic in Baltimore. Meyer defined mental disorder as a form of "maladjustment" or a state in which individuals were no longer able to respond

[65] Ibid., 69.

[66] See: Greta Jones, *Social Hygiene in Twentieth-Century Britain* (London: Croom Helm, 1986).

[67] AU, "The National Council for Mental Hygiene," *Journal of Mental Science* July (1923): 404.

[68] Mathew Thomson, "Mental Hygiene in Britain During the First Half of the Twentieth Century: The Limits of International Influence," in *International Relations in Psychiatry: Britain, Germany, and the United States to World War II,* eds Volker Roelcke, Paul J. Weindling, and Louise Westwood (Rochester, NY: University of Rochester Press, 2010), 134–5.

[69] Henderson, "Nineteenth Maudsley Lecture," 12.

adequately to their environment.[70] His approach was holistic. He believed that the causes of mental disorder could be physiological, psychological, social or environmental. He used the term "psychobiology" to describe this approach, which enabled him to overcome the great divide between "mind" and "body", or between an organicist and a psychological approach to mental illness.[71] His views on the causation, treatment and prevention of mental disorder informed the International Mental Hygiene Movement and influenced the approaches taken by national mental hygiene organisations. The model of treatment advocated by Mental Hygiene Movement included the establishment of outpatient clinics, where patients could receive treatment such as psychotherapy without admission to a mental hospital; the provision of social services for the support of patients both in hospital and at home; child guidance clinics; and facilities for the voluntary treatment of patients at the early stages of their illness (or "open" services, as they were known in France). This model of treatment, and Meyer's teaching, also informed the English Macmillan Report of 1926. This report maintained that "there is no clear line of demarcation between mental and physical illness".[72] Physical illnesses could have a "mental concomitant" just as mental disorders could have a "physical concomitant", and in many cases it was hard to ascertain whether mental or physical symptoms predominated.[73] The notion that mental disorder could be the result of a "medley" of different causes became known as the "continuity of mental disorder", a phrase coined by Edward Mapother of the Maudsley Hospital.[74]

Meyer's influence was apparent in Paris, where the League of Mental Hygiene was founded in 1920 by psychiatrist Édouard Toulouse, supported by Joseph Briand and Georges Génil-Perrin.[75] Toulouse developed a personal friendship with Clifford Beers after reading his book, *A Mind*

[70] Hans Pols, "'Beyond the Clinical Frontiers': The American Hygiene Movement 1910–1945," in *International Relations in Psychiatry: Britain, Germany and the United States to World War II*, eds Volker Roelcke, Paul Weindling, and Louise Westwood (Rochester: University of Rochester Press, 2010), 113–4.

[71] Susan Lamb, "Social Skills: Adolf Meyer's Revision of Clinical Skill for the New Psychiatry of the Twentieth Century," *Medical History* 59 no. 3 (2015): 446.

[72] Cited in Kathleen Jones, *Asylums and After: A Revised History of the Mental Health Services: From the Early 18th Century to the 1990s* (London: Athlone Press, 1993), 131.

[73] Ibid.

[74] Hayward, *The Transformation*, 70.

[75] Annick Ohayon, *L'Impossible Rencontre, Psychologie et psychanalyse en France (1919–1969)* (Paris: Éditions La Découverte, 1999), 32.

That Found Itself, published in 1908, and became greatly interested in the new approach taken by Meyer at the Henry Phipps clinic.[76] Toulouse has been described as an "*organiciste tempéré*".[77] At the beginning of his career, Toulouse was an organicist, but the more he focused on the preventative agenda of the Mental Hygiene Movement the more he became convinced of the role played by social factors in causing mental disorder. Toulouse maintained in 1926 that, "For historical reasons, [psychiatrists] have remained organicists for too long … it must be remembered that mental illness is also affected by the social character of the individual".[78] He maintained that the organicists did not pay sufficient attention to problems of "maladaptation" to a patient's social situation and was adamant that social factors should not be overlooked in the genesis of psychosis.[79] His views were far more in tune with those of Adolf Meyer than they were with French psychiatrists outside Paris, most of whom dismissed his theories. Like Meyer, Toulouse and his Parisian colleagues took a holistic view of mental disorder. One such colleague, Ernest Dupré of the Faculty Clinic at Ste Anne, concluded in 1919 that "mental illnesses are diseases of the personality", suggesting a much broader concept of mental disorder than that held by most French psychiatrists.[80] Toulouse developed a similar concept to Meyer's "psychobiology" that he termed "*la biocratie*".[81] For Toulouse, it was still essential to conduct thorough investigations into potential physical factors, such as the metabolism of the brain, or anatomical changes within the nervous system, but the life-style and character of an individual were equally important considerations.

The British National Council for Mental Hygiene (NCMH), founded in 1922, was not driven by one man's vision in quite the same way as the French League. The British NCMH included a cross-section of founder members, including eminent psychiatrists Sir Frederick Mott, Hubert Bond, Hugh Crichton-Miller and Alfred Tredgold; psychologists Charles Myers and W.H.R. Rivers; neurologist Henry Head; and philanthropist

[76] Michel Huteau, *Psychologie, Psychiatrie et Société Sous La Troisième République, La biocratie d'Édouard Toulouse (1865–1947)* (Paris: L'Harmattan, 2002), 186.

[77] Ibid., 10.

[78] Ibid., 210.

[79] Ibid.

[80] Postel and Quétel, *Nouvelle Histoire*, 343.

[81] Huteau, *Psychologie, Psychiatrie et Société*, 9.

and businessman Sir Courtauld Thomson, the Council's first president.[82] As members of the Council of the Eugenics Education Society, Mott and Tredgold were in favour of segregating the intellectually impaired and of pursuing the policies set out by the Mental Deficiency Act of 1913. Henry Head, like Toulouse, was concerned by industrial fatigue and worker efficiency, maintaining that "much industrial unrest was due to the worry and fatigue induced by unsatisfactory working conditions".[83] Hubert Bond was an active campaigner for voluntary treatment and President of the Association of Occupational Therapists from 1937, while Hugh Crichton-Miller's wartime experiences of treating shell-shock led him to found the Tavistock Clinic for nervous disorders in 1920.[84]

Psychologist Charles Myers, after serving as consultant psychologist to the British Army in France during World War I, founded and became the director of the National Institute of Industrial Psychology in 1922.[85] He was committed to the Mental Hygiene principle that "a man should have pleasure in his work and a feeling that it is worthwhile" and believed that psychiatrists could assist in the "difficult task of fitting men to the jobs they can best do, and jobs to the men they need".[86] W.H.R. Rivers' speciality was applied psychology; an advocate of psychotherapy and psychoanalysis, his treatment of traumatic experience were influential during World War I. Although the movement lacked the drive and passion of its forceful proponent in France, the NCMH members' varied interests and spheres of influence ensured that the principles of the Mental Hygiene Movement had a broader reach in England than in France. The Mental Hygiene Movement effectively took psychiatry out of the confines of the asylum and into the community, thereby bringing to an end the era during which psychiatrists were only concerned with asylum cases, and individuals

[82] AU, "National Council for Mental Hygiene," *British Medical Journal* May 13 (1922): 766.

[83] Ibid., 767.

[84] Aubrey Lewis and Nick Hervey, "Bond, Sir (Charles) Hubert (1870–1945), psychiatrist and mental health administrator," *Oxford Dictionary of National Biography*, 23/09/2004; K. Loughlin, "Miller, Hugh Crichton (1877–1959), psychotherapist," *Oxford Dictionary of National Biography*, 23/09/2004, https://www-oxfordnb-com (accessed 30/04/2022).

[85] F.C. Bartlett, "Obituary: Dr Charles S. Myers CBE, FRS," *Nature* 4019 (1946): 658.

[86] W.J.A. Erskine, "Epitome of Current Literature: Mental Hygiene," *Journal of Mental Science* April (1923): 241–243.

were considered either 'sane' or 'insane'.[87] Fundamental to mental hygiene principles was belief in the curability of mental disorder.

Transformation of the Asile Clinique

Plans to transform the Asile Clinique into a hospital specialising in curable cases had been mooted at intervals since the asylum opened in 1867. These plans were resurrected in 1918. It was proposed that the Asile Clinique should provide the most up-to-date psychiatric treatment and care by doctors at the peak of their profession.[88] Patients admitted to the Asile Clinique would be those defined as "acute". In his 1927 report Dr Marie, head of the Admission Service at Ste Anne's, provided a summary of what was meant by "acute".[89] First and foremost were patients considered "curable" on examination by doctors within the Admissions Service, such as those at the early stage of their illness, preferably not more than six months (or twelve at the outside) from the onset of symptoms, since the evidence suggested that treatment had the greatest chance of being effective during these first few months. This included any case of mental illness (whether confused, toxic or infectious), as long as the case had not been clinically diagnosed as incurable, and patients suffering from emotional states such as phobias or obsessions. At the Asile Clinique, "acute" also included incurable patients who were in need of significant care, even if they were not receiving active treatment, as well as those exhibiting episodes of extreme agitation or depression who needed active medical treatment, or who were dangerous, such as those in the early stages of dementia praecox or chronic delusional states.[90] Incurable cases who were excluded from the Asile Clinique were those whose conditions would not benefit from active treatment, such as later-stage dementia praecox cases; melancholics and maniacs who were not considered dangerous; "organic cases" such as the senile, those suffering from tertiary syphilis; the intellectually impaired; and patients whose condition had deteriorated into chronicity.[91]

[87] Henderson, "Nineteenth Maudsley Lecture," 14.

[88] Louis Dausset, Rapport Général au nom de la 3e Commission sur les propositions budgétaires pour le service des Aliénés (budget de 1919) (Paris: Conseil Général de la Seine, 1918), 20. ADP/D.10K3/27/20.

[89] Dr Auguste Marie, Rapport de la Service d'Admission, Asile Clinique, 1927 (Paris: Préfecture du Département de la Seine, 1927), 49–50. ADP/PER-566-5.

[90] Ibid.

[91] Ibid.

The project to transform the Asile Clinique into an acute service was finally approved in 1923, although the transformation was not fully completed for another five years. Chronic and incurable patients were gradually transferred to the colonies or other asylums in the Seine department or the provinces during 1927 and 1928 to make way for acute cases. The men's and women's divisions were each divided into two sections, with their own medical and nursing teams, to provide the more intensive care required by acute patients. The discharge rates after 1928 indicate a marked change in the movement of patients, demonstrating that the Asile Clinique was indeed operating as a service for acute, curable cases, as Dr Truelle observed in 1931.[92] At the Asile Clinique, "alienism" had given way to psychiatry, although the brand of active treatment practised here was biological, rather than psychosocial.

The effect of the transformation of the Asile Clinique on patient occupation was marked. Most French psychiatrists did not consider work or any form of occupation an appropriate treatment for acute patients. Work was for calm, chronic and convalescent patients; these were the patients who were transferred out of the Asile Clinique. The numbers of patients provided with work in the new treatment divisions of the Asile Clinique dwindled following the transformation to a hospital for acute patients. Patients with acute conditions were initially prescribed bed-rest, sometimes for several weeks, according to the teaching of the Dr Valentin Magnan (1835–1916), the long-serving head of the Admissions Service at Ste Anne's before World War I, and then treated biologically.[93] The financial implications of this policy for the Asile Clinique, which lost its chronic and incurable patients and therefore the majority of its patient workers, are explored further in Chap. 5.

Voluntary Treatment

Separating the curable and incurable allowed a greater focus on treating the curable, but, as had been recognised since the early nineteenth century, the mentally ill needed to receive treatment as soon as possible after the onset of their symptoms for it to be effective. Undergoing the

[92] Dr Truelle, Rapport de la 2e Section (Hommes), Asile Clinique, 1931 (Paris: Préfecture du Département de la Seine, 1931), 13. ADP/PER-566-5.

[93] Magnan worked at Ste Anne's from its opening in 1867 until his retirement in 1912; he was an ardent supporter of degeneration theory and his views were held in high regard.

time-consuming and stigmatising process of legal commitment to an asylum or mental hospital delayed treatment and jeopardised a patient's chance of recovery.[94] In England, calls for the provision of treatment for incipient cases of insanity on a voluntary basis had been ongoing throughout the nineteenth century, but they intensified at the turn of the twentieth. Parliamentary bills were introduced in 1899, 1900, 1904 and 1905, but these had been withdrawn through lack of debating time.[95] Henry Maudsley, a "sharp and persistent critic" of the English system of lunacy legislation and incarceration throughout his long career, was particularly critical of the delay to treatment caused by the certification process.[96] The lack of legislative progress to address this issue led him to make a bequest in 1907 for the establishment of a hospital to focus on the "early treatment of cases of acute mental disorder". This gave the London County Council the impetus, and half the funds, to build an institution that would enable the voluntary admission of poor patients.[97]

Maudsley and his collaborator, Frederick Mott, believed that only by studying mental disorder at its early, acute, yet curable stage could knowledge be generated about its causes. They planned a hospital, named after its benefactor, with facilities for postgraduate training based on Kraepelin's clinic in Munich.[98] Parliament granted the Maudsley Hospital special administrative freedoms to avoid the difficulties surrounding certification. The London County Council (Parks etc) Act of 1915 allowed the Maudsley to "receive and lodge as a boarder and maintain and treat … any person suffering from incipient insanity or mental infirmity". [99] The Maudsley, however, only opened to civilians in 1923 since as soon as it was ready to receive patients in 1916, following various delays due to planning and construction issues, it was taken over by the War Office as a clearing hospital for shell-shocked soldiers, and then by the Ministry of Pensions to treat veterans with severe neuroses.[100]

[94] Thomas, *Treating the Trauma of the Great War*, 149; Michael Collie, "Introduction," in *Henry Maudsley, Victorian Psychiatrist: A Bibliographical Study*, ed. Michael Collie (Winchester: St Paul's Bibliographies, 1988), 6–7.

[95] Sixty-third Annual Report of the Commissioners in Lunacy (London: HMSO, 1909), 32.

[96] Collie, "Introduction," in *Henry Maudsley, Victorian Psychiatrist*, ed. Collie, 6.

[97] Edgar Jones, Shahina Rahman, and Robin Woolven, "The Maudsley Hospital: Design and Strategic Direction, 1923–1939," *Medical History* 51 no. 3 (2007): 358.

[98] Ibid., 359–60.

[99] Ibid., 363.

[100] Ibid., 357.

Admission to the French public asylum system, established by the law of 1838, was by official committal, a system known as *placement d'office*. It was restricted to those who posed a danger to themselves or others, or who threatened public order. The system was administered by the local police, who sent individuals to a special infirmary, usually located next to the police cells, for assessment, certification and referral to an asylum. The process was lengthy, complicated and degrading. In the department of the Seine, there was another, allegedly more humanitarian admission procedure known as *placement volontaire*, that had been introduced in 1876.[101] Individuals could be brought by their families directly to the Admissions Office at Ste Anne's for assessment and referral to one of the Seine asylums. The term *volontaire* is misleading, however, since, once certified, the patient had no rights and was not considered capable of informed consent.[102] Édouard Toulouse sought to create a truly voluntary system, or "open" service, that would be available to all without certification, enabling the early treatment of individuals with acute symptoms or those at the early, mild stage of mental illness, before their conditions deteriorated.[103] After witnessing the Scottish "open door" system in 1897, he attempted to persuade the General Council of the Seine of its merits in lengthy reports written in 1898 and 1913.[104] A revision to the law requiring that patients had to be certified to gain admission to public asylums was proposed in 1914, but World War I broke out before this could be debated.[105]

[101] Patricia Prestwich, "Family Strategies and Medical Power: 'voluntary' committal in a Parisian asylum, 1876–1914," *Journal of Social History* 22 June (1994): 1. https://www.thefreelibrary.com/Family+strategies+and+medical+power%3a+%27voluntary%27+committal+in+a...-a016108112 (accessed 2 April 2021).

[102] Ibid.

[103] Anne-Laure Simonnot, *Hygiénisme et Eugénisme au XXe Siècle: à travers la psychiatrie française* (Paris: Éditions Seli Arslan, 1999), 56.

[104] Edouard Toulouse, Rapport au Conseil Général de la Seine au nom de la commission chargée d'étudier l'assistance des aliénés en Angleterre et en Ecosse (Paris: Conseil Général de la Seine, 1898), ADP/D.10K3/12; Rapport de la division des femmes, 1er section, Asile de Villejuif, 1913 (Paris: Conseil Général de la Seine, 1913), PDS.ADP/PER-566-4. Scottish and English legislation regarding certification of the mentally ill differed until after World War II; in Scotland, the mentally ill had considerably more freedom than English patients prior to 1930.

[105] Henri Colin, "Mental Hygiene and Prophylaxis in France," *Journal of Mental Science* Oct (1921): 459.

The war bolstered Toulouse's arguments for reform by highlighting the need to strengthen the mental health of war survivors, many of whom (both military and civilian) were traumatised by the experience of war and by the economic instability of its aftermath.[106] The French League of Mental Hygiene provided a useful launch pad for Toulouse's campaign to establish an "open" psychiatric service in Paris. He obtained the support of several politicians, notably Councillor Henri Rousselle and Health Minister Justin Godard, but many psychiatrists remained opposed to the project.[107] Despite opposition from colleagues, the *Service Libre de Prophylaxie Mentale* (later named the Henri Rousselle Hospital) opened in 1922 in the grounds of Ste Anne's.[108] The service comprised an outpatients facility (or dispensary); a hospital (where patients were at liberty to discharge themselves voluntarily); various research laboratories; and a social services department. For the first time, poor patients could access free psychiatric treatment, as "in-patients" or "out-patients", without having to be certified.[109] Toulouse recognised the benefits of psychiatric social work and its role in helping patients to secure employment after leaving hospital; this comprised an essential service at the Henri Rousselle Hospital.[110]

The opening of the Henri Rousselle Hospital in Paris in 1922 and the Maudsley Hospital in London in 1923 represented a major reform to treatment of the mentally ill poor in England and France. These two institutions were based on the model of the general hospital and were not subject to the same legislation as asylums. The hospital model was focused on the active treatment of patients who were admitted, and could discharge themselves, voluntarily, in the same way as patients admitted to a general hospital, suffering from a physical condition. Both institutions accepted all classes of patient, including paupers. This meant that the poorest patients could be admitted at the early, curable stage of their

[106] Gregory M. Thomas, "Open Psychiatric Services in Interwar France," *History of Psychiatry* 15 no. 2 (2004).

[107] Ohayon, *L'impossible Rencontre*, 32.

[108] The *Service Libre de Prophylaxie Mentale* became known as the Henri Rousselle Hospital in 1926, named after its most ardent supporter on the General Council, who died that year.

[109] Henri Rousselle, Rapport au nom de la 3e Commission sur le Service départmental de prophylaxie mentale et les service annexes (Paris: Conseil Général de la Seine, 1923), 3–4. ADP/D.10K3/34/87.

[110] Ibid.

mental illness, a facility hitherto denied them. But in 1922/3 these were the only public institutions in England and France where such freedoms existed, and therefore served only a tiny proportion of their respective populations. Outside these two institutions, only those who could afford the fees of private institutions (or in England, those who were eligible for admission to the registered Bethlem Royal Hospital) could access treatment at the early, potentially curable, stage of their illness.

In England after the war, the NCMH and the Board of Control continued the campaign for the extension of voluntary treatment to all public mental institutions. Calls for voluntary treatment peppered the Board of Control's annual reports throughout the 1920s.[111] Their case was boosted by the findings of the Macmillan Report of 1926, produced following a Royal Commission on Lunacy and Mental Disorder established by the Ministry of Health in 1924. This radical and influential report declared that no satisfactory distinction between mental and physical illnesses could be identified and concluded that all illnesses should be managed similarly. It stressed the importance of early treatment and stated that "certification was to be the last resort in treatment, not [its] prerequisite".[112] The aim of psychiatry should no longer be the containment and isolation of individuals with the most severe conditions. Anyone suffering from a mental illness, whether severe or mild, should have access to treatment, either at home, at an outpatients' clinic, in a general hospital or in a mental hospital, depending on the patient's needs and the availability of facilities.[113] This paved the way for an opening up of psychiatric facilities and for the care in the community measures that would be introduced after World War II.

The Mental Treatment Act was eventually passed in 1930. It provided for the admission of uncertified, early-stage, acute cases to all public mental hospitals on a voluntary or temporary basis, without the need for the lengthy process of certification that delayed the commencement of

[111] Phil Fennell, *Treatment without Consent: Law, Psychiatry and the Treatment of Mentally Disordered People since 1845* (London: Routledge, 1996), 114.

[112] Ibid., 116.

[113] Ibid.

treatment.[114] Medical superintendents were encouraged to transfer their chronic patients to Public Assistance Institutions,[115] or specialist institutions for the "mentally deficient" to allow acute cases to have "first call" on mental hospital beds, where they would receive active treatment.[116] The Board of Control described the Act as "the outstanding event of the year".[117] As Mathew Thomson put it, the Act "promised to turn the asylum into a place of treatment and cure, rather than of long-term custody".[118] By 1938, nearly 38% of all admissions to English public mental hospitals were voluntary.[119] This meant that a greater proportion of patients were admitted at the early, curable stage of their illness, thereby increasing the need for effective treatment. In terms of occupation, this called for carefully supervised occupational therapy rather than the traditional, unstructured and routinised institutional work given to chronic, incurable, and convalescent patients. This in turn, required well-trained, specialist staff capable of supervising and delivering occupational therapy.

In France there was no equivalent to the English Mental Treatment Act. The only "open" public services, other than those at the Henri Rousselle, remained the faculty clinics attached to medical schools, but places were extremely limited. It was only in 1937 that French Health Minister Marc Rucart, issued a Circular in which he proposed the reorganisation of care for the mentally ill, along the lines already established by the Henri Rousselle Hospital in Paris. Acknowledging that measures to combat mental illness had not been pursued as vigorously as those taken to fight other social scourges (such as tuberculosis), Rucart set out similar proposals for reform to those put forward by the English authorities in the Macmillan Report over a decade earlier. He stressed the "therapeutic, economic and social importance" of early treatment and maintained that

[114] Prior to 1930, the only hospital in England offering free voluntary care was the Maudsley Hospital in London, which opened in 1923. Bethlem Royal Hospital, a registered hospital, also offered voluntary admission, but only to the "deserving poor" (those who worked) and not to rate-aided patients.

[115] Public Assistance Institutions replaced Poor Law Institutions, formerly known as workhouses.

[116] Nineteenth Annual Report of the Board of Control (London: HMSO, 1931), 4.

[117] Sixteenth Annual Report of the Board of Control (London: HMSO, 1930), 1.

[118] Mathew Thomson, "Mental Hygiene in Britain during the First Half of the Twentieth Century," in *International Relations in Psychiatry: Britain, Germany and the United States to World War II*, eds Volker Roelcke, Paul J. Weindling; Louise Westwood (Rochester: University of Rochester Press, 2010), 143.

[119] Ibid.

pauper mental patients, "easily curable" at the start of their illness, became a danger to themselves and others if their condition was left untreated.[120] By the time their condition had deteriorated to the point where the law intervened, they faced the prospect of long-term internment in an asylum, perhaps for life, for which there was a heavy cost to society, both in terms of the patients' care and their lack of productivity.[121] Rucart wanted to see the provision of "open services" for the voluntary admission of those with mild symptoms, outpatient facilities, social services and assistance for "abnormal children" in all departments.[122] These recommendations were followed by another ministerial Circular, issued in 1938, which attempted to modify (but did not supplant) the regulations set out in 1857. They sought to re-orientate the care of patients in "closed", provincial asylums towards a focus on treatment rather than custodial care.[123] Henceforth asylums were to be known as "psychiatric hospitals" to emphasise this new focus, although as one psychiatrist observed, the change was in name only.[124] Most French provincial institutions lacked either the finances or the will to instigate the proposals set out by Rucart before the outbreak of World War II in 1939, and the new law was not enforced.[125]

This lack of reform within provincial institutions led historians Postel and Quetel to observe that the further French asylums were from cities the more they remained locked into psychiatric conservatism (or alienism). They point to a cleavage between the progressive developments in the area of mental hygiene taking place in Paris, under Toulouse's influence (including the establishment of "open" services, outpatient clinics, social services and research facilities), and other institutions outside the

[120] Marc Rucart, Ministre de la Santé publique, Circulaire du 13/10/1937 relative à la ré-organisation de l'Assistance psychiatrique dans le cadre départementale, https://www.ascodocpsy.org/wp-content/uploads/textes_officiels/Circulaire_13octobre1937.pdf (accessed 20/11/2018).

[121] Ibid.

[122] Ibid.

[123] Marc Rucart, Ministre de la Santé publique, Circulaire du 5/02/1938, Règlement modèle du Service Intérieur des Hôpitaux psychiatriques, 39. https://www.ascodocpsy.org/wp-content/uploads/textes_officiels/Circulaire_5fevrier1938.pdf (accessed 03/12/2017).

[124] Postel and Quétel, *Nouvelle Histoire*, 351. The Asile Clinique became a "treatment hospital" in 1927, but only officially changed its name in 1938 to the Hôpital Psychiatrique de Ste Anne.

[125] Isabelle von Bueltzingsloewen, *L'Hécatombe des Fous : La Famine dans les Hôpitaux Psychiatriques Français, 1914–1940* (Paris : Éditions Flammarion, 2007), 282.

metropolis that were effectively left behind.[126] Their distance from the capital and the reformist agenda of the Mental Hygiene Movement, together with their isolation from the rest of medicine, meant that psychiatrists in French provincial asylums, such as the Asile de la Sarthe, continued to run their establishments according to the custodial model of care associated with alienism. The new approach to psychiatric care taken by Toulouse and his colleagues at the Henri Rousselle Hospital was not welcomed by these provincial psychiatrists for whom it represented a threat.[127] Routinised work continued to characterise occupation for calm, chronic and incurable patients in these provincial institutions, while acute and severely disturbed patients were sedated or isolated.

The ongoing cleavage between alienism and psychiatry, which remained marked in France outside the capital, attracted the attention of Aubrey Lewis (who joined the Maudsley Hospital staff in 1929 and was appointed Clinical Director in 1936) during his tour of European psychiatric institutions in 1937. His subsequent report offers a useful insight into the approach of French interwar institutions. While he was in Paris, Lewis observed that "the gulf between the '*médecins des hôpitaux*' and the '*médecins des asiles*' [was] wide".[128] By this he meant that hospital doctors, such as those practising at the Henri Rousselle Hospital and Henri Claude's Faculty Clinic, took a more holistic view of mental illness and were engaged in active treatment, including psychotherapy, while asylum doctors were more likely to hold a traditional organicist stance and to offer only custodial care. Lewis maintained that French asylum medicine drew on its heritage of neurology, hysteria and neurosis, and remained intent on identifying physical causes of mental illness.[129] Furthermore, Lewis observed that there did not appear to be much communication between the various sectors of French psychiatry.

Although psychotherapy and psychoanalysis had begun to make inroads into French psychiatry in the 1920s, it was very much "on tiptoe".[130]

[126] Postel and Quétel, *Nouvelle Histoire*, 352. See also: Aude Fauvel, "Aliénistes contre psychiatres: la médecine en crise (1890–1914)," *Psychologie Clinique* no. 17 (2004): 61–76.

[127] Postel and Quétel, *Nouvelle Histoire*, 351.

[128] Aubrey Lewis, "Report on Visits to Psychiatric Centres in Europe 1937," in *European Psychiatry on the Eve of War: Aubrey Lewis, the Maudsley Hospital and the Rockefeller Foundation in the 1930s*, eds Katherine Angel, Edgar Jones, and Michael Neve (London: Wellcome Trust Centre, 2003), 81.

[129] Ibid., 79.

[130] Ibid.; Postel and Quétel, *Nouvelle Histoire*, 350.

Organisations such as the *Société de l'Évolution Psychiatrique*, the *Mouvement Psychanalytique Français* had emerged and the *Revue Française de Psychanalyse* appeared in 1926, but the numbers involved were small and all were based in Paris.[131] Lewis noted that the "more progressive people" were associated with Eugène Minkowski's *Évolution Psychiatrique* group (also Parisian), whose main interest was psychopathology.[132] Rather than trying to explain mental disorder in terms of brain lesions, psychopathologists concentrated on the clinical observation of symptoms, putting them at odds with the neurologists, who regarded the *Évolution Psychiatrique* group somewhat contemptuously.[133] The group mainly comprised hospital doctors (many were attached to the Henri Rousselle, including Georges Heuyer, or to Professor Claude's Faculty Clinic at Ste Anne's, such as Jacques Lacan and Henri Ey) and psychiatrists working outside the asylum system. Most asylum doctors, concluded Lewis, had "very little opportunity" to experience "the more psychological side of therapy".[134] As French historian of psychiatry Jean Guyotat observes, psychoanalysis did not impinge on neuro-psychiatry, but it had considerably more influence at the psychiatric hospital level, thereby reinforcing the division between the "resolutely" organicist stance of the asylums and the psychosocial psychiatry associated with psychoanalysis, of psychiatrists working in private practice and at the Parisian public mental hospitals.[135] The organicism of French asylum medicine was at odds with Lewis' own views; he maintained the psychobiological perspective of most of the Maudsley psychiatrists, who were heavily influenced by Adolf Meyer.

Conclusion

The fact that French psychiatry was slow to develop as a separate discipline, independent of neurology, impeded development of the holistic interpretation of mental disorder that was the prerequisite for the adoption of new methods of occupational therapy. Psychiatrists adopting a holistic approach were more likely to embrace the idea of treating of acute

[131] Ibid.

[132] Psychopathology was defined by Bernard Hart in 1912 as "the science which attempts to explain the problems of mental disorder by psychological principles" (Bernard Hart, *The Psychology of Insanity*, Fourth ed., Cambridge University Press: 1930, xv).

[133] Lewis, "Report on Visits to Psychiatric Centres in Europe 1937," 80.

[134] Ibid., 81.

[135] Jean Guyotat, "La Formation Des Psychiatres En France," 70.

cases with occupational therapy.[136] It was only in Paris, and more specifically, at the Henri Rousselle Hospital and the Faculty Clinic, that more psychological or psychobiological interpretations developed. When mental disorder was interpreted holistically, psychotherapy and re-educative methods, including occupational therapy, were considered beneficial for acute, curable patients. This interpretation was limited to Paris until after World War II because of the organicist stance of most psychiatrists, who perceived mental disorder in purely neurological terms. Sedation for the agitated and routinised work for the calm were the standard means of controlling the large numbers of incurable and chronically ill patients. For the latter groups, work was a useful distraction. This custodial approach, associated with the term "alienism", continued in provincial asylums, such as the Asile de la Sarthe, for far longer than those based in the larger cities, such as Paris and Lyon. At the Asile Clinique, where acute cases received active treatment from the late 1920s onwards, the treatment was biological rather than psychological; acute cases were not prescribed work or occupational therapy. At the Henri Rousselle Hospital, on the other hand, Toulouse considered occupation an important aspect of treatment.

In England, psychiatry was more firmly established as an independent discipline. Campaigning by reformists such as Henry Maudsley; the emergence of psychology; the experience of dealing with shell shocked soldiers during World War I; the influence of the Mental Hygiene Movement; and the findings of the Macmillan Report all contributed to bringing about an ideological shift amongst English psychiatrists. Mental disorder was regarded holistically by a greater proportion of English psychiatrists during the interwar period than before 1914. For psychiatrists, rather than alienists, an individual was no longer regarded as simply sane (without physiological disease) or insane (suffering from a physiological lesion of the brain or central nervous system). Varying degrees of mental disorder placed an individual somewhere on a spectrum of normality and pathology and took account of physiological, psychological, social and environmental factors that could all affect an individual's mental state.[137] This erosion of the traditional boundary between "sanity" and "insanity", referred to as the "continuity" of mental disorder, led to a greater focus on active treatment in mental hospitals.[138] This was facilitated by the separation of

[136] This approach, based on the American model, would develop after World War II.

[137] Hayward, *Transformation of the Psyche*, 71.

[138] Ibid.

curable and incurable cases, and by the passage of the Mental Treatment Act in 1930 that radically changed the rules surrounding admission to public mental hospitals. The move away from a custodial approach to caring for the mentally disordered, where work was routinised and only prescribed for incurable, chronic and convalescent patients, and towards a medicalised approach typical of hospitals, created an environment in which new methods of occupational therapy could gain acceptance as a curative treatment for acute, curable patients.

Bibliography

Secondary Sources

Bartlett, F.C. "Obituary: Dr Charles S. Myers CBE, FRS." *Nature* 158 no. 4019 (1946): 657–8.

Bewley, Thomas. *Madness to Mental Illness.* London: The Royal College of Psychiatrists, 2008.

Bogousslavsky, Julien and Laurent Tatu. "French Neuropsychiatry in the Great War: Between moral support and electricity." *Journal of the History of the Neurosciences,* 22 (2013): 144–54.

Clarac, F. and F. Boller. "History of Neurology in France." In *Handbook of Clinical Neurology* edited by S. Finger, F. Boller, and K.L. Tyler, 629 56. Amsterdam: Elsevier B.V., 2010.

Colas, M.D., F. Benslama and M. Daudin. "Les "Gueules Cassées", d'une génération à une autre: approche sociologique et psychopathologique des blessures de guerre." *Annales Médico-psychologiques,* 170 no. 4 (2012): 238–43.

Collie, Michael. "Introduction." In *Henry Maudsley, Victorian Psychiatrist: A Biographical Study,* edited by Michael Collie, 1–8. Winchester: St Paul's Bibliographies, 1988.

Crammer, John. "Training and Education in British Psychiatry 1770–1970." In *150 Years of British Psychiatry, Volume II: The Aftermath,* edited by Hugh Freeman and German E. Berrios, 209–42. London: Athlone Press, 1996.

Crocq, Louis. "La Psychiatrie de la Première Guerre Mondiale. Tableaux Cliniques, Options Pathogéniques, Doctrines Thérapeutiques." *Annales Médico-psychologiques,* 163 no. 3 (2005): 269–89.

Delaporte, S. *Les Gueules Cassées.* Paris: Noesis, 1996.

Ellenberger, Henri F. *The Discovery of the Unconscious: The History and Evolution of Dynamic Psychiatry.* London: Fontana, 1994.

Erskine, W.J.A. "Epitome of Current Literature: Mental Hygiene." *Journal of Mental Science,* 69 no. 285 (1923): 241–243.

Fauvel, Aude. "Aliénistes contre psychiatres: la médecine en crise (1890–1914)," *Psychologie Clinique* no. 17 (2004): 61–76.

Fennell, Phil. *Treatment without Consent: Law, Psychiatry and the Treatment of Mentally Disordered People since 1845*. London: Routledge, 1996.

Feudtner, Chris. "'Minds the dead have ravished': Shell-shock, the History and Ecology of Disease Systems." *History of Science*, 31 no. 4 (1993): 337–420.

Fitzgerald, Michael. "Why did Sigmund Freud refuse to see Pierre Janet? Origins of Psychoanalysis: Janet, Freud or Both?" *History of Psychiatry*, 28 no. 3 (2017): 358–64.

Freeman, Hugh. "Psychiatry in Britain, *c*.1900." *History of Psychiatry*, 21 no. 3 (2010): 312–24.

Gelfand, Toby. "Neurologist or Psychiatrist? The Public and Private Domains of Jean-Martin Charcot." *Journal of the History of the Behavioural Sciences*, 46 no. 3 (2000): 215–29.

Guillemain, Hervé and Stéphanie Tison. *Du Front à L'Asile, 1914–1918*. Paris : Alma éditeur, 2013.

Guyotat, Jean. "La Formation des Psychiatres en France." *Raison Présente* no. 83 (1987): 69–81.

Hart, Bernard. *The Psychology of Insanity*. Cambridge: Cambridge University Press, 1912.

Hayward, Rhodri. *The Transformation of the Psyche in British Primary Care, 1880–1970*. London and New York: Bloomsbury, 2014.

Henderson, David K. "Nineteenth Maudsley Lecture: A Revaluation of Psychiatry." *Journal of Mental Science*, 85 no. 354 (1939): 1–21.

Huteau, Michel. *Psychologie, Psychiatrie et Société Sous La Troisième République, La biocratie d'Édouard Toulouse (1865–1947)*. Paris: L'Harmattan, 2002.

Jones, Edgar and Simon Wessely. *Shell Shock to PTSD: Military Psychiatry from 1900 to the Gulf War*. Hove and New York: Psychology Press, 2005.

Jones, Edgar, Shahina Rahman and Robin Woolven. "The Maudsley Hospital: Design and Strategic Direction, 1923–1939." *Medical History*, 51 no. 3 (2007): 357–78.

Jones, Greta. *Social Hygiene in Twentieth-Century Britain*. London: Croom Helm, 1986.

Jones, Kathleen. *Asylums and After. A Revised History of the Mental Health Services: From the Early 18th Century to the 1990s*. London: Athlone Press, 1993.

Kuhn, Philip. *Psychoanalysis in Britain 1893–1913: Histories and Historiography*. New York: Lexington Books, 2016.

Lamb, Susan. "Social Skills: Adolf Meyer's Revision of Clinical Skills for the New Psychiatry of the Twentieth Century." *Medical History*, 59 no. 3 (2015): 443–64.

Le Centre Neurologique de Montpellier; Dr Villaret; Référence B 189. https://imagesdefense.gouv.fr/fr/le-centre-neurologique-de-montpellier-docteur-villaret.html#. Accessed 24/04/2022.

Leese, Peter. *Shell Shock: Traumatic Neurosis and the British Soldiers of the First World War*. Basingstoke: Palgrave Macmillan, 2002.

Lewis, Aubrey and Nick Hervey. "Bond, Sir (Charles) Hubert (1870–1945), psychiatrist and mental health administrator." *Oxford Dictionary of National Biography*, 23/09/2004. https://www.oxfordnb.com. Accessed 30/04/2022.

Lewis, Aubrey. "Report on Visits to Psychiatric Centres in Europe 1937." In *European Psychiatry on the Eve of War: Aubrey Lewis, the Maudsley Hospital and the Rockefeller Foundation of the 1930s*, edited by Katherine Angel, Edgar Jones and Michael Neve, 64–147. London: Wellcome Trust Centre, 2003.

Loughlin, K. "Miller, Hugh Crichton (1877–1959), psychotherapist." *Oxford Dictionary of National Biography*, 23/09/2004. https://www.oxfordnb.com. Accessed 30/04/2022.

Ohayon, Annick. *L'impossible rencontre, Psychologie et Psychanalyse en France (1919–1969)*. Paris: Éditions La Découverte, 1999.

Pols, Hans. "'Beyond the Clinical Frontiers': The American Hygiene Movement 1910–1945." In *International Relations in Psychiatry: Britain, Germany, and the United States to World War II*, edited by Volker Roelcke, Paul J. Weindling, and Louise Westwood, 111–33. Rochester, NY: University of Rochester Press, 2010.

Porter, Roy. "Two Cheers for Psychiatry! The Social History of Mental Disorder in Twentieth-Century Britain." In *150 Years of British Psychiatry, Volume II: The Aftermath*, edited by Hugh Freeman and German E. Berrios, 383–406. London: Athlone Press, 1996.

Postel, Jacques and Claude Quétel. *Nouvelle Histoire de la Psychiatrie*. Paris: Dunod, 2012.

Prestwich, Patricia. "Family Strategies and Medical Power: 'voluntary' committal in a Parisian asylum, 1876–1914." *Journal of Social History,* 22 June (1994). Accessed 2 April 2021 from *The Free Library*: https://www.thefreelibrary.com/Family+strategies+and+medical+power%3a+%27voluntary%27+committal+in+a...-a016108112.

Roudebush, Marc. "A Battle of Nerves: Hysteria and its Treatment in France during World War I." In *Traumatic Pasts: History, Psychiatry and Trauma in the Modern Age, 1870–1930*, edited by Paul Lerner and Mark Micale, 253–79. Cambridge: Cambridge University Press, 2001.

Simonnot, Anne-Laure. *Hygiénisme et Eugénisme au XXe Siècle: à travers la psychiatrie française*. Paris: Éditions Seli Arslan, 1999.

Thomas, Gregory M. *Treating the Trauma of the Great War: Soldiers, Civilians and Psychiatrists in France, 1914–1940*. Baton Rouge: Louisiana State University Press, 2009.

Thomas, Gregory M. "Open Psychiatric Services in Interwar France." *History of Psychiatry*, 15 no. 2 (2004): 131–53.

Thomson, Mathew. "Mental Hygiene in Britain During the First Half of the Twentieth Century: The Limits of International Influence." In *International Relations in Psychiatry: Britain, Germany, and the United States to World War II,* edited by Volker Roelcke, Paul J. Weindling, and Louise Westwood, 134–55. Rochester, NY: University of Rochester Press, 2010.

Von Bueltzingsloewen, Isabelle. *L'Hécatombe des Fous: La Famine dans les Hôpitaux Psychiatriques Français sous l'Occupation.* Paris: Éditions Flammarion, 2007.

Archival Sources

Author Unknown. "National Council for Mental Hygiene." *British Medical Journal,* May 13 (1922): 766–7.

Autor Unknown. "The National Council for Mental Hygiene." *Journal of Mental Science.* 69 no. 286 (1923): 403–4.

Babinski, Joseph. *Hystérie-pithiatisme et troubles nerveux d'ordre réflexe* (Paris: Masson, 1917).

Colin, Henri. "Mental Hygiene and Prophylaxis in France." *Journal of Mental Science,* 67 no. 279 (1921): 459–70.

Dausset, Louis. Rapport Général au nom de la 3e Commission sur les propositions budgétaires pour le service des Aliénés (budget de 1919). Paris : Conseil Général de la Seine, 1918. ADP/D.10K3/27/20.

Marie, Auguste. Rapport de la Service d'Admission, Asile Clinique, 1927. Paris : Préfecture du Département de la Seine, 1927. ADP/PER-566-5.

Mercier, Charles. *Conduct and Its Disorders: Biologically Considered.* London: Macmillan & Co., 1911.

Mott, Frederick. *War Neuroses and Shell Shock.* London: Henry Frowde and Hodder & Stoughton, 1919.

Nineteenth Annual Report of the Board of Control. London: HMSO, 1931.

Report of the War Office Committee Enquiry into "Shell Shock". London: HMSO, 1922.

Rousselle, Henri. Rapport au nom de la 3e Commission sur le Service départmental de prophylaxie mentale et les service annexes. Paris : Conseil Général de la Seine, 1923, 3–4. ADP/D.10K3/34, Report 87.

Rucart, Marc. Ministre de la Santé publique. Circulaire du 13/10/1937 relative à la ré-organisation de l'Assistance psychiatrique dans le cadre départementale. https://www.ascodocpsy.org/wp-content/uploads/textes_officiels/Circulaire_13octobre1937.pdf. Accessed 20/11/2018.

Rucart, Marc. Ministre de la Santé publique. Circulaire du 5/02/1938, Règlement modèle du Service Intérieur des Hôpitaux psychiatriques, 39. https://www.ascodocpsy.org/wp-content/uploads/textes_officiels/Circulaire_5fevrier1938.pdf. Accessed 03/12/2017.

Salmon, Thomas W. *The Care and Treatment of Mental Diseases and War Neuroses ("shell shock") in the British Army.* New York: War Work Committee of the National Committee for Mental Hygiene, 1917.

Sixteenth Annual Report of the Board of Control. London: HMSO, 1930.

Sixty-third Report of the Commissioners in Lunacy, 1909. London: HMSO, 1910.

Smith, G. Elliot and T.H. Pear. *Shell Shock and its Lessons*, Second Edition. Manchester: Manchester University Press, 1917.

Toulouse, Édouard. Rapport au Conseil Général de la Seine au nom de la commission chargée d'étudier l'assistance des aliénés en Angleterre et en Ecosse. Paris : Conseil Général de la Seine, 1898. ADP/D.10K3/12.

Toulouse, Édouard. Rapport de la division des femmes, 1er section, Asile de Villejuif, 1913. Paris: Préfecture du Département de la Seine, 1913. PDS. ADP/PER-566-4.

Truelle, Dr. Rapport de la 2e Section (Hommes), Asile Clinique, 1931. Paris : Préfecture du Département de la Seine, 1931. ADP/PER-566-5.

CHAPTER 4

New Approaches to Patient Work and Occupation

The concept of work as a therapy for the mentally disordered (as revealed in chapter two) was originally conceived in the context of moral treatment, a psychological mode of treatment that emerged in the late eighteenth and early nineteenth centuries. Originally tailored to suit individual patient's needs, the nature of work underwent subtle changes, becoming more routinised and formulaic, as psychiatry moved from a psychological to a physiological paradigm in the late nineteenth century. A subsequent ideological shift within psychiatry, stimulated by World War I towards a more holistic interpretation of mental disorder in England and Paris (analysed in Chap. 3), paved the way for a re-assessment of the therapeutic value of patient work. This chapter examines the new theories regarding patient occupation that emerged in the USA and Germany just before World War I. What were these new ideas, what prompted their development and how did French and English psychiatrists react to them? Did patient occupation in French and English mental hospitals change as a result of the new ideas? What were the medical and non-medical factors that either impeded or encouraged their adoption? This chapter seeks to answer these questions and to ascertain just how new the ideas were, since as Gerald Grob, Marijke Gijswijt-Hofstra and Harry Oosterhuis have pointed out, both the American and the German theories of "occupational therapy" bore a significant resemblance to early

J. Freebody, *Work and Occupation in French and English Mental Hospitals, c.1918–1939*, Mental Health in Historical Perspective,
https://doi.org/10.1007/978-3-031-13105-9_4

nineteenth-century moral treatment.[1] The similarities and differences between moral treatment, late nineteenth-century patient work and the two "new" types of occupational therapy are assessed.

Criticisms of Patient Work After World War I

Psychiatrists Julian Raynier (1888–1936) and Henri Beaudouin (1885–1968)[2] expressed their frustration with how patient work was organised in France.[3] In their influential book that became known as the "bible" for asylum doctors, they claimed that work was not being used to its full therapeutic potential.[4] Chief medical officers should not need reminding of work's beneficial effects on a patient's mental and physical well-being, they claimed, because this had been accepted since the early nineteenth century.[5] Work, especially farm work, was beneficial for all patients, and the authors lamented its absence from many asylums.[6] Contrary to the views of most contemporary French psychiatrists, who insisted on bed-rest for acute cases, they maintained that work accelerated convalescence from acute psychosis, as well as providing an excellent distraction for incurable and chronic patients. Among the benefits, patients gained "social dignity" through productive work, even if they were only

[1] Marijke Gijswijt-Hofstra and Harry Oosterhuis, "Introduction: Comparing National Cultures of Psychiatry," in *Psychiatric Cultures Compared: Psychiatry and Mental Health Care in the Twentieth Century: Comparisons and Approaches*, eds Marijke Gijswijt-Hofstra, et al. (Amsterdam: Amsterdam University Press, 2005), 46. Gerald N. Grob, Review: Pathologist of the Mind: Adolf Meyer and the Origins of American Psychiatry by S. D. Lamb, *Bulletin of the History of Medicine* 89 no. 3 (2015): 618.

[2] Raynier served on several professional commissions of the Superior Council of Public Hygiene, and was appointed inspector general of administrative services for the Ministry of the Interior in 1935; Beaudouin, renowned for his clinical skills and reformist agenda, was chief medical officer of the Parisian Maison Blanche asylum for women from 1926 to 1955.

[3] J. Raynier and H. Beaudouin, *L'Aliéné et les Asiles d'Aliénés au point de vue administratif et juridique, assistance et législation* (Paris: Librairie Le François, 1924), 298. A second edition of their work appeared in 1930 and became the acknowledged reference work for all those working in the asylum system.

[4] Jane Freebody, "'The Root of All Evil is Inactivity': The Response of French Psychiatrists to New Approaches to Patient Work and Occupation, 1918–1939," in *Voices in the History of Madness: Personal and Professional Perspectives on Mental Health and Illness*, eds Robert Ellis, Sarah Kendall, Steven J. Taylor (Cham: Palgrave Macmillan, 2021), 75.

[5] Raynier and Beaudouin, *L'Aliéné,*312.

[6] Ibid.

able to achieve minimal results.[7] The authors urged chief medical officers to do whatever was required to facilitate the provision of work for patients, such as buying or renting land, establishing indoor workshops, or instructing workshop managers in how to supervise patients.[8]

Psychiatrists Charles Ladame (Swiss) and Georges Demay (French) agreed that medical thinking on the suitability of work for patients at the acute stage of their illness had evolved.[9] In their 1926 work, *La thérapeutique des maladies mentales par le travail*, they opined that work was now considered appropriate for acute-stage patients. It was no longer considered necessary to wait until the agitation of delirious patients had completely disappeared since work could focus their attention, channel their energy and lead to a change in habits.[10] Ladame and Demay illustrated the benefit of work by citing the case of a patient suffering from delusions of persecution who remained calm when working in the fields for six days out of seven, but his delirium returned, accompanied by noisy monologues and gesticulations, on Sundays when patients did not work.[11] Work was beneficial for melancholic patients, either encouraging them to engage with their surroundings, or acting as a refuge and distraction. Some dementia patients were also capable of work, including those suffering from dementia praecox, for whom it was particularly beneficial, as Swiss psychiatrist Eugen Bleuler (1857–1939) had observed.[12]

Acute-stage patients needed medical surveillance, requiring workshops for these patients to be located inside the patient quarters.[13] Such workshops could provide the kind of work found in the local area, with which patients were already familiar, such as lace-making, glove-making or weaving.[14] The authors highlighted the Villejuif Asylum's Third Section for criminal and dangerous patients, established by Henri Colin (1860–1930) in 1910, where even the most challenging patients worked. This was unusual in French asylums; an enquiry revealed that only two out of the

[7] Ibid., 313.

[8] Ibid.

[9] C. Ladame and G. Demay, *La Thérapeutique des Maladies Mentales par le Travail* (Paris: Masson et Cie, 1926).

[10] Ibid, 22.

[11] Ibid.

[12] Ibid., 24.

[13] Ibid.

[14] Ibid.

25 asylums surveyed had interior workshops.[15] All Third Section patients were expected to engage in productive work, since criminals had to contribute to the costs of their maintenance.[16] Colin claimed that many of his criminally insane patients were excellent workers.[17] These patients had been unable to attend workshops in the hospital grounds in the past because they required continual surveillance. Situating the workshops inside the patients' quarters enabled them to work, supervised by nurses experienced in manual labour, thereby improving patient behaviour without putting other inmates or staff at risk.[18] The benefits of the Third Section's interior workshops were emphasised by Dr. Calmels at the Congress of French Alienists and Neurologists in Geneva in 1926. He highlighted how patients who were normally forced to remain in their quarters with nothing to do, despite being capable of simple work, including those suffering from chronic delirium or GPI, could be given employment.[19] The boredom and sadness suffered by patients forced to be idle, could be replaced by purposeful activity and an atmosphere of contentment.[20]

In England, debate concerning the way employment was organised in mental hospitals was stimulated by an incendiary book by Dr. Montagu Lomax, entitled *Experiences of an Asylum Doctor*, published in 1921.[21] The book, based on the two years Lomax spent working at Prestwich Mental Hospital in Lancashire between 1917 and 1919, criticised many aspects of asylum administration, including the provision of patient work, recreation, and exercise. The book caused a public outcry and prompted

[15] Ibid., 7.

[16] The disciplinary and therapeutic aims of work for patients in the Third Section are the subject of research by Fau-Vincenti, "Valeur du Travail à la Troisième Section de L'hôpital De Villejuif: Entre Thérapie et Instrument de Discipline." *Savoirs, Politiques et Pratiques de l'exécution des peines en France au XXe siecle*, 2014," http://criminocorpus.revues.org/2788 (accessed 15/04/2016).

[17] Henri Colin, "Le Quartier de Sureté de Villejuif (Aliénés criminels, vicieux, difficiles, habitués des asiles)," *Annales Médico-psychologiques* November (1912): 370–391, December (1912): 540–548; January (1913): 36–65, February (1913): 170–177.

[18] Colin, "Le Quartier de Sureté," *Annales Médico-psychologiques* February (1913): 171–2.

[19] M. Calmels, "Le Travail par petits ateliers à la troisième section de l'Asile de Villejuif," *Annales Médico-psychologiques* January (1927): 283.

[20] Ibid.

[21] Montagu Lomax, *Experiences of an Asylum Doctor: With suggestions for asylum and lunacy law reform* (London: G.Allen and Unwin Ltd., 1921).

an inquiry by the Ministry of Health, led by Sir Cyril Cobb.[22] Lomax deplored the type of work given to patients, the conditions of work and the fact that patients were not paid for their efforts, accusing the authorities of exploitation. He criticised the job of coir-picking: "It is unpleasant, unhealthy work, reminiscent of oakum-picking to those who have been in jail or worked as 'casuals' in workhouses".[23] Lomax remarked that this type of work was "very useful" for the asylum authorities as it saved "much expense", although the dust it generated was dangerous for patients suffering from bronchitis or other chest conditions..[24] It is worth noting here that in 1887 "hair-picking" (a similar task) was introduced at Littlemore where respiratory ailments, including tuberculosis and pneumonia, were rife.[25] At Prestwich, the "closet-barrow gang" were allocated the unenviable task of emptying the asylum's commodes, described by Lomax as the "most unpleasant and unhealthy work of all".[26] It would not have been so bad if the patient workers had been "well fed, well clothed and properly compensated", but this was not the case.[27]

Following these allegations, the Cobb Report, published in 1922, concluded that with regard to Prestwich Hospital "there was room for considerable development in organising the occupation of patients, both as regards the number of patients employed and the variety of work undertaken".[28] Commenting on patient work more generally, Cobb noted that 57% of mental hospital patients were "usefully employed" but that over half were engaged in ward work.[29] This was clearly an area for improvement, particularly when compared with the "remarkable range of work" undertaken at some of the newer mental hospitals, which were equipped with modern workshops.[30] Cobb agreed that patients should receive some sort of remuneration, which would offer an "incentive to

[22] Kathleen Jones, *Asylums and After. A Revised History of the Mental Health Services: From the Early 18th Century to the 1990s* (London: Athlone, 1993), 128.

[23] Lomax, *Experiences of an Asylum Doctor*, 104.

[24] Ibid.

[25] Report of the Medical Superintendent, Annual Reports of the Littlemore Asylum, Oxford, 1887.

[26] Lomax, *Experiences of an Asylum Doctor*, 106.

[27] Ibid.

[28] Cyril Cobb, Ministry of Health, Report of the Committee on Administration of Public Mental Hospitals (London: HMSO, 1922), 51–2.

[29] Ibid., 52.

[30] Ibid.

work" and would stimulate the patient's self-respect.[31] He also agreed that more should be done to "promote social life" than was currently afforded by the weekly entertainments, and that patients should be provided with more opportunities for parole and exercise.[32]

Prompted by the Cobb Report, the Board of Control noted in its annual report of 1923 that "the organisation of occupations in most hospitals is not altogether satisfactory" and that "the number of patients of both sexes whose only employment is ward work is noteworthy".[33] The commissioners remarked, "We attach so much importance to occupation as a curative agent and as a means of promoting the contentment and well-being of patients that we should like to see the organisation of industries placed upon a better footing, possibly by the appointment of an occupations officer."[34] They observed that an Occupations Officer had been appointed at the recently-opened Maudsley Hospital, where provision had been made for occupation and recreation, such as carpentry, gardening and tennis.[35] A Royal Commission on Lunacy and Mental Disorder, ordered to further investigate Lomax's claims, conducted its inquiries between 1924 and 1926.[36] The resulting Macmillan Report of 1926, which advocated the active treatment of all curable patients, indicated that further facilities for occupation and amusement should be provided and that the appointment of an occupations officer should be considered.[37] The findings of the Cobb and Macmillan Reports coincided with the arrival in the UK of new ideas regarding patient occupation from the USA, and would ultimately lead to a divergence in approaches to occupation in England and France.

New Ideas Regarding Patient Occupation

A new approach to using occupation therapeutically was developed in the USA in the 1910s. The National Society for the Promotion of Occupational Therapy (NSPOT) was formed in 1917, on the eve of the USA's entry

[31] Ibid.

[32] Ibid., 55.

[33] Tenth Report of the Board of Control (London: HMSO, 1923), 10.

[34] Ibid.

[35] Ibid, 29.

[36] Jones, *Asylums and After*, 130.

[37] AU, "Royal Commission on Lunacy and Mental Disorder: Summary of Report," *The British Journal of Nursing* (September 1926): 201.

into World War I. The American version of occupational therapy, referred to henceforth as American Occupational Therapy or AOT, grew out of the collective vision of individuals from a variety of backgrounds including psychiatry, psychology, nursing, social reform, and the arts, who shared a belief in the therapeutic value of occupation. Influences included the Arts and Crafts Movement, the philosophy of pragmatism, the Work Cure, and the Mental Hygiene Movement, as Catherine Paterson and Ann Wilcock have outlined.[38] The emergence of AOT coincided with a "backlash" against the "extreme somaticism" of late nineteenth-century psychiatry (discussed in chap. 2). By the 1880s, as Ben Harris explains, American psychiatrists were under increasing pressure from neurologists to become more scientific in their approach to mental disease.[39] As a result, American psychiatrists took a far greater interest in pathology, physiology, and pharmacology, and in surgical and endocrinological treatments, and far less interest in the psychologically-oriented moral treatment. A resurgence in enthusiasm for psychological methods followed in the 1910s, associated with the rise of psychotherapy in the USA at that time.[40]

Dr. Adolf Meyer developed what he called a "psychobiological" approach to psychiatry, which took account of both psychological and biological factors. He believed that "psyche" (mind) and "soma" (body) should be considered as different dimensions of the same entity. In other words, he believed in the "continuity" of mental disorder. For Meyer, mental illness was not a structural defect of the mind or body, but a lowering of an individual's capacity to function or to adapt to his social situation. Differences between normality and abnormality, between psychosis and neurosis were not absolute, but shades of grey.[41] Meyer recognised that occupation could be used to help individuals solve problems of adaptation to their environment, which he regarded as one of the main causes

[38] Catherine F. Paterson, *Opportunities not Prescriptions: The Development of Occupational Therapy in Scotland 1900–1960* (Aberdeen: Aberdeen History of Medicine Publications, 2010), 27–44; Ann A. Wilcock, *Occupation for Health, Vol II* (London: British Association and College of Occupational Therapists, 2002).

[39] Ben Harris, "Therapeutic work and mental illness in America, *c.*1830–1970," in *Work, Psychiatry and Society, c.1750–2015* ed. Waltraud Ernst (Manchester: Manchester University Press, 2016), 64.

[40] Ibid.

[41] Edgar Jones, "Aubrey Lewis, Edward Mapother and the Maudsley," *Medical History Supplement* 22 (2003): 14.

of mental disease.[42] He also believed that "the proper use of time in some helpful and gratifying activity" was fundamental to the treatment of the psychiatric patient.[43] Meyer observed that work around the asylum did little to stimulate a patient's interest or enthusiasm. Conversely, he noted how readily patients responded to a simple programme of craft activities, taking "a pleasure in achievement, a real pleasure in the use of and activity of [their] hands and muscles, and a happy appreciation of time".[44] Through engaging in arts and crafts, patients learned how to organise their time and make the best use of the opportunities available to achieve their goals.[45]

The Arts and Crafts Movement of the late nineteenth century encouraged the creation of hand-made goods and offered an alternative to the perceived harshness of late nineteenth-century industrialism. It emphasised the spiritually uplifting nature of quality work and craftsmanship, attributes that resonated with the aims of occupational therapy.[46] Like the British founders of the Arts and Crafts Movement, John Ruskin (1819–1900) and William Morris (1834–96), Meyer was critical of the industrial processes of production. He maintained that "Our industrialism has created the false, because one-sided, idea of success in *production* to the point of overproduction, bringing with it a kind of nausea to the worker" and a loss of "the capacity and pride of workmanship".[47] Echoes of Karl Marx (1818–83) can be detected here. Marx regarded the capitalist mode of production as "alienating" to the worker, who was deprived of control over the product, and of engaging in psychologically satisfying activity.[48] There was no dignity in the work. Marx regarded "purposeful activity" as necessary for the "realisation of the full humanity of the individual".[49] Meyer recognised the potential of arts and crafts to enable patients to express their creativity, experiment with different materials and "give the satisfaction of completion and achievement", thereby boosting a

[42] Adolf Meyer, "The Philosophy of Occupational Therapy," in *The Collected Papers of Adolf Meyer, Volume IV: Mental Hygiene*, ed. Eunice Winters (Baltimore: Johns Hopkins Press, 1952), 86.

[43] Ibid., 87.

[44] Ibid., 86.

[45] Ibid., 90–2.

[46] Rosalind Blakesley, *The Arts and Crafts Movement* (London and New York: Phaidon, 2006), 8.

[47] Meyer, "Philosophy," 90.

[48] Paterson, *Opportunities not Prescriptions*, 35.

[49] John Dupré and Regenia Gagnier, "A Brief History of Work," *Journal of Economic Issues* 30 no. 2 (1996): 554.

patient's self-esteem.[50] Meyer believed that "the main advance of the new scheme was the blending of work and pleasure" and the fact that the activities were organised according to their appropriateness for individual patients, rather than part of an overall, centralised scheme of work.[51]

Meyer introduced one of the earliest programmes of AOT at the Henry Phipps Clinic, part of the Johns Hopkins Hospital, in 1913. Overseen by one of the founders of NSPOT, Eleanor Clarke Slagle, the daily routine was structured to resemble "normal" everyday life. On being woken up, patients were encouraged to bathe and dress. They were expected to do this themselves; it was considered "pointless" to try and force them to dress. This was a different approach to that adopted in most asylums where it was the nurses' or attendants' responsibility to dress the patients.[52] After breakfast, during which conversation was encouraged, patients went to work in the occupation rooms. Here they were given classes in clay modelling, painting, weaving, bookbinding, knitting, leatherwork and basket weaving. These handicrafts were aimed at igniting impulses of self-interest and the desire for satisfaction essential for efficient adaptation.[53] In the carpentry workshops, male patients made wooden trays, tables and bookshelves, while in the needlework room, female patients made slippers, shawls and tablecloths. Patients were encouraged to send their achievements home as gifts for their families and friends.[54] Lunch was followed by rest or quiet activities on the wards. These might include reading, card-playing, letter-writing or domestic tasks, such as clearing away meal trays, sweeping, bedmaking or polishing brass fixtures. Meyer maintained that help around the wards, provided it was voluntary and pleasant and gave a sense of "helpful enjoyment", acted as an "instrument of biological adaptation".[55] This contrasted with the obligatory, often exploitative, labour of patients in the large public institutions, but was nonetheless focused on cultivating productivity and usefulness.

The concept of usefulness was key to the German method of occupational therapy. This also involved the re-education of the asylum patient through the establishment of a regular routine of occupation, rest and

[50] Meyer, "Philosophy," 90.

[51] Ibid., 88.

[52] S. D. Lamb, *Pathologist of the Mind: Adolf Meyer and the Origins of American Psychiatry* (Baltimore: Johns Hopkins University Press, 2014), 169.

[53] Ibid., 170.

[54] Ibid.

[55] Ibid., 180.

recreation. It differed from AOT in that patients were engaged in the "real work" required to run the institution, rather than in arts and crafts. This was considered important because it allowed patients to feel that their work had a useful purpose. Unlike AOT, the German method was developed by just one individual, the Dr. Hermann Simon (1867–1947). When Simon took up the position of medical director of the newly built Warstein Asylum in 1905, he discovered that the landscaping of the grounds had not been completed. With staff in short supply, Simon directed his patients to complete the work.[56] Initially he only selected those patients who were considered fit enough to work, but soon involved those who were agitated or in bed.[57] He observed that as more patients were involved, a general improvement in behaviour took place.[58] Patients became much calmer and more orderly, reducing the need for sedative drugs and isolation cells.[59] Within nine years, Simon was able to occupy *c.*90% of his patients. The whole atmosphere of the institution changed into one of purposeful activity, and patients took a renewed interest in their surroundings.[60] Simon realised that regular, serious activity was part of normal, everyday life and that scheduling some sort of work (however limited) into the daily routine, made the adjustment to institutional life easier for newly admitted patients.[61]

Simon used his experiences at Warstein to develop "More Active Therapy" (*aktivere Krankenbehandlung*), henceforth known as MAT. Simon instituted MAT at the Gütersloh Asylum in Westphalia, Germany, where he became medical director in 1919. His theory was published in a German psychiatric journal in 1927 and in book-form in 1929.[62] Simon believed that every patient, even those at the acute stage of their illness, should be set to work on admission to hospital.[63] Patients should feel that the work assigned to them had a purpose, since this was essential to maintaining their engagement and interest in the activity. Simon main-

[56] Frederic Wertham, "Progress in Psychiatry II: The Active Therapy of Dr. Simon," *Archives of Neurology and Psychiatry* 24 no. 1 (1930): 151.

[57] Monika Ankele, "The Patient's View of Work Therapy: The Mental Hospital Hamburg-Langenhorn during the Weimar Republic." In *Work, Psychiatry and Society, c.1750–2015*, ed. Waltraud Ernst (Manchester: Manchester University Press, 2016), 242.

[58] Wertham, "Progress", 151.

[59] Ankele, "The Patient's View of Work Therapy," 242.

[60] Wertham, "Progress," 151.

[61] Ibid., 152.

[62] Hermann Simon, "Aktivere Krankenbehandlung in der Irrenanstadt," *Allegemeine Zeitschrift fur Psychiatrie* 87(1927): 97–145.

[63] Ankele, "The Patient's View of Work Therapy," 242.

tained that "the work allocated to a patient should be real and serious" such as work in the laundry, kitchen, grounds, poultry house or offices. The patient should be paid for this work, no matter how small the amount.[64] Simon insisted that "Whatever task can be done by someone who is sick is not to be done by someone who is healthy."[65] He also insisted that the economic value of the work was not its main goal, but that the focus should be on the benefit to the patient. Simon believed that the purpose of treatment was to "re-introduce a healthy logic into the life and mental world of the patient".[66] Idleness, according to Simon, was not only the "root of all evil… but also of impending idiocy".[67] In line with the principles of moral therapy, work was to be balanced by social and recreational activities. Like Ladame and Demay, Simon believed it was necessary to schedule some form of "activity" every day including Sundays.[68]

Simon acknowledged that it required considerable education to habituate patients to work. The physician had to assess the mental state of the patient when allocating work and deciding on the training method. Simon divided the work into five stages of increasing difficulty, which he compared to the grades of a school.[69] New patients were initially assigned simple tasks—such as helping to carry a basket or dusting furniture—and as their performance improved, they were given ever more demanding assignments.[70] The final stage, representing the normal work capacity of a healthy individual, might involve work outside the hospital, thereby strengthening a patient's sense of responsibility and independence.[71] Whilst a patient should not be allowed to become overtired, it was important to "push a patient to the upper limits of their abilities" in order that they made progress. Striking this balance was a challenge for the physician.[72] The aim was to educate patients to take responsibility for their actions and to play a useful role in the community. This ethos, as Monica

[64] Wertham, "Progress," 153.

[65] Hermann Simon, "Active Therapy in the Lunatic Facility (1929)," in *From Madness to Mental Health: Psychiatric Disorder and Its Treatment in Western Civilisation*, ed. Greg Eghigian (New Brunswick: Rutgers University Press, 2010), 273.

[66] Ibid.

[67] Ibid., 272.

[68] Wertham, "Progress," 153.

[69] Ibid., 152.

[70] Simon, "Active Therapy in the Lunatic Facility (1929)," 273–5.

[71] Ibid., 275.

[72] Ibid., 274.

Ankele emphasises, resonated with that of other German welfare institutions during the Weimar period, such as alms-houses, youth centres and prisons, where work played a central role.[73]

The Reception of New Approaches to Patient Occupation in England and France

England

The "new" approaches to occupational therapy began to influence how patients were occupied in England in the early 1920s, but they failed to gain traction in France outside Paris. Initially, the practices associated with AOT, rather than MAT, influenced patient occupation in England. AOT was introduced to Britain in the early 1920s, while MAT only began to attract attention in Britain in the late 1920s and early 1930s. The Scottish psychiatrist, Dr. (later "Sir") David Henderson, who worked for Meyer in the USA before World War I, is believed to be the first psychiatrist to bring AOT to Britain.[74] Henderson had learned Meyer's methods of identifying patients' reaction types, analysing the significance of their social backgrounds, taking meticulous case notes, and prescribing AOT.[75] On becoming Physician-Superintendent of the Glasgow Royal Lunatic Asylum at Gartnavel, Scotland, in 1921, Henderson applied Meyer's psychobiological approach to treatment and introduced occupational therapy.[76] In the Gartnavel Annual Report of 1922, Henderson maintained that "attempts should be made to cultivate good habits in both mind and body, to stimulate interests, and attempt in every way to reconstruct the personality".[77] Papers from the 1924 conference on occupational therapy, organised by Henderson, were published in the *Journal of Mental Science* in 1925. At the conference, Henderson expressed the belief that through AOT "many recoveries are hastened, many improvements are effected, good habits are substituted for bad ones, physical and mental deterioration are retarded,

[73] Ankele, "The Patient's View of Work Therapy," 242.

[74] Henderson worked for Meyer in New York from 1908 to 1910, and at the Henry Phipps Clinic in Baltimore from 1912 to 1915 before serving in the RAMC during World War I.

[75] Hazel Morrison, "Henderson and Meyer in Correspondence: A Transatlantic History of Dynamic Psychiatry, 1908–1929," *History of Psychiatry* 28 no. 1 (2016): 75–6.

[76] Ibid., 82.

[77] Cited in Paterson, *Opportunities Not Prescriptions*, 30.

and life is made more endurable for the great bulk of our permanent population".[78]

Henderson's commitment to Meyer's holistic approach to psychiatry and his advocacy of occupational therapy were re-affirmed in his *Textbook of Psychiatry* published in 1927, written with his colleague R.D. Gillespie and dedicated to Meyer. The volume became the standard textbook for postgraduate students of psychiatry, running to several editions until finally being withdrawn in the late 1970s.[79] In the chapter devoted to occupational therapy, Henderson declared that every mental hospital should have an occupational therapy department, "varied to suit the individual needs of the institution, private or parochial, rural or urban".[80] He believed that guiding patients into "satisfactory work channels" could accomplish more than "all the drugs in the pharmacopoeia".[81] Noting that in the past, patient work frequently constituted "mere drudgery" that must have "antagonised" rather than helped many patients, he maintained that properly organised occupational therapy could inspire in patients a "spirit of hopefulness and of happiness".[82]

Having made few remarks on patient occupation since the war, the Board of Control began to advocate occupational therapy as a means of engaging mentally ill patients who refused, or were unable, to perform hospital maintenance work. In 1928, the Board admitted that employment had been restricted to "those patients whose readiness to work was spontaneous or needed only the urge of some small reward".[83] As Henderson highlighted, many patients who had been considered unemployable could be occupied by "staff trained in teaching handicrafts" if sufficient care was taken to select an activity that appealed to the patient and was appropriate for their mental condition.[84] The Board noted in 1931 that where a "skilled occupation officer" had been appointed, as at Barming Heath, this had resulted in the "employment of types of recent

[78] Cited in: Ibid., 31.

[79] Michael Gelder, "Adolf Meyer and His Influence on British Psychiatry," in *150 Years of British Psychiatry 1841–1991*, eds G.E. Berrios and Hugh Freeman (London: Gaskell, 1991), 431.

[80] D.K. Henderson and R.D. Gillespie, *A Text-Book of Psychiatry for Students and Practitioners*, 5th ed. (London: Oxford University Press, 1940), 600.

[81] Ibid.

[82] Ibid., 593, 600.

[83] Fifteenth Annual Report of the Board of Control (London: HMSO, 1928), 4–5.

[84] Ibid.

and progressing cases who a few years ago would have been regarded as incapable of benefiting by such treatment".[85]

This interest in occupational therapy prompted research visits to Holland and Germany to see MAT in action. These were convened by the Royal Medico-Psychological Association (RMPA)[86] between 1928 and 1933. Visitors to the Santpoort Asylum, where MAT had been instigated in 1926, were impressed by the "silence and stillness" of the wards and by the numbers of patients engaged in activities. Only 10% of patients were unoccupied and this was mainly due to physical sickness.[87] Dr. van der Scheer, medical director of the Santpoort Asylum, explained that his patients' mental disorder was due to educational and environmental factors. His regime was aimed at the re-education of patients through the "acquisition of new experiences" and the generation of "new conditional reflexes".[88] Scheer, like Simon, believed that "every patient is able to do certain kinds of work" including "imbecile, demented [and] maniacal" patients.[89] Visitors noted the various types of work undertaken by patients, which varied in complexity, and the fact that other occupations included "reading, games and dancing".[90] They also noted that "this system of occupation requires a high proportion of staff, who must co-operate in the treatment with the exercise of much patience and intelligence".[91] The Board of Control used the research from such study visits, together with existing knowledge of AOT, to produce a *Memorandum on Occupation Therapy for Mental Patients* in 1933. The 27-page booklet, published to encourage the introduction of occupational therapy into mental hospitals, constituted "a significant early policy document in the history of occupation", according to John Hall.[92]

The Board of Control's booklet defined occupational therapy as "the treatment, under medical direction, of physical or mental disorders by the

[85] Eighteenth Annual Report of the Board of Control (London: HMSO, 1931), 9–10.

[86] The MPA became a "Royal" association in 1926.

[87] AU, "Study Tour and Post-Graduate Educational Information Sub-Committee: Tour of Dutch Mental Hospitals and Clinics," *Journal of Mental Science* Jan (1929): 204.

[88] Ibid.

[89] Ibid.

[90] Ibid.

[91] Ibid.

[92] John Hall, "From work and occupation to occupational therapy: The policies of professionalisation in English mental hospitals from 1919 to 1959," in *Work, Psychiatry and Society, c.1750*–2015, ed. Waltraud Ernst (Manchester: Manchester University Press, 2016), 320.

application of occupation and recreation with the object of promoting recovery, of creating new habits, and of preventing deterioration".[93] Outlining the methods used in the USA, Germany and Holland and how these might be adapted for use in English mental hospitals, it also advised on staff training and the financial implications of introducing occupational therapy. The *Memorandum* downplayed the economic value of occupation to the hospital, stating that "the object of occupation is primarily therapeutic" and should not be viewed as a means of "providing commodities for use in the hospital at a low cost".[94] The Board of Control were keen to make a sharp distinction between occupational *therapy* and mere occupation; for occupation to be therapeutic it had to be prescribed by a doctor and delivered by specialist staff trained in this "branch of *medical* therapy".[95] This emphasised both the medicalisation and professionalisation of occupational therapy, and set it apart from the routinised, systematic application of work that had characterised patient occupation since the mid-nineteenth century.

The Board of Control's *Memorandum* was followed in 1936 and 1938 by the publication of two further guides to providing occupation for mental patients. The first was by Dr. Richard Eager (1881–1947), former medical superintendent of the Devon Mental Hospital.[96] Eager had successfully instigated occupational therapy in Devon, following a visit to Dutch mental hospitals in 1932, where he witnessed Hermann Simon's MAT in action. Eager maintained that where a system of occupation was introduced on "intensive lines", as it had been in Holland, a change in the "general atmosphere" of the hospital could be detected: "the hospital which had formerly been a refuge for idlers, becomes a hive of industry".[97] However, considerable staff cooperation was required, involving the Matron, head Male Nurse and their respective teams, all of whom had a role to play in supervising patients in various tasks. Medical staff needed to devote a "considerable amount of time in allotting patients to suitable

[93] Board of Control, *Memorandum on Occupation Therapy for Mental Patients* (London, HMSO: 1933), 3.

[94] Ibid., 7.

[95] Ibid., 15. [italics my own]

[96] Richard Eager, *Aids to Mental Health: The Benefits of Occupation, Recreation and Amusement* (Exeter: W.V. Cole and Sons, 1936). Interestingly, Eager had been a medical officer at the Littlemore Asylum during World War I (when it was known as the Ashhurst Hospital), working alongside medical superintendent Thomas Saxty Good.

[97] Eager, *Aids to Mental Health*, 6.

classes" since the prescription of occupation was as important—if not more so—than the prescribing of drugs.[98] He continued, "one might even go further, and suggest that the latter might be largely dispensed with, if more attention were given to the former".[99] The introduction of occupational therapy, Eager warned, was not a way of reducing staff numbers or a form of cheap labour. Echoing the Board's *Memorandum*, he emphasised that occupational therapy was a "definite *treatment*, and has no relationship to the value of the work done".[100] Because Eager had instigated MAT most of the occupations he discussed were related to work around the hospital, but like Simon, he also advocated amusements, recreations, and social activities such as cricket, football, Swedish drill, singing, dances, ward parties and games. It was important to engage patients in these activities which warranted just as much thought and planning as work-related activities.[101] He also thought it was important that nurses learned basic craft activities that they could teach to patients, a practice that had been adopted at the Littlemore. Eager's booklet was made available to medical superintendents, public health officials and members of the public.

The second publication was written by Dr. John Ivison Russell (1888–1970), medical superintendent of the Clifton Hospital, York, where he established a very active occupational therapy department. Russell was described in the *British Medical Journal* as "one of the foremost psychiatrists of his generation".[102] His book indicates the ongoing enthusiasm for occupational therapy in England despite the introduction of the new shock treatments by the time the book was published in 1938.[103] Russell's recommendations were based on AOT, with a whole section of the book devoted to the execution of various arts and crafts, including woodwork, basketry, brush-making, bookbinding, matmaking, work with plaster, concrete and stone, needlework, papier maché and weaving. Russell also addressed the rationale for occupational therapy, appropriate occupations for the various psychological types of patients, how to organise the provision of occupational therapy, suggested routines for habit training and

[98] Ibid.

[99] Ibid.

[100] Ibid., 8.

[101] Ibid. 28.

[102] J.M.F., "Obituary: J. Ivison Russell," *The British Medical Journal* 4 no. 5738 (1970): 808.

[103] John Ivison Russell, *The Occupational Treatment of Mental Illness* (London: Ballière, Tindall and Cox, 1938).

the doctor's role in prescribing occupational therapy. The book included a specimen occupational therapy prescription for doctors to use.

The first school of occupational therapy in England, the Dorset House School of Occupational Therapy, was founded by the English physician, Elizabeth Casson, in Bristol in 1930 (see Fig. 4.1). Inspired by Henderson's 1925 article in the *Journal of Mental Science*, Casson visited the occupation department of Bloomingdale Hospital in New York and the Boston School of Occupational Therapy in 1926. She wanted to find out more about occupational therapy, having been troubled by the "atmosphere of bored idleness" she encountered on the wards of the psychiatric unit where she was working.[104] The curriculum at Dorset House was based on AOT. Fifty per cent of the teaching was devoted to arts and crafts, which supplemented classes in anatomy, physiology and psychiatry. Casson had developed an affinity for arts and crafts whilst growing up in "a family with more than average dramatic and musical talent" and she herself had "considerable gifts in drawing and painting".[105] Her tastes and accomplishments were typical of what Lauren Goodlad has referred to as a strong British, middle-class commitment to productive occupation.[106] Casson had been exposed to the ethos of the Arts and Crafts Movement, whilst working for Octavia Hill (1838–1912) at one of her social housing projects, before embarking on her medical training. Hill's mentor had been John Ruskin, co-founder of the Arts and Crafts Movement. Ruskin had encouraged her to bring art, beauty and nature into the lives of her working-class tenants. This made a "deep impression" on Casson according to the first principle of Dorset House, Constance Owens.[107]

The Dorset House School was initially staffed by teachers who had trained in the USA and it remained committed to AOT. This inevitably gave AOT a firm foothold in England since most of the occupational therapists that populated English interwar mental hospitals either trained at Dorset House or were trained by those who had, such as those who were

[104] Elizabeth Casson, "Forward," in *The Story of Dorset House School of Occupational Therapy, 1930–1986*, ed. Betty Collins (Oxford: unpublished), 3.

[105] Esther Reed, "Dr Casson's Early Life – by her sister," *Occupational Therapy* 18 no. 3 (1955): 87.

[106] Lauren E. Goodlad, *Victorian Literature and the Victorian State: Character and Governance in a Liberal Society* (Baltimore and London: The Johns Hopkins University Press, 2003), 35.

[107] Constance Owens, "Obituary – Elizabeth Casson OBE, MD, DPM," *Occupational Therapy* 1 no. 18 (1955): 4.

Fig. 4.1 Student occupational therapists perfecting their weaving skills at the Dorset House School of Occupational Therapy, Bristol, in the 1930s. (*Oxford Brookes University Special Collections, Dorset House School of Occupational Therapy Collection, DH/3/1/ Vol. 1*)

employed at the Maudsley School, which opened in 1932. Dorset House's influence, and thus the prevalence of AOT, was extended when occupational therapy's governing body was established by a group of former Dorset House students in 1936, supported by Casson, and chaired by Constance Owens. This Association of Occupational Therapists was responsible for setting the occupational therapy syllabus, curriculum and examinations. Arts and crafts remained an integral aspect of occupational therapy in England until the 1960s. As John Hall has observed, psychiatrists Wilhelm Mayer-Gross, Eliot Slater and Martin Roth characterised occupational therapy in Britain as "being more of *pastimes and hobbies* than of rough manual work, as on the European continent" in their new psychiatric textbook, *Clinical Psychiatry*, published in 1954.[108]

Occupational therapy was provided as soon as the Maudsley Hospital opened in 1923. The Maudsley's medical superintendent Edward Mapother, and his successor Aubrey Lewis, both embraced the

[108] Hall, "From Work and Occupation to Occupational Therapy," 322.

psychobiological approach of Adolf Meyer.[109] They recognised the benefits of psychological modes of treatment and believed that occupation had a role to play in helping patients re-adjust to their environments. The Maudsley's commitment to occupational therapy was demonstrated by the establishment of the hospital's own school of occupational therapy in 1932. The experience of treating war neuroses during World War I had introduced the Littlemore's medical superintendent, Thomas Saxty Good, to the benefits of psychotherapy and occupation. Good began prescribing occupational therapy (taught by nurses rather than occupational therapists) when the Littlemore re-opened for civilians in 1923. John Porter-Phillips, physician superintendent of the Bethlem Royal Hospital, was more of an organicist and it was his long-established practice to sedate acute-stage patients, rather than occupy them. An occupational therapy department did not open at Bethlem until 1932. It took a letter from the hospital's architect, Alfred Cheston, to the Treasurer of the Board of Governors, Lionel Faudel-Phillips, in which he stated that occupational therapy was "considered by most modern practitioners as [an] essential curative agency", together with complaints from patients about being bored, to bring about provision of occupational therapy.[110]

France

Following attempts by Beaudouin, Raynier, Ladame and Demay to improve the use of patient work in France, French psychiatrists became aware of MAT. The theory and methods of "more active therapy" (MAT) were explained in detail to an audience of French psychiatrists attending the Congress of French-speaking alienists and neurologists at Anvers in July 1928. After the Congress, Parisian psychiatrist Paul Courbon was amongst a group of delegates invited to visit the Dutch asylum at Santpoort, where Hermann Simon's MAT had been introduced two years earlier.[111] Courbon praised the method in the *Annales Médico-psychologiques* in November 1928.[112] Simon's articles in the German psychiatric journal,

[109] Hayward, *The Transformation of the Psyche*, 69.

[110] Jonathan Andrews et al., *The History of Bethlem* (Abingdon and New York, Routledge), 689.

[111] This visit to Santpoort must have just pre-dated the study visits paid by the English representatives of the RMPA to the same establishment.

[112] P. Courbon, "Un voyage d'étude dans les asiles de Hollande," *Annales Médico-psychologiques* no. 2 (1928): 289–306 and 385–405.

the *Allgemeine Zeitschrift* (1927) and his book, *Aktivere Krankenbehandlung in der Irrenanstalt* (1929) were reviewed positively by Parisian psychiatrists, Paul Schiff and G. Halberstadt, in 1929, and by Jacques Vié in 1934.[113] Porot, writing in *L'Hygiène mentale* in 1929, was also complimentary about the way patient work was organised in Dutch asylums, noting how unfavourably work in French asylums compared with the system at Santpoort.[114] Simon shared his views directly with a French audience in 1933, also writing in *L'Hygiène mentale*, where he claimed that work was one of the best means of combatting agitation, impulsivity and a tendency to violence. Whilst work could not cure organic lesions, it could nonetheless strengthen a patient's physical and mental faculties.[115] A report on patient work presented to the Superior Council for Public Assistance in 1934 added to the growing list of recommendations for the adoption of MAT in French asylums.[116]

The arguments appeared persuasive. There were very few conditions that would prevent patients from participating in MAT. When Paul Courbon visited Santpoort, he noted that out of 1420 patients, 1273 were working. Out of the 147 unoccupied patients, 112 were too physically weak to work, leaving just 35 whose mental conditions prevented them from working. Bed-rest quarters were greatly reduced and were reserved for "organic" cases or the very agitated who were being treated with the sedative, somnifene.[117] Porot observed that hydrotherapy equipment had been abandoned because it was no longer required to soothe agitated patients, who were now occupied. There were no patients huddled on benches, nor any crying out or shouting. Night-time agitation had practically disappeared, and isolation cells were no longer in use.[118] All the available workforce was being used, with 85–90% of patients systematically

[113] P. Schiff, "Le movement d'hygiène mentale en Allemagne," *L'Hygiène mentale* no. 8 (1929): 232–233; G. Halberstadt, "A propos de l'Ergothérapie," *Annales Médico-psychologiques* March (1929): 193–198; J. Vié, "Le traitement des malades mentaux par le travail: les idées et les réalisations de H. Simon," *L'Aliéniste français* no. 10 (1934): 589–598.

[114] A. Porot, "L'Assistance par le travail dans les Asiles Hollandais," *L'Hygiène mentale* no. 2 (1929): 41–54.

[115] Hermann Simon, "La Psychothérapie de l'Asile," *L'Hygiène mentale* no. 1 (1933): 20.

[116] D-M. Legrain and G. Demay, "Le travail des aliénés convalescents," *L'Aliéniste Français* no. 6 (1936): 283.

[117] P. Courbon, "Voyage d'études dans les asiles de Hollande," *Annales Médico-psychologiques* no. 2 (1928): 395.

[118] Porot, "L'Assistance par le travail," 45–7.

working—either in bed, in their quarters, in the workshops, or in the grounds.[119]

Like his English colleagues who visited Santpoort, Paul Courbon was deeply impressed by the silence. During the six hours they spent working, patients were forbidden from speaking.[120] The moment patients became agitated they were removed from the work room for half an hour to rest and then brought back again. If they re-offended, they were removed for a longer period; if this proved insufficient, the patients were given the sedative somnifene. Courbon found the results of this method were "extraordinary", while for Porot they were "revolutionary".[121] Courbon described the method of treatment as "*ergothérapie*" or occupational therapy, which was not a term in regular use in France prior to this period. It consisted, he explained, of re-educating the automatic responses of patients and helping them to readapt to social life.[122] The application of MAT totally transformed the atmosphere of asylums; everywhere there was silence, concentration and an impressive discipline.[123] Porot believed that the methods he had witnessed at Santpoort and other Dutch asylums was not simply a way of filling the time and occupying idle hands, but a means of active therapy that countered agitation and other symptoms. It was, as Courbon claimed, "*une véritable rééducation*".[124]

Simon's methods of MAT generated considerable interest amongst a specific cohort of French psychiatrists between 1928 and 1939.[125] These psychiatrists, mainly (although not exclusively) from Paris, shared their views nationally through the various professional journals. They recognised the benefits of a re-educative approach to treating mental disorder, based on Simon's psychological methods and incorporating his type of occupational therapy, MAT. They were pessimistic, however, about the practicalities of introducing MAT into French asylums. There were too many impediments inherent in the French system including overcrowding; lack of funds; management issues; staff quality and training; and the ratio of staff to patients. As A. Walk commented in the English *Journal of*

[119] Ibid., 52.

[120] Courbon, "Voyage d'études," 396.

[121] Ibid., 397; Porot, "L'Assistance par le travail," 52.

[122] Courbon, "Voyage d'études," 397.

[123] Porot, "L'Assistance par le travail," 52.

[124] Ibid.

[125] Although according to Hervé Guillemain, Simon's methods were not widely known in France until after World War II.

Mental Science, the "fully-developed Simon system [MAT] … involves a transformation of the hospital régime" that could only be compared with the asylum reforms introduced in the early nineteenth century.[126] Under Simon's system, Walk maintained, "the entire institution becomes one vast occupational centre" and that the "therapeutic application of work, rest and recreation" had to be the "foremost concern" of all the asylum staff working in collaboration.[127] MAT clearly involved a fundamental change not only in the way the asylum was run, but in the attitudes and skills of staff. This represented a daunting undertaking even for an institution without the problems suffered by French asylums.

An additional factor that could have contributed to a rejection of MAT by French psychiatrists was its German origin. The French had been traditionally hostile to everything German since the defeat of France by Germany in the Franco-Prussian War of 1870–1871.[128] This defeat intensified the French sense of inferiority with regard to Germany, initiated by the loss of French scientific and medical supremacy to Germany earlier in the century.[129] The university chair in psychiatry established in Germany in 1863 demonstrated German psychiatry's scholarly and scientific legitimacy, while the ground-breaking work of Emil Kraepelin, whose influential *Textbook of Psychiatry* was published in eight editions between 1883 and 1915, sealed the German victory over French psychiatry.[130] Postgraduate students flocked to Kraepelin's clinic in Munich, rather than Paris, to study psychiatry under the "new master".[131] Those Parisian psychiatrists who advocated MAT may have been attempting to keep pace with German developments, while others rejected Simon's methods on the grounds of anti-German sentiment.

Édouard Toulouse, on the other hand, was influenced by the American model of occupational therapy, as a result of his contact with American psychiatry through Clifford Beers (with whom he developed a close

[126] A. Walk, "Occupation Therapy and 'Active Therapy'," *Journal of Mental Science* Oct (1933): 759.

[127] Ibid.

[128] Postel and Quétel, *Nouvelle Histoire De La Psychiatrie*, 350.

[129] George Weisz, "Reform and Conflict in French Medical Education, 1870–1914," in *The Organisation of Science and Technology in France 1808–1914*, eds Robert Fox and George Weisz (Cambridge: Cambridge University Press, 1980), 64–5.

[130] Jan Goldstein, *Console and Classify: The French psychiatric profession in the nineteenth century* (Chicago and London: University of Chicago Press, 2001), 350; Paul Lerner, *Hysterical Men War: Psychiatry and the Politics of Trauma in Germany, 1890–1930* (Ithaca, NY: Cornell University Press, 2003), 16.

[131] Ibid.

personal friendship) and the Mental Hygiene Movement.[132] At the Henri Rousselle Hospital, patients were encouraged to keep busy. Art classes, sewing and music were among the activities provided. However, Toulouse himself was a divisive figure who did not "win over" his more neurologically oriented colleagues. They regarded his establishment of the Henri Rousselle Hospital as a threat to their *modus operandi* which revolved around the "closed" asylum.[133] Toulouse's methods, and those of Adolf Meyer, including the prescription of occupational therapy, would eventually be adopted more widely in France after World War II.[134]

The main barrier to the introduction of either AOT or MAT was the organicist stance of most French psychiatrists. The new concepts of patient work and occupation were associated with a psychological or psychobiological approach, rather than an organicist approach to psychiatry. The French psychopathological and psychoanalytical groups highlighted in Aubrey Lewis's 1937 report "remained strictly Parisian" (i.e. Paris-based) during the interwar period and had little impact on asylums outside the capital.[135] New ideas regarding patient occupation were therefore unlikely to find favour outside Paris. Dr. Henry Christy of the provincial Asile de la Sarthe was quite clear that his treatment priorities for curable patients were actively biological, while work remained a useful means of occupying incurable and chronic patients.[136] He was adamant that "It is biology, it is neurology that has generated progress within mental medicine".[137] Like most of his colleagues outside Paris, Christy was wedded to a rigidly organicist interpretation of mental disorder and preferred biological methods of treating acute, curable patients. The Parisian Asile Clinique also remained heavily influenced by neurology. The more psychologically oriented, younger members of the medical staff did not have a significant impact until after World War II, which helps to explain the small numbers of patient workers within the treatment sections of the Asile Clinique.

[132] Michel Huteau, *Psychologie, Psychiatrie et Société*, 186–187.

[133] Postel and Quétel, *Nouvelle Histoire*, 352.

[134] Ibid.

[135] Jean-Christophe Coffin, "'Misery and Revolution' the organisation of French psychiatry, 1900–1980," in *Psychiatric Cutlures Compared: Psychiatry and Mental Health Care in the Twentieth Century. Comparisons and Approaches*, edited by Marijke Gijswijt-Hofstra, Harry Oosterhuis, Joost Vijselaar and Hugh Freeman (Amsterdam: Amsterdam University Press, 2005), 231.

[136] Rapport du Médecin en Chef, l'Asile de la Sarthe, 1935 (Le Mans, 1936), 11. ADS-1X964/5.

[137] Rapport du Médecin en Chef, l'Asile de la Sarthe, 1936 (Le Mans, 1937), 4. ADS-1X964/5.

Moral Treatment, Late Nineteenth-Century Patient Work, and Occupational Therapy

The similarities between the tenets of occupational therapy, as set out in the Board of Control's 1933 *Memorandum*, and early nineteenth-century moral treatment, are striking. Both were based on a psychological approach to treatment and on the belief that occupation could be curative if carefully selected, supervised and monitored. Both were aimed at modifying behaviour by engaging the patient in satisfying activities. Focusing the patient's attention on an appropriate task distracted her from her troubles, helped her develop self-control and concentration, and boosted self-esteem. Occupations were to be selected according to patient preference, existing skills and the severity of symptoms. The occupations formed part of a daily routine that included regular hours for work, meals, rest and recreation, thereby helping to re-establish a "normal" pattern of behaviour. This was known as habit training in the early twentieth century, but the principles of establishing a routine were the same as in the early nineteenth century. Pragmatism—or what proved to be effective—informed both moral and occupational therapy. These similarities between moral and occupational therapy set both apart from the bureaucratic system of patient work that had evolved in the late nineteenth century.

One of the main practical differences between moral and occupational therapy was the grading, according to different levels of complexity, of the tasks associated with occupational therapy. This was particularly apparent with MAT. Patients were guided through five different levels of difficulty, with the earliest, easiest stage (such as litter-picking) reserved for newly admitted, acute patients, while work at the final stage, designed to be as similar as possible to normal work outside the hospital, was prescribed to patients nearly ready for discharge. This ensured that the patient was continually challenged without being overstretched. The idea of graded activities, prescribed to suit the level and abilities of the patient, had originated in the treatment of tuberculosis. The English physician Marcus Paterson, superintendent of the Brompton Hospital Sanatorium, developed a programme of graduated labour, comprising six levels of work designed to build up a patient's stamina, in the early 1900s.[138] Paterson noticed that the programme resulted in psychological as well as physical improve-

[138] Cynthia Creighton, "Graded Activity: Legacy of the Sanatorium," *American Journal of Occupational Therapy* 47 no. 8 (1993): 746.

ment.[139] Simon may have been aware of this method of treatment for tuberculosis.

The tasks comprising AOT were based on arts and crafts, while with MAT the tasks comprised work around the hospital. That said, the line between what comprised work that contributed to the institution and what constituted arts and crafts could be blurred, as in the cases of woodwork, rug making or basketry. It really depended on the intended purpose of the items being made, whether they were for use in the institution or for the patients themselves. Within the context of AOT, activities based on arts and crafts were aimed at enabling the patient to produce an item that was aesthetically pleasing. The process of making the item by hand should be satisfying and the finished product should instil pride. The skill of the therapist was to inspire the patient, to identify a craft for which the patient had an affinity and that was not too difficult (or too simple) for their ability and condition. The rationale for MAT, on the other hand, was to give patients a sense of genuine purpose—a feeling that what they were doing was useful and contributed to their community.

There was nothing intrinsically different about the *type* of work outlined in Simon's programme and the work performed by patients in the late nineteenth century, since both were based on the work required to maintain the institution. The difference lay in the way it was prescribed, such as the grading of tasks, the meticulous supervision by nursing and medical staff, and the condition of the patients to whom it was allocated. Almost all patients, including the acutely ill, were given some sort of productive work, according to the principles of MAT. Simon believed that even patients at the acute stage of their illness should be occupied. Late nineteenth-century patient work was limited to patients who required little supervision, namely calm, incurable and chronic patients and convalescents. Patients were allocated tasks according to where their labour was needed, rather than on the basis of their needs. Patient work, while still provided within the framework of moral treatment, no longer took account of the patient's individual preferences, aptitudes and skills, and the patient's progress was not closely monitored by a doctor. Late nineteenth-century psychiatrists, while continuing to believe in the benefits of occupation as a distraction, no longer regarded it as curative, and occupation ceased to be considered suitable for patients at the acute stage of their illness.

MAT required total commitment from the chief medical officer; the existence of medical officers to assist with supervision; sufficient numbers

[139] Ibid.

of well-trained, competent nurses; and enough flexibility within the system to allow for a re-organisation of the hospital regime. In most French mental hospitals during the interwar period, the above requirements were lacking. In particular, support for occupational therapy was less likely to be forthcoming from French chief medical officers than their English counterparts either because of French psychiatrists' preference for biological modes of treatment or because they remained wedded to a custodial model of care. In both cases an adherence to an organicist interpretation of mental disorder led to a rejection of occupation for acute cases or for patients who required significant supervision. For all of these reasons, MAT was not adopted in French mental hospitals during the interwar period, despite the recommendations of certain Parisian psychiatrists.

In England, where a psychobiological rather than a neurological approach prevailed, occupational therapy was used curatively and was considered appropriate for acute patients, including those who were still confined to bed. Arts and crafts were also used to engage patients who refused, or were unable, to perform hospital maintenance work. As Henderson had noted, under the late nineteenth-century system of work, many patients remained unemployed, either because of "inefficiency, helplessness, or poor general state of health", while the new system of occupational therapy was designed to appeal to those who had never been employed and to "stimulate anew" those who had failed.[140] It was noted by the Board of Control that many patients who had been considered unemployable could be occupied by "staff trained in teaching handicrafts" if sufficient care was taken to select an activity that appealed to patients and was appropriate for their mental condition.[141] Psychiatrist A. Walk agreed that the employment of an occupational therapist made a significant difference to the conditions of patients who had previously been idle.[142]

Walk suggested that the introduction of AOT only benefited a certain number of patients and left "untouched the general character of the hospital". Hospitals that adopted MAT, on the other hand, were "transformed" into hives of productive activity, with almost all patients engaged in supervised occupation of some kind.[143] The recommendations of the Board of Control's *Memorandum*, however, indicated that all occupa-

[140] Henderson and Gillespie, *A Text-Book of Psychiatry for Students and Practitioners*, 594.
[141] Ibid.
[142] Walk, "Occupation Therapy and 'Active' Therapy," 758.
[143] Ibid.

tions, whether arts and crafts or work around the hospital, should be focused on the therapeutic benefits to the patient, rather than on the maintenance requirements of the institution. In a departure from the stance taken before World War I, the financial benefits of work around the hospital (discussed in chap. 5) were downplayed by the Board of Control. The introduction of occupational therapy for acute patients, and for those with severe symptoms who had previously been idle, marked a significant change in the way these patients were treated in English mental hospitals. If much of the work around an institution serving a mixed clientele continued to be performed by incurable and chronic and convalescent patients, the primary aim of this work—in theory at least—was to benefit the patient.

Conclusion

From this analysis, it appears that occupational therapy represented a return to the principles of moral treatment. Occupational therapy, whether based on arts and crafts or on work around the hospital, can be seen as a more sophisticated form of the individualised work programmes that comprised moral treatment. Formalised in medical practice and through the training and professional qualifications of its practitioners, occupational therapy was a medicalised and professsionalised version of moral treatment. As such, occupational therapy provides an example of a therapeutic method that was not entirely new, but a re-envisaged form of a pre-existing approach to treating the mentally ill. The curative use of occupation followed in the wake of changes within psychiatric ideology from psychological, to physiological, and to psychobiological perspectives, and from therapeutic optimism to pessimism and back to optimism. In England, and at the Henri Rousselle Hospital in Paris, occupational therapy became a recognised form of treatment for acute cases by psychiatrists who had embraced Adolf Meyer's psychobiological stance and who believed in the curability of mental disorder. In provincial France, where most psychiatrists continued to adhere to a physiological interpretation of mental disorder, patient work remained in the style of the late nineteenth century. Here, the benefits of work to the institution were at least as important as those to the patient, not least because the provision of work was limited to incurable, chronic and convalescent patients. The financial implications of patient work and occupational therapy are discussed in chap. 5.

Bibliography

Secondary Sources

Andrews, Jonathan, Asa Briggs, Roy Porter, Penny Tucker, and Keir Waddington. *The History of Bethlem*. Abingdon and New York: Routledge, 1997.

Ankele, Monika. "The Patient's View of Work Therapy: The Mental Hospital Hamburg-Langenhorn during the Weimar Republic." In *Work, Psychiatry and Society, c.1750–2015*, edited by Waltraud Ernst, 238–61. Manchester: Manchester University Press, 2016.

Blakesley, Rosalind. *The Arts and Crafts Movement*. London and New York: Phaidon, 2006.

Casson, Elizabeth. "Forward." In *The Story of Dorset House School of Occupational Therapy, 1930–1986,* edited by Betty Collins, 3–6. Oxford: unpublished, 1986.

Coffin, Jean-Christophe. "'Misery and Revolution': The Organisation of French Psychiatry, 1900–1980." In *Psychiatric Cultures Compared: Psychiatry and Mental Health Care in the Twentieth Century: Comparisons and Approaches*, edited by Marijke Gijswijt-Hofstra, Harry Oosterhuis, Joost Vijselaar and Hugh Freeman, 225–48. Amsterdam: Amsterdam University Press, 2005.

Creighton, Cynthia. "Graded Activity: Legacy of the Sanatorium." *American Journal of Occupational Therapy* 47, no. 8 (1993): 745–8.

Dupré, John and Regenia Gagnier. "A Brief History of Work." *Journal of Economic Issues*, 30 no. 2 (1996): 553–9.

Fau-Vincenti, Véronique. "Valeur du Travail à la Troisième Section de l'Hôpital de Villejuif: Entre Thérapie et Instrument de Discipline." *Savoirs, Politiques et Pratiques de l'exécution des peines en France au XXe siecle*, 2014. Accessed 15/04/2016. http://criminocorpus.revues.org/2788.

Freebody, Jane. "'The Root of All Evil is Inactivity': The Response of French Psychiatrists to New Approaches to Patient Work and Occupation, 1918–1939." In *Voices in the History of Madness: Personal and Professional Perspectives on Mental Health and Illness,* edited by Robert Ellis, Sarah Kendall, Steven J Taylor, 71–94. Cham: Palgrave Macmillan, 2021.

Gelder, Michael. "Adolf Meyer and his Influence on British Psychiatry." In *150 Years of British Psychiatry 1841–1991*, edited by German E. Berrios and Hugh Freeman, 419–35. London: Gaskell, 1991.

Gijswijt-Hofstra, Marijke and Harry Oosterhuis. "Introduction: Comparing National Cultures of Psychiatry." In *Psychiatric Cultures Compared: Psychiatry and Mental Health Care in the Twentieth Century: Comparisons and Approaches*, edited by Marijke Gijswijt-Hofstra, Harry Oosterhuis, Joost Vijselaar and Hugh Freeman, 9–34. Amsterdam: Amsterdam University Press, 2005.

Goldstein, Jan. *Console and Classify: The French psychiatric profession in the nineteenth century*. Chicago and London: University of Chicago Press, 2001.

Goodlad, Lauren E. *Victorian Literature and the Victorian State: Character and Governance in a Liberal Society*. Baltimore and London: The Johns Hopkins University Press, 2003.

Grob, Gerald N. Review of *Pathologist of the Mind: Adolf Meyer and the Origins of American Psychiatry*, by S.D. Lamb. *Bulletin of the History of Medicine,* 89 no. 3 (2015): 617–18.

Hall, John. "From Work and Occupation to Occupational Therapy: the Policies of Professionalisation in English Mental Hospitals from 1919 to 1959." In *Work, Psychiatry and Society, c.1750–2015*, edited by Waltraud Ernst, 314–33. Manchester: Manchester University Press, 2016.

Harris, Ben. "Therapeutic work and mental illness in America, *c*.1830–1970." In *Work, Psychiatry and Society, c.1750–2015*, edited by Waltraud Ernst, 55–74. Manchester: Manchester University Press, 2016.

Hayward, Rhodri. *The Transformation of the Psyche in British Primary Care, 1880–1970*. London and New York: Bloomsbury, 2014.

Jones, Edgar. "Aubrey Lewis, Edward Mapother and the Maudsley," *Medical History Supplement* 22 (2003): 3–38.

Jones, Kathleen. *Asylums and After. A Revised History of the Mental Health Services: From the Early 18th Century to the 1990s*. London: Athlone, 1993.

Lamb, S.D. *Pathologist of the Mind: Adolf Meyer and the Origins of American Psychiatry*. Baltimore: Johns Hopkins University Press, 2014.

Lerner, Paul. *Hysterical Men War: Psychiatry and the Politics of Trauma in Germany, 1890–1930*. Ithaca, NY: Cornell University Press, 2003.

Morrison, Hazel. "Henderson and Meyer in Correspondence: A Transatlantic History of Dynamic Psychiatry, 1908–1929." *History of Psychiatry* 28, no. 1 (2016): 72–86.

Owens, Constance. "Obituary—Elizabeth Casson OBE, MD, DPM." *Occupational Therapy*, 1 no. 18 (1955): 3–6.

Paterson, Catherine F. *Opportunities not Prescriptions: The Development of Occupational Therapy in Scotland 1900-1960*. Aberdeen: Aberdeen History of Medicine Publications, 2010.

Postel, Jacques and Claude Quétel. *Nouvelle Histoire de la Psychiatrie*. Paris: Dunod, 2012.

Reed, Esther. "Dr Casson's Early Life—by her sister." *Occupational Therapy*, 18 no.3 (1955): 87–90.

Weisz, George. "Reform and Conflict in French Medical Education, 1870–1914." In *The Organisation of Science and Technology in France 1808–1914*, edited by Robert Fox and George Weisz, 61–94. Cambridge: Cambridge University Press, 1980.

Wilcock, Ann A. *Occupation for Health, Volume II: A Journey from Prescription to Self Health*. London: British Association and College of Occupational Therapists, 2002.

Archival Sources

Author Unknown. "Royal Commission on Lunacy and Mental Disorder: Summary of Report." *The British Journal of Nursing,* September (1926): 201.

Author Unknown. "Study Tour and Post-Graduate Educational Information Sub-Committee: Tour of Dutch Mental Hospitals and Clinics." *Journal of Mental Science,* Jan (1929): 192–207.

Board of Control. *Memorandum on Occupation Therapy for Mental Patients.* London: HMSO, 1933.

Calmels, M. "Le Travail par petits ateliers à la troisième section de l'Asile de Villejuif," *Annales Médico-psychologiques,* no. 1 (1927): 277–283.

Cobb, Cyril. Report of the Committee on Administration of Public Mental Hospitals, Ministry of Health. London: HMSO, 1922.

Colin, Henri. "Le Quartier de Sureté de Villejuif (Aliénés criminels, vicieux, difficiles, habitués des asiles)." *Annales Médico-psychologiques* novembre (1912): 370–391; décembre(1912): 540–548; janvier (1913): 36–65; février (1913): 170–177.

Courbon, P. "Un voyage d'étude dans les asiles de Hollande." *Annales Médico-psychologiques* no. 2 (1928): 289–306 and 385–405.

Eager, Richard. *Aids to Mental Health: The Benefits of Occupation, Recreation and Amusement.* Exeter: W.V. Cole and Sons, 1936.

Eighteenth Annual Report of the Board of Control. London: HMSO, 1931.

Fifteenth Annual Report of the Board of Control. London: HMSO, 1928.

Halberstadt, G. "A propos de l'Ergothérapie." *Annales Médico-psychologiques,* mars (1929): 193–8.

Henderson, D.K. and R.D. Gillespie. *A Text-Book of Psychiatry for Students and Practitioners,* 5th edition. London: Oxford University Press, 1940.

J.M.F. "Obituary: J. Ivison Russell." *The British Medical Journal* 4 no. 5738 (1970): 808.

Ladame, C. and G. Demay, *La Thérapeutique des Maladies Mentales par le Travail.* Paris: Masson et Cie, 1926.

Legrain, D-M. and G. Demay, "Le travail des aliénés convalescents: rapport présenté à la 4e Section du Conseil supérieur de l'Assistance publique en 1934." *L'Aliéniste français,* no. 2 (1936): 281–97.

Lomax, Montagu. *Experiences of an Asylum Doctor: With suggestions for asylum and lunacy law reform.* London: G. Allen and Unwin Ltd, 1921.

Medical Superintendent's Report. Annual Reports of the Littlemore Asylum, 1887. OHA-L1/A1/40.

Meyer, Adolf. "The Philosophy of Occupational Therapy." In *The Collected Papers of Adolf Meyer, Volume IV: Mental Hygiene,* edited by Eunice Winters, 86–92. Baltimore: Johns Hopkins Press, 1952.

Porot, A. "L'Assistance par le travail dans les Asiles Hollandais," *L'Hygiène mentale* no. 2 (1929): 41–54.

Rapport du Médecin en Chef. L'Asile de la Sarthe, 1935. Le Mans, 1936. ADS-1X964/5.

Rapport du Médecin en Chef. L'Asile de la Sarthe, 1936. Le Mans, 1937. ADS-1X964/5.

Raynier, J. and H. Beaudouin. *L'Aliéné et les Asiles d'Aliénés au point de vue administratif et juridique, assistance et législation.* Paris: Librairie Le Français, 1924.

Russell, John Ivison. *The Occupational Treatment of Mental Illness.* London: Ballière, Tindall and Cox, 1938.

Schiff, P. "Le movement d'hygiène mentale en Allemagne." *L'Hygiène mentale* no. 8 (1929): 232–3.

Simon, Hermann. "La Psychothérapie de l'Asile." *L'Hygiène mentale* no. 1 (1933): 16–28.

Simon, Hermann. "Active Therapy in the Lunatic Facility (1929)." In *From Madness to Mental Health: Psychiatric Disorder and Its Treatment in Western Civilisation*, edited by Greg Eghigian, 271–75. New Brunswick: Rutgers University Press, 2010.

Simon, Hermann. "Aktivere Krankenbehandlung in der Irrenanstadt." *Allegemeine Zeitschrift fur Psychiatrie*, 87 (1927): 97–145.

Tenth Annual Report of the Board of Control. London: HMSO, 1923.

Vié, J. "Le traitement des malades mentaux par le travail: les idées et les réalisations de H. Simon." *L'Aliéniste français* no. 10 (1934): 589–98.

Walk, A. "Occupation Therapy and 'Active Therapy'." *Journal of Mental Science* Oct (1933): 758–63.

Wertham, Frederick. "Progress in Psychiatry II: The Active Therapy of Dr Simon." *Archives of Neurology and Psychiatry*, 24 no. 1 (1930): 150–60.

BY

CHAPTER 5

Money and Management

Patient labour saved asylums money. Those in charge of institutions for the mentally ill recognised that the work carried out by patients, originally conceived as therapy by the moral therapists, could also contribute to institutional running costs. By the late nineteenth century, the economic rationale for patient work had begun to take precedence over (although not supplant) its perceived value as therapy. As Geoffrey Reaume observed, asylum work programmes had become cost-saving measures created "under the guise of moral therapy".[1] The work carried out by calm, chronically ill and convalescent patients (those who required minimal supervision) around the asylum, its grounds and workshops, and on the asylum farm reduced the requirement for paid staff, thereby making savings on the budget for personnel. Furthermore, much of the fresh food and other essential items, such as clothing and bedding, were produced *in situ* using patient labour instead of needing to be purchased. Before World War I, the asylum authorities in both England and France praised the cost savings generated by patient labour. The straitened financial circumstances brought about by the war rendered the economic contribution made by patient work even more valuable to those trying to balance the asylum accounts, particularly in France where the post-war financial situation was

[1] Geoffrey Reaume, "A Wall's Heritage: Making Mad People's History Public," *Public Disability History* 1 (2016): 20.

J. Freebody, *Work and Occupation in French and English Mental Hospitals, c.1918–1939*, Mental Health in Historical Perspective,
https://doi.org/10.1007/978-3-031-13105-9_5

especially challenging. But during the 1920s, attitudes towards patient occupation began to change as asylums were accused of exploiting patients. This chapter examines the response of those concerned with asylum management in England and France to the financial challenges of the interwar period in the light of such criticisms and the emergence of new approaches to occupation. Could asylums survive if patient work was re-organised in a way that might compromise its economic contribution to institutional budgets? How did a patient's perceived curability, and the management structure of institutions affect decisions regarding the occupation of patients?

The Financial Situation After World War I

Financial pressures following the war meant that hospital budgets were extremely tight, while post-war inflation caused prices to rise unpredictably for essential items such as food. French losses during the war were particularly heavy; the industrial north-east of the country was occupied by enemy forces who laid waste to factories, towns and agricultural land as they retreated. 1.31 million Frenchmen were killed and 1.1 million were severely wounded with permanent work incapacity. France lost 7.2% of its human capital, 25% of its domestic assets and 49% of its overseas assets and 31% of its GDP was spent on the war.[2] Britain avoided the level of devastation experienced by France, although 715,000 British servicemen died and more than twice that number were wounded. Britain lost 3.6% of its human capital; 10% of its domestic and 24% of its overseas assets were destroyed; and over 25% of its GDP was spent on the war effort between 1915 and 1918.[3] Whilst healthcare budgets were strained in Britain, the severity of the French economic devastation resulted in even greater pressure on public finances. Budgetary restrictions not only influenced attitudes towards patient work but also affected the resources available for entertainment and improvements to the material comforts of patients.

Asylums in both France and England were in state of disarray following the cessation of hostilities in 1918. Many patients had been transferred to other asylums following the requisition of their original institutions by the

[2] Jari Eloranta and Mark Harrison, "War and Disintegration, 1914–1950," in *The Cambridge Economic History of Modern Europe*, eds Stephen Broadberry and Kevin O'Rourke (Cambridge: Cambridge University Press, 2010), 149.

[3] Ibid.

military.[4] This had resulted in overcrowding, poor nutrition, fewer staff, lower standards of hygiene and the increased incidence of disease (such as tuberculosis and dysentery) in the establishments receiving the transferred patients. The consequent increase in patient mortality led to a reduction in numbers in both French and English asylums during the war.[5] The situation was exacerbated by the influenza epidemic of 1918–1919, which in France killed over 2500 asylum patients in the Seine department alone.[6] Despite this reduction in patient numbers, a return to pre-war levels of overcrowding was predicted by Parisian psychiatrist Édouard Toulouse. He envisaged an increase in the incidence of mental illness amongst the general population whose resistance had been eroded by years of wartime privation and anxiety.[7] After the war, expenditure on health and welfare had to be tightly controlled as the French and English economies struggled to recover. While the Board of Control's report of 1918 indicates that the post-war condition of English asylums was poor, the General Council of the Seine's report suggests that French asylums were in an even worse state. It might be expected, therefore, that the contribution made by patient work to asylum finances became even more of a priority than it had been during the late nineteenth century.

[4] By the end of 1918, 17 out of the 97 public asylums in England and Wales were in use as war hospitals, including the Littlemore Asylum in Oxford and the Maudsley Hospital in London. Three out of the nine asylums of the Seine Department were requisitioned by the military (but not Ste. Anne's).

[5] The number of individuals in care in England and Wales fell from 140,456 on 31/12/1914 to 116,703 on 31/12/1918, a reduction of 17%. In the Department of the Seine, patient numbers fell from 15,919 on 31/12/1914 to 12,989 on 31/12/1918, a reduction of 18%. See: Dausset, "Rapport Générale," 1918, 20; Fifth Report of the Board of Control 1918 (London, HMSO: 1919), 24–5.

[6] Isabelle von Bueltzingsloewen, *L'Hécatombe des Fous: La Famine dans les Hôpitaux Psychiatriques Français sous l'Occupation* (Paris: Éditions Flammarion, 2007), 252.

[7] Louis Dausset, Rapport Général au nom de la 3e Commission sur les propositions budgétaires pour le Service des Aliénés (budget de 1919) (Paris: Conseil Général de la Seine, 1918), 26. ADP/D.10 K3/27/20.

The Financial Contribution of Patient Work in France and England

France

In France, this contribution remained an essential aspect of asylum finances during the interwar period. Because of the management structure of French asylums, the financial imperative of patient work made it an area of potential conflict between the asylum director and chief medical officer.[8] According to the legislation of 1838/1857, asylum management was split between a chief medical officer, responsible for all medical decisions, and an asylum director, responsible for all administrative and financial issues.[9] The asylum director was at pains to keep asylum maintenance costs down and deliver the cost savings demanded by the local prefecture, while the chief medical officer's priority was the treatment of patients. Pressure from the asylum director to maximise the economic advantages of patient work could compromise the wishes of the chief medical officer to insist on a primarily therapeutic agenda for patient work. Toulouse was anxious to avoid this type of conflict and insisted on being appointed "medical director" of the Henri Rousselle Hospital, which put him in charge of both medical and administrative matters. This meant he had overall control of patient work.

Although finances were tight, there was intense pressure on French asylums to cure as many patients as possible so that they could be returned to the labour market. The country needed individuals to resume their duty as productive citizens since France's "human capital was so diminished" by the war.[10] Successful treatment of patients was hampered, however by the state of provision for the mentally ill in the aftermath of the war. Psychiatrists complained that services were far behind those of "other great countries" and asylums were embarrassingly ill-equipped.[11] They described the post-war state of French asylums as "deplorable", with poor general hygiene; cramped, undifferentiated quarters; a lack of outdoor

[8] This issue is discussed further in chapter six.

[9] Jane Freebody, "The Root of All Evil is Inactivity": The response of French psychiatrists to new approaches to patient work and occupation, 1918–1939," in *Voices in the History of Madness: Personal and Professional Perspectives on Mental Health and Illness*, eds Robert Ellis, Sarah Kendal and Steven J. Taylor (Cham: Palgrave Macmillan, 2021), 84.

[10] Dausset, Rapport Général, 1918, 26.

[11] Ibid., 28.

space; and poor facilities. Entertainments, sport and other forms of recreation, considered so important for patients whose conditions were beginning to improve, were practically non-existent.[12] Treatment for curable patients was compromised by the mixing of curable and incurable patients in the same quarters, and by ratios of over 400 patients per doctor, resulting in most patients being merely "overseen" rather than receiving "modern" treatment. Furthermore, most asylums lacked facilities for laboratory testing and the equipment necessary for delivering modern treatments, such as hydrotherapy, UV-ray treatment, electrotherapy and radiography. Many patients, who, in more favourable conditions, might be cured or improved, remained in asylums far longer than was necessary, adding to the costs of their care and depriving the nation of their labour.[13]

The French Asylums' Medical Society, led by Toulouse, put forward proposals for post-war reforms to services for the mentally ill, including the separation of acute, recent cases of mental illness and incurable or chronic patients; an admission's system for recent or incipient cases of mental illness to asylums or mental hospitals that avoided the usual legal formalities; and improvements to support for recently discharged patients in order to avoid a relapse of their symptoms.[14] It was understood by the French Asylums' Medical Society that the reforms would be costly, but the provision of separate care for incurable and chronic patients, who did not require intensive medical treatment or specialist facilities, would save money in the longer term. General Councillor Dausset suggested that more could be done to enhance the economic value of patient work by ensuring that the work carried out by patients was "really productive," and by engaging more patients in the process.[15]

While Dausset accepted that patient work was primarily a means of therapy (as the law stipulated), he felt that its therapeutic benefits should not be allowed to overshadow the economic advantages. He quoted the nineteenth-century moral therapist Jean-Baptiste-Maximien Parchappe (1800–1866), who had declared, "I do not believe that it is against the principles of humanity, or morality, to expect the work of patients to benefit the establishments that offer them refuge".[16] Parchappe believed that it was

[12] Ibid., 27.

[13] Ibid.

[14] Ibid., 31–41; Fifth Annual Report of the Board of Control, 1918, 1–2.

[15] Dausset, Rapport Général, 1918, 43.

[16] Ibid., 44.

perfectly legitimate to organise the work in "the double interest of patients and establishments".[17] But by 1918, patient work in Parisian asylums was not sufficiently well organised to fulfil either objective. Dausset observed that "a few women were employed in the laundry, in the ironing and sewing rooms, or in the kitchen, and that some men were engaged in cultivation or gardening, but the great majority of patients remained unoccupied in their quarters".[18] This was not the case in the service for "difficult" patients organised by Henri Colin at the Villejuif asylum, however, where work was organised in a methodical, rational manner. Nor was it the case in provincial asylums, such as the Asile de la Sarthe in Le Mans, from whom Dausset felt much could be learned.[19] Here, he claimed, one found both asylum farms and workshops that were genuinely productive. Ironically, he added that many of the best workers in those workshops were calm, chronic and incurable patients who had been transferred to the provinces from the Seine.[20] The Asile de la Sarthe, where the maintenance charge was cheaper than that of the Parisian asylums, received around six patients each year from the Seine, part of an arrangement with several provincial institutions designed to relieve overcrowding in the metropolitan asylums.

Dausset recommended that the Seine asylums extended their use of market gardening, which required a much smaller area than agriculture. Market gardening had the potential to employ a large and easily surveyed patient workforce. From the economic point of view, it was very profitable because the products harvested could be consumed by the asylum population. While few Parisians were used to such work, Dausset maintained (perhaps disingenuously) that being able to grow their own fruit, vegetables and flowers was a "dream" for many.[21] At the provincial Asile de la Sarthe the asylum director made frequent references to the cost savings made by "agricultural exploitation". In 1920, the asylum director commented that farming had given good results that year, saving the asylum considerable sums. Farming was the second largest employer of patient labour, surpassed only by housework. Produce from the market gardens, piggery and poultry farm were evaluated at 85,971F, which, after deducting costs of 47,253F, generated a "profit" of 38,718F.[22] The products,

[17] Ibid.
[18] Ibid., 45.
[19] Ibid.
[20] Ibid.
[21] Ibid.
[22] Rapport du Directeur, Asile de la Sarthe, 1920 (Le Mans, 1921), 15. ADS-1X964.

including bacon, eggs, chicken, fruit and vegetables, were all consumed on the premises. Nothing was wasted. The asylum director—echoing Parchappe—saw the cultivation of food stuffs as a means of improving the condition of patients, from which the establishment could also benefit.[23] The ongoing importance of this food production to the asylum's finances was demonstrated in 1936 when the asylum director commented that there was no question of building additional accommodation on the land devoted to market gardening, despite overcrowding. The four hectares of land under cultivation was barely enough to supply the required quantities of fresh produce for the swelling asylum population; losing this land would have serious financial consequences.[24]

The price fluctuations of essential goods, such as food, in France in the mid-1920s resulted in economic difficulties for asylums in both urban and provincial asylums. In Le Mans, the asylum director had warned the prefect in 1924 of the asylum's precarious financial circumstances caused by rising prices.[25] It is perhaps no coincidence that 1925 saw the highest proportion of patient workers indicated for the entire interwar period.[26] In the same year, the escalating costs of maintaining the Seine's mentally ill population,[27] forced the General Council to look at ways of reducing the numbers of paid asylum employees.[28] The work provided by patients was crucial, serving to "lighten the maintenance costs which weighed so heavily on the collective purse".[29] This helps explain the hostile reaction of the General Council to a proposal to replace patients working in the kitchen with three paid kitchen assistants at the Asile Clinique in 1925. Following an accident, the kitchen was deemed too dangerous for patients by medical staff. The Council argued that patients should be given tasks that were not

[23] Ibid.

[24] Rapport du Directeur, Asile de la Sarthe, 1936 (Le Mans, 1937), 4. ADS-1X964.

[25] Rapport du Directeur et Rapport Médical, Asile de la Sarthe, 1924 (Le Mans, 1925), 11. ADS-1X964.

[26] Rapport du Directeur, Asile de la Sarthe, 1925 (Le Mans, 1926), 3. ADS-1X964.

[27] Between 1924 and 1925 the maintenance costs rose by 11 million Francs and were set to rise by another 6.3 million in 1926.

[28] M.E. Chausse, Rapport Général au nom de la 3e Commission sur le compte du Service des aliénés pour 1924, le budget rectificatif de 1925 et le projet de budget pour 1926 (Paris: Conseil Général de la Seine, 1925), 7. ADP/D.10 K3/38/83.

[29] François Latour, Rapport au nom de la 6e Commission sur les comptes du directeur de l'Asile Clinique et du receveur des asiles relatifs à la gestion de cet établissement en 1924 (Paris: Conseil Général de la Seine, 1925), 20–1. ADP/D.10 K3/38/75.

dangerous, leaving the riskier work to existing employees.[30] If every Seine asylum decided not to use patient labour in their kitchens, the cost implications of employing staff for them all would be huge (an estimated 170,000F).[31] Furthermore, patients were not to be denied the opportunity to work since it was a means of therapy. In the interests of both patients and budgets, it was recommended that kitchen staffing levels remained as they were, and that more effort was made to employ patients safely.[32]

Proposals to transform the Asile Clinique from an asylum for all types of patient into a "hospital" for acute, curable cases had been discussed at intervals since the establishment of the institution in 1867. Specialising in acute cases meant incurring the loss of incurable and chronic cases and thus the majority of the patient workforce (made up almost exclusively of calm, incurable and chronic patients). The anticipated financial consequences of this loss had resulted in the repeated postponement of the transformation from asylum to hospital. A feasibility study had been requested by the General Council in 1913, but the war had intervened and approval for the project was delayed until 1923.[33] The process of gradually transferring incurable and chronic patients out of the Asile Clinique to colonies or other asylums to make room for the acute cases, began in 1927.[34] Replacing the incurable and chronic patient workers with paid members of staff was not financially feasible, and there were unlikely to be sufficient numbers of convalescent patients to carry out the work. The decision was taken to create a special division for working patients within the Asile Clinique, separate from the treatment sections. Two pavilions were constructed for the 120 male and 50 female incurable and

[30] The safety of the work allocated to mental patients is a topic discussed by Lee-Ann Monk, "Exploiting patient labour at Kew Cottages, Australia, 1887–1950," *British Journal of Learning Disabilities* 38 (2010): 86–94.

[31] M. Béquet, Rapport au nom de la 3e Commission sur l'Asile Clinique: budget additionnel de 1926 et budget de 1927 (Paris: Conseil Général de la Seine, 1926), 35. ADP/D.10 K3/40/105.

[32] Ibid.

[33] Henri Rousselle, Rapport au nom de la 3e Commission sur les budgets et comptes de l'Asile Clinique (Paris: Conseil Général de la Seine, 1915), 3. ADP/D.10 K3/25/20; Procès Verbal de la Visite de la Commission, 21–4 Avril 1923 (Paris: Commission de Surveillance des Asiles Publiques d'Aliénés de la Seine, 1923), 122. ADP/D1X3–44/6.

[34] Colonies for the Seine's chronic patients had been established at Dun-sur-Auron in 1892 and at Ainay-le-Château in 1897. Colonies, which facilitated the separation of care for acute, curable patients and chronic, and incurable patients, were not established elsewhere in France during the interwar period.

chronic patient workers who would be able to fulfil the tasks around the hospital that acute patients were deemed incapable of performing.[35] The transformation of the service was completed in 1928, with all patients in the treatment quarters at the acute stage and therefore presumed curable, while the patients in the separate workers' pavilions, who were assumed to be incurable, did not receive active treatment. The economic contribution made by these patient workers became even more important than before the transformation because of the additional personnel and pharmaceutical costs associated with caring for acute-stage patients. Two additional chief medical officers, four additional interns and 34 extra nurses were employed, whose salaries added to the maintenance charge.[36]

The Great Depression put renewed pressure on asylum finances, rendering the economic contribution made by patient workers even more important to the Asile Clinique. While the impact of the Depression on France was minimal at the outset, its arrival was merely delayed.[37] The crisis reached France in 1931 and its effects persisted until 1938, continuing despite the upturn of the world economy in 1935. French economic activity during the interwar period peaked in 1929, but from 1931 France was engulfed by economic, political, social and moral malaise.[38] The Seine's departmental budget was particularly stretched as the cost of unemployment benefit rose from 1.5 million Francs in 1931 to 191 million in 1932 and was set to rise again to 210 million in 1933. Councillor Fiancette called for a reduction of 3 million Francs in the budget for maintaining the mentally ill of the Seine.[39] This demand was extremely challenging since the asylum population expanded during the early 1930s, adding to the costs of care.[40] A fall in the discharge rate was blamed on the economic crisis. Doctors hesitated to discharge recovered patients, know-

[35] François Latour, Rapport au nom de la 6e Commission sur les comptes du directeur de l'Asile Clinique en 1929 (Paris: Conseil Général de la Seine, 1930), 13. ADP/D.10 K3/46/76.

[36] M.E. Chausse, Rapport au nom de la 3e Commission sur un projet de modification des cadres du personnel des asiles des aliénés (Paris: Conseil Général de la Seine, 1928), 2. ADP/D.10 K3/44/63.

[37] James F. McMillan, *Twentieth-Century France: Politics and Society 1898–1991* (London: Arnold, 1992), 101.

[38] Ibid., 102.

[39] M. Fiancette, Rapport Générale presenté au nom du Comité du budget, du compte et du contrôle (Paris: Conseil Général de la Seine, 1933); 6. ADP/D.10 K3/49/64.

[40] The number of individuals passing through the Ste. Anne's Admissions Service increased from 4856 to 4942, while the number of discharges from Seine asylums fell from 2938 to 2170 between October 1932 and 16,888 in October 1933.

ing that they would struggle to support themselves in the exceptionally difficult economic climate of the early 1930s.[41] The economic crisis was therefore contributing to overcrowding in the Seine's asylums.[42] Furthermore, as psychiatrist Aubrey Lewis noted with regard to England and historian Richard Warner observed in relation to the USA, cases of mental disorder rose as a result of the anxiety caused by unemployment and financial hardship.[43] The increased costs of care, coupled with demands for further cost-savings, increased the importance of productive patient work to the Asile Clinique budget.

French patients, unlike their English counterparts, were incentivised to work by the prospect of a nominal daily wage, known as a *pécule,* paid to all pauper patient workers, as set out in the Law of 1857.[44] The payment of French patients underscores the essential nature of their work to the asylum, even though the amount paid was a fraction of what an ordinary worker would receive outside the asylum. The precise amount varied according to the gender of the patient, the type of work and, in Paris, the skill of the labourer. The *pécule* paid to women was less than that paid to men, reflecting pay structures outside the asylum. In 1927, for example, male and female workers in the Asile Clinique's laundry undertook similar work but men were paid 1.20F per day while women earned 1F per day. Male patients working in the tailors' workshop earned 0.80F per day, while women in the sewing rooms earned between 0.63F and 0.70F per day. The male patient assisting in the mechanics' workshop was highly skilled and received the maximum *pécule* of 2.5F per day.[45] At the Asile de la Sarthe, there were just two rates of pay. Most tasks, including working in the laundry or knitting, were attributed a value of 1F per day, while more skilled work, such as clerical work, stonemasonry, painting, carpentry, locksmithing, and shoemaking, was evaluated at 1.50F per day.[46]

[41] M.E. Chausse, Rapport Général au nom de la 3e Commission sur le compte du Service des Aliénés pour 1932 (Paris: Conseil Général de la Seine, 1933), 4. ADP/D.10 K3/50/110.

[42] Ibid., 4.

[43] Richard Warner, *Recovery from Schizophrenia: Psychiatry and Political Economy* (London and New York: Routledge and Kegan Paul, 1985), 269.

[44] Article 153, Bulletin Officiel du Ministère de l'Intérieur, Circulaire No. 7, 20/03/1857, 76. www.gallica.bnf.fr / (accessed 03/12/2017).

[45] François Latour, "Comptes d'Ateliers," Rapport au nom de la 6e Commission sur les comptes du directeur de l'Asile Clinique et du receveur des asiles relatifs à la gestion de cet établissement en 1927 (Paris: Conseil Général de la Seine, 1928), 12–17. ADP/D.10 K3/44/64.

[46] These evaluations had not changed since before WW1.

Women were assigned less skilled roles in the lower wage category. The work performed by patients, in terms of days or half-days, and the type of work performed were carefully recorded by the asylum bursar. This record enabled the bursar to work out not only how much each patient should be paid, but also the value of the contribution made by patient work for inclusion in the asylum accounts.

The annual value of patient work to the asylum (referred to as the *produit du travail des aliénés*) was calculated at three times the total amount paid to patients. The value of the produce harvested from the farm or market gardens, or made by patients, appeared in the accounts as the *produits recoltés*. For example, at the Asile de la Sarthe, the items made in the sewing room during 1923 included 190 pillowcases, 470 sheets, 67 mattress covers, 360 men's shirts, 499 women's blouses, 238 dresses, 152 pairs of trousers, and 1622 handkerchiefs, each of which was assigned a monetary value.[47] Dresses were valued at 4F, pillowcases at 0.50F and handkerchiefs at 0.10F each. The *produit du travail* and the *produits recoltés* constituted the total financial contribution made by patients to the asylum maintenance budgets. At the Asile de la Sarthe this amount represented between 5% and nearly 12% of the total maintenance budget [see Table 5.1]. The Asile Clinique's city centre location meant that the land available for cultivation was limited, so receipts from the *produits recoltés* were modest compared with those of the provincial asylums, such as the Asile de la Sarthe [see Table 5.2], and those of other Seine asylums located on the outskirts of Paris. In 1930, the value of vegetables and flowers grown at the Asile Clinique was estimated at 59,278F, while the value of produce from the Maison Blanche was 262,404F.[48]

At the Henri Rousselle Hospital, patient work was organised rather differently to that at the Asile Clinique and the Asile de la Sarthe. All patients at the Henri Rousselle Hospital were in the unique position (in France) of being admitted *voluntarily* to a public institution at the early, acute stage of their illness, and able to leave the hospital at any time (on giving 72 hours' notice), as were patients at the Maudsley Hospital in London. Patients were not expected to work to contribute to the costs of their care. They were advised to keep busy and were paid a *pécule* for their labour if they chose to work, but this was optional. Patient work was not evaluated and did not appear in the hospital accounts as it did at the other French

[47] État des Produits en Nature, Compte Administratif, Asile de la Sarthe, 1923 (Le Mans, 1924). ADS-1X964.

[48] Procès Verbal, Séance du 28/11/1931, Asile de la Sarthe (Le Mans, 1931), 223.

Table 5.1 Table to show the value, in French Francs (F), of goods produced or harvested; the value of patient labour; and the total value of goods and labour expressed as a percentage of total expenditure at the Asile de la Sarthe, Le Mans, 1919–1939

Year	*Value of goods produced/ harvested (F)*	*Value of Patient Labour (F)*	*Total Contribution (F)*	*Total Asylum expenditure (F)*	*% of Total Expenditure*
1919	50,265	33,078	83,343	923,894	9%
1920	60,731	50,270	111,001	1345,689	8.2%
1921	87,204	55,630	142,834	1,511,191	9.5%
1922	95,289	60,999	156,288	1,619,019	9.7%
1923	106,304	66,007	172,311	1,529,616	11%
1924	116,521	60,960	177,481	1,633,652	10.9%
1925	121,994	67,158	189,152	1,610,620	11.7%
1926	135,529	64,780	200,309	2,042,720	9.8%
1927	153,498	60,736	218,309	2,352,401	9.3%
1928	142,564	63,810	206,374	2,370,067	8.7%
1929	147,063	67,486	214,549	2,932,546	7.3%
1930	139,338	63,942	203,280	2,797,749	7.3%
1931	148,013	66,163	214,176	3,187,855	6.7%
1932	138,308	69,320	207,628	2,776,108	7.5%
1933	137,579	69,473	207,052	2,836,554	7.3%
1934	176,051	75,779	251,830	2,855,111	8.8%
1935	165,875	71,311	237,186	2,932,788	8%
1936	187,590	78,272	265,862	2,792,421	9.5%
1937	239,029	77,629	316,658	7,072,670	4.5%
1938	308,954	77,514	386,468	5,694,461	6.8%
1939	271,924	80,234	352,158	6,353,049	5.5%

Source: Reports of the Asylum Director, Asile de la Sarthe, 1919–1939. ADS-1X964/5

F = French francs

institutions. This different approach to patient work was made possible by the fact that there was just one person in overall charge of the hospital, the medical director. Toulouse was able to instil a therapeutic priority for occupation because he did not have to comply with an asylum director's financial demands from patient work.

England

In England, there were also different levels of expectation regarding the financial contribution made by patient work, and towards the costs of occupational therapy. The Littlemore was typical of the traditional English

Table 5.2 Table to show the value, in French Francs, of goods produced or harvested, and of patient labour, at the Asile Clinique, Paris, 1920–1939

Year	*Value of goods produced/ harvested in Francs*	*Value of patient labour in Francs*	*Total contribution made by patient work in Francs*
1920	56,428	108,843	165,271
1921	59,182	117,400	176,582
1922	42,529	113,526	156,055
1923	41,659[a]	113,818	155,477
1924	44,567	121,435	166,002
1925	53,158	111,135	164,293
1926	65,200	112,284	177,284
1927	57,854	207,609	265,463
1928	50,467[b]	221,250	271,717
1929	57,077	250,320	307,397
1930	59,278	263,777	323,055
1931	55,662	267,935	323,597
1933	71,208	275,037	346,245
1935	111,418	273,426	384,844
1936	134,680	276,100	410,780
1937	150,955	274,822	425,777
1938	149,878	275,072	424,950
1939	206,093	285,007	491,100

Source: Reports to the General Council, 1920–1939. ADP-D.10K3 28-59

[a]The summer of 1923 was exceptionally dry, which adversely affected Market Gardening

[b]The figure for ***produits récoltés*** was low in 1928 while the new pavilions for patient workers were being completed. The workers arrived towards the end of 1928

asylum, in that it catered for a mix of curable, incurable and chronic cases, with pauper patients' maintenance fees paid out of the local rates. Bethlem, a registered institution, was England's only mental hospital specialising in acute cases until the state funded Maudsley Hospital opened in 1923. Bethlem was financed charitably through a mix of charitable donations and income from investments. The physician superintendent had to answer to a Board of Governors regarding the maintenance charge rather than a local authority. Bethlem catered for the "deserving poor", working people on low incomes or who found themselves in straitened circumstances without being paupers. Some of the Bethlem's patients were received "gratuitously" and did not have to pay for their care, while others paid fees, depending on their circumstances. The proportion of Bethlem's patients who paid fees steadily increased during the interwar period.

Manual labour was an anathema for this increasingly middle-class clientele.[49] The Maudsley was a public hospital that accepted both pauper and paying patients (*c*.25% of total admissions paid for their care). Like the Henri Rousselle Hospital in Paris, the Maudsley was in the unique position in England of receiving pauper patients on a voluntary basis, facilitating the treatment of the poorest patients at the onset of their illness without the delays caused by the lengthy process of certification.[50]

Before 1913, English public asylums, which at that time cared for a mix of both chronic, incurable and acute, curable patients (with the exception of Bethlem which only admitted acute, curable patients), were expected to ensure that a high proportion of patients were "usefully employed".[51] As in France, it was the calm, incurable and chronic, and convalescent patients who performed "useful" work around the asylum that offset institutional running costs. This was the case at the Littlemore before the outbreak of World War I. As discussed in chap. 2, the Mental Deficiency Act of 1913 was intended to provide for the separate accommodation of the intellectually impaired and chronic patients in colonies, enabling asylums to concentrate their efforts on treating patients who were presumed curable. Plans to build colonies had to be scaled back as a result of the financial challenges presented by World War I, so the Act was not as far-reaching as had been originally intended.[52] Nonetheless, the planned separation of the curable and incurable led to the divergence of rationales for patient occupation.

From 1913, the Board of Control reports were divided into two sections, one focusing on institutions specialising in "mental deficiency" (the colonies) and the other on those specialising in "lunacy" (the asylums, or mental hospitals as they became known). The two sections of the Board's reports revealed very different aims for the occupation of patients. In the

[49] Jonathan Andrews, Asa Briggs, Roy Porter, Penny Tucker and Keir Waddington, *The History of Bethlem* (Abingdon and New York: Routledge, 1997), 689. The class of patients and its influence on occupation is addressed in chapter eight.

[50] The Maudsley remained the only public English hospital to admit patients on a voluntary basis until 1930 when the Mental Treatment Act was passed that allowed all public mental hospitals to do so. Similar legislation was not passed in France until 1938 and was not effective until after World War II.

[51] Sarah Chaney, "Useful members of society, or motiveless malingerers? Occupation and malingering in British asylum psychiatry, 1870–1914," in *Work, Psychiatry and Society, c.1750–2015*, ed. Waltraud Ernst (Manchester: Manchester University Press, 2016), 277–97.

[52] Institutions for "mental defectives" were originally intended to house 94,000 individuals; in 1937, 129 such institutions housed c.32,600 individuals. See: Mathew Thomson, *The Problem of Mental Deficiency: Eugenics, Democracy and Social Policy in Britain, c.1870–1959* (Oxford: Oxford University Press, 1998), 130.

"mental deficiency" section, the Board was keen to emphasise self-sufficiency and the productive work of inmates in the colonies. In the "lunacy" section, the emphasis was on therapy for curable patients. In reality, however, the mental hospitals continued to house many chronic and incurable cases simply because there were insufficient places for incurable patients in the colonies.[53] Some mental hospitals attempted to retain their incurable and chronic working patients (rather than recommending their transfer to a colony) in order to keep maintenance costs down, although this attracted the Board's disapproval.[54] The result was that within the same institution, there could be different rationales for occupation, depending on whether the patient was curable or incurable, as Mary Macdonald, principal of the Dorset House School of Occupational Therapy, recognised.[55] She maintained that as well as keeping the incurable patient healthy and "as near normal as possible", work "enables those who would otherwise be a burden on society to contribute in some measure to their maintenance". In her view, occupational therapy should be aimed at the curable cases who were expected to make a recovery.[56]

The emphasis placed on the financial contribution made by patient work in mental hospitals began to be downplayed, while it remained an important consideration in the colonies. The economic viability of the Caterham Imbecile Asylum, for example, depended on patient work and was expected to be a "self-sufficient site".[57] The Board of Control's 1917 report emphasised that a significant proportion of "mentally deficient" patients in the colonies could be "trained to contribute towards their own support" instead of being a "dead weight upon the community".[58] The phrase "usefully employed" gradually disappeared from the lexicon of the annual and inspection reports of mental hospitals, which emphasised therapeutic occupation rather than work that was "useful" to the institution. At Bethlem, for example, inspectors recorded the numbers of "usefully employed" patients for the last time in 1928. In the long-term it was

[53] Kathleen Jones, *Asylums and After: A Revised History of the Mental Health Services: From the Early 18th Century to the 1990s* (London and Atlantic Highlands, NJ: Athlone Press, 1993), 123.

[54] Nineteenth Annual Report of the Board of Control, 1932, 4.

[55] Mary Macdonald and Norah Haworth, *Theory of Occupational Therapy for Students and Nurses* (London: Bailliere, Tindall and Cox, 1940), 8.

[56] Ibid.

[57] Stef Eastoe, *Idiocy, Imbecility and Insanity in Victorian Society: Caterham Asylum, 1867–1911* (Cham: Palgrave Macmillan, 2020), 106.

[58] Fourth Annual Report of the Board of Control, 1917, 66.

deemed more cost-effective to provide curable patients with effective therapeutic occupation that would expedite their recovery and increase discharge rates.[59] This would enable greater numbers of patients to spend shorter periods in hospital, thereby saving money and freeing up beds for others. The recovered patients would then be able to re-join the labour market outside hospital.

The negative publicity surrounding the exploitative nature of patient work generated by publication of Montagu Lomax's *The Experiences of an Asylum Doctor* (1921) and the subsequent government enquiry prompted a re-appraisal of the financial benefits of patient work. The Board was keen to distance itself from the pre-war pre-occupation with its monetary value and emphasised a commitment to patient occupation "as a curative agent and as a means of promoting the contentment and well-being of [hospital] patients" in its annual report of 1923.[60] The Macmillan Report of 1926 reinforced the need to deliver active treatment to curable patients; occupational therapy formed an important aspect of such treatment.[61] The Board began to criticise the provision of occupations for patients in most hospitals, noting that for many, the only available employment was ward work. The commissioners recommended the appointment of an "occupations officer" who would ensure that patient occupation was organised therapeutically.[62] Such an appointment would obviously add to the maintenance charge, but the Board considered it money well spent. Occupations officers were appointed at the Maudsley in 1924 and at Bethlem in 1932. At the Littlemore, the medical superintendent chose not to employ a specialist occupations officer; instead Littlemore nurses were expected to learn a craft that they could teach to patients.[63] In 1928, the Board highlighted the "gratifying results" obtained in some hospitals where occupational therapy had been developed, and admitted that, "in the past there has been a tendency to concentrate on the employment of those patients

[59] Fifteenth Annual Report of the Board of Control, 1928, 5.

[60] Tenth Annual Report of the Board of Control, 1923, 10. Following the 1913 Mental Deficiency Act, the Board's annual reports were divided into two sections, one dealing with "Lunacy" (or mental illness) and the other dealing with "Mental Deficiency" (or intellectual impairment). Work was discussed in much more detail in the latter section.

[61] AU, "Royal Commission on Lunacy and Mental Disorder: Summary of Report," *The British Journal of Nursing* September (1926): 201.

[62] Tenth Annual Report of the Board of Control, 1923, 10.

[63] Thomas Saxty Good, "The History and Progress of the Littlemore Hospital," *Journal of Mental Science* Oct (1930): 614.

who readiness to work was spontaneous … and upon work which was of some economic value to the institution".[64] The Board's *Memorandum on Occupation Therapy* (1933) emphasised that "the economic value of occupation need not be stressed" and that "the object of occupation is primarily therapeutic".[65] The aim of occupational therapy was "not … a means of providing commodities for use in the hospital at a low cost".[66] There should not be "too much stress … laid on output and finish".[67] Patients were not to be expected to produce "articles of high artistic merit or of commercial value".[68] As Adolf Meyer had maintained, the most effective aspect of occupational therapy was the "actual doing, actual practice" (that is, the process of making something) rather than the finished item.[69]

Despite the Board's exhortations to the contrary, some items for institutional use were produced by the occupational therapy departments, but the purpose of the activities was primarily therapeutic. At the Littlemore, brickmaking was introduced in 1926. Dr. Good observed that this activity provided effective "assistance in treatment" and the bricks were used to build a new mess room. The Board's inspectors were enthusiastic about the new activity and wondered if patients might be taught how to make concrete curbing for the paths in the ward gardens.[70] Littlemore's female patients were given the task of making cotton-wool swabs for use at the Radcliffe Infirmary. Male patients made a complete set of folding tables for the Littlemore's Recreation Hall.[71] At the Maudsley, wireless sets, bedside tables, lampshades and mortuary trolleys were among the items destined for hospital use produced by patients undergoing occupational therapy. These activities, although they saved the hospitals money, were not evaluated financially. The one aspect of productive patient work that was ascribed a monetary value was the food produced by the Littlemore's farm. The sums raised by the sale of farm produce, and the value of food

[64] Fifteenth Annual Report of the Board of Control, 1928, 4–5.

[65] Board of Control, *Memorandum on Occupation Therapy for Mental Patients* (London: HMSO, 1933), 4.

[66] Ibid., 7.

[67] Ibid.

[68] Ibid. 25.

[69] Adolf Meyer, "The Philosophy of Occupational Therapy," in *The Collected Papers of Adolf Meyer, Volume IV: Mental Hygiene*, ed. Eunice Winters (Baltimore: Johns Hopkins Press, 1952), 89.

[70] Board of Control Report, Annual Reports of the Littlemore Hospital, 1925/1926, 14–15.

[71] Medical Superintendent's Report, Annual Reports of the Littlemore Hospital, 1938/1939, 8.

supplied to the hospital, featured in the hospital accounts, and offset maintenance costs [see Table 5.3]. The value of patient labour involved in the food production was not evaluated, however.

While occupational therapy was not intended as a means of raising funds, hospitals were expected to keep the net costs of the craft activities that comprised occupational therapy to a minimum, particularly during the testing economic climate of the Depression. The Board maintained that "the best work can often be done with the simplest and cheapest materials", so "no great outlay" was required. They recommended the use of waste materials, claiming that "a little ingenuity ... solves many problems".[72] The cost of materials for handicrafts was usually covered by sales of the finished goods within the institution. This was the case at

Table 5.3 Table to show the value, in GBP, of farm goods supplied to the hospital; the income generated from the sale of surplus farm goods; and the total value of the goods expressed as a percentage of total expenditure at the Littlemore Hospital, 1923–1939

Year	*Value of goods supplied to the hospital*	*Income from goods sold*	*Total contribution*	*Total hospital expenditure*	*Value of goods as % of total expenditure*
1923	£1387	£434	£1821	£48,946	3.7%
1925	£1345	£598	£1943	£62,281	3.1%
1926	£1410	£1208	£2618	£69,011	3.8%
1927	£1282	£755	£2037	£69,515	2.9%
1928	£1354	£924	£2278	£70,580	3.2%
1929	£1128	£713	£1841	£70,237	2.6%
1930	£1371	£521	£1892	£69,762	2.7%
1931	£1368	£417	£1785	£68,311	2.6%
1932	£1317	£490	£1807	£68,554	2.6%
1933	£1404	£685	£2089	£49,949	4.2%
1934	£1424	£481	£1905	£61,702	3.1%
1935	£1205	£362	£1567	£52,841	3%
1936	£1224	£16	£1240	£55,732	1.8%
1937	£1539	£377	£1916	£58,767	2.7%
1938	£1394	£637	£2031	£59,286	2.9%
1939	£1624	£797	£2421	£60,132	3.5%

Source: Littlemore Hospital Annual Reports, 1923–1939. OHA-L1/A2/1-17

[72] Ibid.

Bethlem, where the physician superintendent was proud to announce in 1933 that the work produced by the occupational therapy department (established the previous year) had "found a ready sale amongst patients and their friends and articles to the value of £70 have been sold".[73] At the Maudsley, many items produced by the occupational therapy department were sold in the hospital canteen, which also sold tobacco, sweets, stationary and toiletries to patients, visitors and staff.[74] The medical superintendent was keen to point out in 1930 that occupational therapy continued to be run on a self-supporting basis; patients who wanted to keep the items they made bought the materials themselves at a reduced price.[75] Reporting on occupational therapy at the Littlemore in 1929, the Board's inspectors remarked that, "It is pleasant to hear that the periodical sales of work are remunerative".[76]

Even though some items were sold, and other items may have been used in the hospital (such as the decorative lampshades made by Maudsley patients), the arts and crafts that comprised occupational therapy could be considered as based on hobbies. This was certainly the view of Wilhelm Mayer-Gross, Eliot Slater and Martin Roth, whose 1954 textbook, *Clinical Psychiatry*, described British occupational therapy as "being more of ***pastimes and hobbies*** than of rough manual work, as on the European continent."[77] As such, occupational therapy fell somewhere between work and recreation. Patients were not rewarded for engaging in occupational therapy (as they were for work around the hospital) and, even if sales of items covered the expense of raw materials, the employment of an occupational therapist represented a cost to the hospital budget.

[73] Physician Superintendent's Report, Annual Reports of the Bethlem Royal Hospital, 1933, 10.

[74] Medical Superintendent's Report, Maudsley Hospital, 1924, 9.

[75] Medical Superintendent's Report, Maudsley Hospital, 1930, 16.

[76] Sixteenth Annual Report of the Board of Control, 1929, 236.

[77] Cited in John Hall, "From Work and Occupation to Occupational Therapy: the Policies of Professionalisation in English Mental Hospitals from 1919 to 1959," in *Work, Psychiatry and Society, c.1750–2015*, ed. Waltraud Ernst (Manchester: Manchester University Press, 2016), 322. [Italics in the original]

Entertainment—Therapy or a Drain on Resources?

The provision of entertainment for patients also represented a cost to French and English institutions. The resources allocated for entertainment varied between France and England and between metropolitan and provincial institutions. As he made plain in 1918, Toulouse was strongly in favour of "*distractions*" or entertainment for patients, which he believed were an essential aspect of treatment.[78] As the medical director of the Henri Rousselle Hospital, he was responsible for deciding how much to spend on these activities. At the Asile Clinique and the Asile de la Sarthe, decisions regarding the amount to be spent and the content of the programme, were taken by the asylum director. The latter also negotiated the costs—or the supply free of charge—of the entertainments. Not all French psychiatrists believed in the therapeutic value of entertainments, but the law of 1857 stipulated that "distractions" and "intellectual occupations" should be provided for patients.[79] Although there were only 80 beds at the Henri Rousselle Hospital, the entertainments budget was increased from 4000F (or 50F per capita) in 1925 to 10,000F (or 125F per capita) in 1928, indicating the priority attributed to this area by Toulouse. A piano, wireless and gramophone were available for patients to use,[80] and creative afternoons were organised with visiting artists.[81] Games, books and writing materials were also provided [see Fig. 5.1].[82]

At the Asile Clinique, where an average of 1100 patients were accommodated, expenditure per capita on entertainments was much lower, increasing from 5788 F (or 5.3 F per capita) in 1921 to 10,100 (or 9.18 F

[78] For analysis of the perceived curative value of amusements, see: Ute Oswald, "'Distraction from Hurtful Thoughts': Recreational Activities as Agents of Healing in Nineteenth-Century British Asylums," *Medizinhistorisches Journal* 56 no. 1–2 (2021): 30–57.

[79] Article 164, Bulletin Officiel du Ministère de l'Intérieur, Circulaire No. 7, 20/03/1857, 77. www.gallica.bnf.fr / (accessed 03/12/2017).

[80] Henri Rousselle, Rapport au nom de la 3e Commission sur l'Asile Clinique et sur le Service libre de prophylaxie mentale (Paris: Conseil Général de la Seine, 1922), 6. ADP/D.10 K3/32/77.

[81] Toulouse and his close colleagues, notably Dr Auguste Marie of the Ste Anne's Admissions Service, were interested in the artistic creations of their patients, using their art and writing to help understand the nature of their condition. Dr Marie had begun collecting patient art whilst practising at the Villejuif Asylum in the 1900s and continued to do so throughout his career. One of his last journal articles was devoted to patient art: Auguste Marie, "L'Art chez les aliénés," *Revue du Médecin* July (1930).

[82] Rousselle, Rapport au nom de la 3e Commission, 1922, 6.

Fig. 5.1 Recreation room at the Henri Rousselle Hospital during the 1930s. (© *Collection Musée d'histoire de la psychiatrie et des neurosciences, GHU Paris, photographie Direction de la communication du GHU*)

per capita) in 1930 [see Table 5.4]. Concerts, such as those given by the ensembles "L'Harmonie de la Préfecture", the brass band "La Sirène, "Musique et poésie à l'Hôpital" and "L'Harmonie Municipale de la Ville de Paris", took place in the asylum gardens in summer and in the *Salle des Fêtes* (recreation hall) in winter approximately once a month.[83] Stars from the Parisian music-hall gave performances at the Asile Clinique free of charge, and in winter there was a fancy-dress ball for the women patients.[84] Music for the latter was provided either by the hospital band, or by phonograph, and therefore did not incur a cost. Patients could play tennis and boules and there was a games room, well used by the working patients, where billiards was particularly appreciated.[85] Christmas celebrations

[83] François Latour, Rapport au nom de la 6e Commission sur les comptes du directeur de l'Asile Clinique en 1929 (Paris: Conseil Général de la Seine, 1930), 12. ADP/D.10 K3/46/76.

[84] Rapport Moral de l'Asile Clinique, 1933 (Paris: Préfecture de la Seine, 1933), 9. ADP/3719-W68-W69.

[85] Rapport Moral de l'Asile Clinique, 1934 (Paris: Préfecture de la Seine, 1934), 10. ADP/3719-W68-W70.

Table 5.4 Table to show the annual expenditure in French Francs (F) on entertainments, and expenditure per patient, by the Asile Clinique 1921–1937

Year	*Patient Population—Asile Clinique*	*Expenditure on Entertainments*	*Expenditure per patient—Asile Clinique*
1921	1109	5788 F	5.22 F
1922	1094	5976 F	5.46 F
1923	1055	5993 F	5.68 F
1924	994	6878 F	6.92 F
1925	1018	7241 F	7.11 F
1927	923	9518 F	10.31 F
1928	1043	9551 F	9.16 F
1929	1077	10,000 (budget)	9.29 F
1930	1083	10,100 (budget)	9.33 F
1932	1631	14,473 F	8.87 F
1933	1085	12,170 F	11.22 F
1934	1092	14,891 F	13.63 F
1935	1094	14,025 F	12.82 F
1936	1090	13,879 F	12.73 F
1937	1084	15,261 F	14.08 F

Source: Reports to the General Council, 1921–1937. ADP-D.10K 29-57

involved a party organised by the asylum director's wife and a Christmas tree from which gifts donated by the managers of large Parisian stores were distributed to patients. In 1932, 450 items of various kinds were donated, including 80 kg of sweetmeats, jewellery, undergarments, smoking apparatus, toiletries and tobacco.[86]

The events, whether paid for or free of charge, attracted criticism from the General Council for their infrequency.[87] Noting the wide variation in the sums allocated to distractions, the Commission de Surveillance recommended that a minimum sum of 15,000F be included in all the Seine asylum budgets, to allow a complete programme of therapeutic entertainments to be offered to patients, including the very popular film showings and the installation of wireless equipment.[88] The Asile Clinique's budget was duly amended [see Table 5.4]. In 1932, two wireless sets were installed in the

[86] Rapport Moral de l'Asile Clinique, 1932 (Paris: Préfecture de la Seine, 1932), 9. ADP/3719-W68-W68.

[87] M. Collavéri, Rapport au nom de la 6e Commission sur les comptes du directeur de l'Asile Clinique en 1931 (Paris: Conseil Général de la Seine, 1932), 12. ADP/D.10 K3/48/54.

[88] Procès Verbal, Séance de 28/11/1931, (Paris: Commission de Surveillance des Asiles Publiques d'Aliénés de la Seine, 1931), 208. CDS des APAS. ADP/D1X3–52.

workers' quarters, which were reported as greatly appreciated.[89] In 1933, apparatus for showing "talkie" films was purchased and film showings were organised each month.[90] What is unclear from the records is the rationale for these events. The provision of entertainments could simply have been to comply with legislation, to avoid accusations of inhumane treatment, to reward the patient workers, to encourage social skills amongst convalescent patients, or the programme could have been perceived as curative by psychiatrists who viewed entertainments in the same light as Toulouse.

At the Asile de la Sarthe, provision of *distractions* was much more limited than in the capital. Before World War I, a programme of entertainments for patients had been itemised in the asylum director's reports. These had included picnics in the countryside, walks into town, concerts, watching the town fireworks on 14 July and celebration of the Fête-Dieu, a Catholic festival that took place 60 days after Easter, organised by the chaplain. These events had not incurred any costs to the asylum; the cost of travelling on the tram to local beauty spots for picnics was deducted from the patients' *pécule*. No such events were mentioned in the post-war reports. This is not to say that they did not occur, but they were not discussed in the annual reports. A budget for "distractions and games for patients" appeared in the asylum accounts for the first time in 1934, and continued to do so until 1939 (expenditure varied between 215F and 377F; see Table 5.5). This indicated that some forms of entertainments

Table 5.5 Table to show expenditure on entertainments, and expenditure per patient, in French Francs (F), by the Asile de la Sarthe, 1934–1939 (no figures appeared in the Asile de la Sarthe accounts for entertainments prior to 1934)

Year	*Patient Population—Asile de la Sarthe*	*Expenditure on Entertainments*	*Expenditure per patient—Asile de la Sarthe*
1934	831	215 F	0.26 F
1935	837	272 F	0.32 F
1936	845	150 F	0.18 F
1937	869	377 F	0.43 F
1938	879	680 F	0.77 F
1939	899	290 F	0.32 F

Source: Reports of the Asylum Director, Asile de la Sarthe, 1934–9. ADS-1X964/5

[89] Rapport Moral de l'Asile Clinique, 1932 (Paris: Préfecture de la Seine, 1932), 9. ADP/3719-W68-W68.

[90] Rapport Moral de l'Asile Clinique, 1934 (Paris: Préfecture de la Seine, 1934), 10. ADP/3719-W68-W70.

were provided, at least from 1934 onwards, although the sums allocated were very small. Prior to 1934, the asylum director may have continued the pre-World War I policy of organising free entertainment.

The three French asylums demonstrate different levels of financial expectation from patient work and different attitudes towards expenditure on entertainments. The acute patients admitted to the Asile Clinique's treatment divisions after 1927 were not expected to work, but this left a "gap" in the hospital's finances. A solution to this problem was the creation of separate quarters for incurable and chronic, working patients who required minimal supervision and custodial care rather than active treatment. Although work was deemed unsuitable for the acute patients, entertainment for them and for the working patients, was encouraged by the General Council. The asylum director committed increasing amounts to the budget for amusements and negotiated the provision of some forms of entertainment and even Christmas gifts at no charge. At the Asile de la Sarthe, where there was a mix of acute, incurable and chronic, pauper and private patients, work continued to be performed by the calm, incurable and chronic pauper patients who were expected to contribute to the costs of their care. Work was organised in much the same way as it had been before World War I. Entertainment was minimal and the budget tiny in comparison to that of the Asile Clinique. Occupation at the Henri Rousselle Hospital was organised on a voluntary basis; no monetary value was attributed to any work carried out by patients; and the budget for amusements was particularly generous here. These varied policies concerning the value of patient work and expenditure on entertainment reflect the different management structures existing at the Henri Rousselle Hospital and the other two asylums; the attitudes towards occupation of those in charge; and the different financial circumstances of metropolitan and provincial institutions. On this last point, certain similarities between the situation in France and England can be seen.

In England, the Board of Control encouraged hospitals to provide "objects of interest and amusement" for patients.[91] The levels of expenditure committed to this area varied considerably. Bethlem had the most lavish budget for entertainment out of the three hospitals [see Table 5.6]. For physician superintendent Dr. John Porter-Phillips (1877–1946), Bethlem's entertainment programme constituted "one of the most potent factors in treatment" and was a more appropriate diversion for his

[91] Sixth Annual Report of the Board of Control, 1919, 73.

Table 5.6 Table to show the annual expenditure on entertainments by the Bethlem Royal Hospital, and expenditure per patient, in GBP, 1919–1922 (the only years for which figures were available)

Year	*Patient Population*	*Expenditure on entertainments*	*Expenditure per patient*
1919	210	£937	£4 4 s 6d
1920	226	£890	£3 9 s 3d
1921	213	£946	£4 1 s 9d
1922	243	£902	£3 7 s 1d

Source: Bethlem Royal Hospital Annual Reports, 1919-22. BMM/BAR-53

middle-class patients than manual work.[92] As studies of both private and public institutions before World War I by Louise Hide and Anne Shepherd have shown, middle-class patients tended to prefer leisurely pursuits over manual work.[93] Bethlem's programme was curtailed during World War I, but by 1923 was back up to speed with a full schedule of dances, theatrical performances, concerts, sewing parties and lectures.[94] The Christmas Eve Fancy Dress Ball was a regular feature, constituting "a prominent landmark in the domain of winter entertainment [that] affords much enjoyment and comfort of mind to those whose mental horizon may be tinged with pessimism or clouded with unhappiness."[95] Visits to the Boat Race and Epsom Races, and picnics in the countryside, were scheduled in the summer months. Patients were taken out for drives or for accompanied walks each week, and those who were well enough could leave the hospital "on parole". Considerably more patients attended the entertainments (an average of 79 patients between 1920 and 1929) than were "usefully employed" (an average of 64 patients between 1919 and 1928).

[92] Physician Superintendent's Report, Annual Reports of the Bethlem Royal Hospital, 1926, 20.

[93] Louise Hide, *Class and Gender in English Asylums, 1890–1914* (Basingstoke: Palgrave Macmillan, 2014), 111; Anne Shepherd, "The Female Patient Experience in Two Late Nineteenth-Century Surrey Asylums," in *Sex and Seclusion, Class and Custody. Perspectives on Gender and Class in the History of British and Irish Psychiatry*, eds Jonathan Andrews and Anne Digby (Amsterdam and New York: Rodopi, 2004), 223–248.

[94] Physician Superintendent's Report, Annual Reports of the Bethlem Royal Hospital, 1923, 17.

[95] Physician Superintendent's Report, Annual Reports of the Bethlem Royal Hospital, 1928, 16–17.

In 1931, Porter-Phillips maintained that "…every form of entertainment and amusement has been encouraged by the Governors for the treatment and happiness of the patients and staff".[96] In 1931, the Board of Control remarked that "probably nothing has done more in recent years to add to the happiness and contentment of the patients than the installation of the cinema".[97] A "movie-tone" apparatus was duly installed at Bethlem in 1932 to enable the showing of new films. It also had a "Radio-gram attachment" so that gramophone records could be played in the Recreation Hall, "thus obviating the necessity of an Orchestra".[98] During an outbreak of chicken pox in 1932, the new technology enabled Christmas celebrations to be broadcast from the Recreation Hall to various wards around the hospital.[99] In 1935, Bethlem's library held around 2000 books, mostly fiction, with an additional music section. A circulating system, organised by a male nurse, ensured that parcels of books were delivered every fortnight to the various wards. The stock was replenished every year by the addition of *c*.50 new books and 60–70 books were repaired annually.[100] Since 1891, Bethlem had published its own magazine, *Under the Dome*, edited by the Chaplain, to which patients, staff and Governors contributed. It was financed by sales to patients, former patients, and patients' families and friends. A new, more up-to-date version, *Orchard Leaves*, appeared in 1934, comprising short articles, quizzes, poems, a crossword, letters, and book reviews as well as details concerning the entertainments programme and sporting events.[101] The Governors noted in 1935 that production costs were not covered by sale of the magazine, but Porter-Phillips persuaded them to allow publication to continue, claiming; "The magazine [was] proving a means of interest and pleasure to all types of patients"; patients enjoyed contributing to it; and it "provide[d] a medium which create[d] and encourage[d] a good-will and fellowship amongst guests and staff alike."[102] In this instance the interests of patients overcame cost considerations.

[96] Physician Superintendent's Report, Annual Reports of the Bethlem Royal Hospital 1931, 18.

[97] Eighteenth Annual Report of the Board of Control, 1931, 10.

[98] Physician Superintendent's Report, Annual Reports of the Bethlem Royal Hospital 1932, 14.

[99] Ibid.

[100] Physician Superintendent's Report, Annual Reports of the Bethlem Royal Hospital, 1935, 16.

[101] The new name of the magazine reflected Bethlem's move to Monks Orchard in Kent.

[102] Physician Superintendent's Report, Annual Reports of the Bethlem Royal Hospital, 1935, 15.

Fig. 5.2 Physician Superintendent John Porter-Phillips (front row, centre) with the Bethlem Royal Hospital's cricket team. (© *By permission of Bethlem Museum of the Mind, HPC-20*)

Porter-Phillips put great emphasis on sport and other activities that kept patients active. The extensive grounds at Bethlem's new premises at Monks Orchard enabled a variety of sports to be played, including cricket, tennis, bowls, hockey and football. Regular matches were played by the hospital teams, which comprised both patients and staff. For example, it was reported in *Orchard Leaves* that during the summer season of 1935, 28 cricket matches were played, of which Bethlem won 12, lost 8 and drew 5, while 3 had to be abandoned due to bad weather [see Fig. 5.2]. Eleven tennis clubs sent teams to play at Bethlem, and the contests were described as "very keen".[103] A new sports pavilion was completed in 1937, providing Bethlem with "what is generally believed to be one of the best sports grounds for many miles around" and facilitating the entertainment

[103]WMA, "Sporting Activities," *Orchard Leaves*, Bethlem Royal Hospital, Winter (1935): 8.

of visiting cricket teams.[104] For the less sportively-inclined, Swedish drill and country dancing were introduced in 1927, providing another "valuable asset in the domain of treatment" from which the patient "unconsciously derives enormous benefit in mind and body".[105] By 1938, various dance and eurythmic movement classes of different levels, even for the most disturbed patients, were held twice weekly for both sexes.[106] Teaching of these classes by specialist instructors represented an addition cost. Bethlem's provision of indoor recreations and outdoor physical activities was described by the Board of Control as "plentiful and varied" in 1938, and they were pleased to note that patients were encouraged to pursue their individual hobbies.[107] Bethlem's programme of entertainment was designed to appeal to a middle-class audience, reflected by its lecture series and the contents of its magazine. The Maudsley's patients included labourers and the unemployed, as well as clergymen and doctors, and therefore the hospital had to cater for a broader range of tastes.

At the Maudsley, sport was limited, owing to the hospital's city location, but there were two billiard tables, tennis and badminton courts, and croquet was set up in the gardens attached to ground-floor wards.[108] A hard tennis court was laid in 1928 enabling play to continue all year round.[109] Fortnightly picnics were organised in neighbouring parts of London in summer, and at least one indoor event, either a concert, dance or whist drive, was held each week. Volunteers contributed to patient entertainment in less formal ways, for example, by singing in the wards.[110] The hospital was well-supplied with pianos, gramophones, wireless sets, cards and draughts, and its library (run by two nurses in their own time) held over 1000 books.[111] Weekly singing and dancing classes were introduced in 1924. "Such classes", maintained Mapother, "not only mitigate the monotony of hospital life and promote cheerfulness but can be made

[104] WMA, "Bethlem Royal Hospital Sports Club," *Orchard Leaves*, Bethlem Royal Hospital, Summer (1937): 8.

[105] Physician Superintendent's Report, Annual Reports of the Bethlem Royal Hospital, 1927, 20.

[106] Physician Superintendent's Report, Annual Reports of the Bethlem Royal Hospital, 1938, 13.

[107] Ibid., 26.

[108] Medical Superintendent's Report, Maudsley Hospital, 1924, 9.

[109] Medical Superintendent's Report, Maudsley Hospital, 1927–1931, 8.

[110] Ibid.

[111] Ibid.

to play a definite part in the re-education of many neurotic patients".[112] A patient choir was established, which gave many successful performances during the winter.[113] The choral instructress, Miss Erhart, gave her services for free. Mapother paid tribute to her in 1926, remarking that "all grades of staff as well as patients are agreed that her work is one of the most useful activities of the hospital".[114] Christmas celebrations included a fancy dress dance and entertainment around the Christmas tree on Boxing Day for child patients or the offspring of patients.[115] Attendance at this event grew each year and "strain[ed] the capacity of the hall".[116]

In terms of audio-visual equipment, "wireless" sets were provided in two of the Littlemore's largest wards, and also in the nurses' quarters, in 1926.[117] The Maudsley was similarly equipped in 1928. Music could be delivered to patients' rooms with headphones provided in every single room, and between each pair of adjoining beds in the dormitories. Loudspeakers were installed at various points throughout the hospital.[118] Both the Maudsley and the Littlemore benefited from the installation of a cinematograph in 1927 and 1928 respectively. The Board of Control urged hospitals to invest in the equipment for showing "talkie" films in 1934 because silent films were going out of circulation.[119] Sound apparatus for showing "talkie" films was introduced at the Littlemore in 1934.[120] At the Maudsley, patients had to wait for completion of the extension to the hospital in 1936, which included the provision of a larger recreation hall and new premises for the library.[121]

At the Littlemore, many of the events organised for patients did not incur any costs. For example, Dr. Good reported in 1925/1926 that concerts were held "throughout the year [performed] by Staff, both male and female, for the amusement of patients". He added that "dances have been

[112] Medical Superintendent's Report, Maudsley Hospital, 1925, 3.

[113] Ibid., 3.

[114] Medical Superintendent's Report, Maudsley Hospital, 1926, 6.

[115] Ibid.

[116] Medical Superintendent's Report, Maudsley Hospital, 1927–1931, 17.

[117] Medical Superintendent's Report, Annual Reports of the Littlemore Hospital 1927/1928, 10; and 1928/1929, 9.

[118] Medical Superintendent's Report, Maudsley Hospital, 1927–1931, 17.

[119] Twenty-first Annual Report of the Board of Control, 1934, 6.

[120] Board of Control Report, Annual Reports of the Littlemore Hospital, 1934/1935, 17.

[121] Medical Superintendent's Report, Maudsley Hospital, 1935, 40.

held in the Entertainment Hall, and in many of the Wards".[122] These free events continued throughout the interwar period, as indicated by the reports of 1926/1927 and 1937/1938.[123] Dr. Armstrong (appointed in 1936) introduced the practice of sending groups of "the better type of working patients" for charabanc outings to some of the local beauty spots. He noted that several of the older patients had not left the grounds for many years and were very appreciative of this innovation.[124] Dr. Penton (the new assistant medical officer) introduced folk dancing classes in 1939 and organised concerts given by the patients.[125] Again, such events organised by staff did not incur additional costs. The Littlemore's budget for entertainment was a fraction of that of Bethlem [see Table 5.7].

Attempts were made at the Littlemore in 1937 to increase opportunities for outdoor exercise.[126] Dr. Armstrong introduced weekly keep fit classes at the Littlemore, conducted by an instructor from the Keep Fit movement, and classes in Physical Training were held regularly for both

Table 5.7 Table to show the annual expenditure on entertainments by the Littlemore Hospital, and expenditure per patient, in GBP, 1923–1932

Year	*Patient Population*	*Expenditure on entertainments*	*Expenditure per patient*
1923	522	£44 18 s	1 s 7d
1925	606	£33 15 s	1 s 11d
1926	667	£79 17 s	2 s 4d
1927	707	£91 6 s	2 s 6d
1928	729	£63 12 s	1 s 7d
1929	743	£95 19 s	2 s 6d
1930	735	£125 14 s	3 s 4d
1931	787	£85 6 s	2 s 2d
1932	786	£98 5 s	2 s 5d

Source: Littlemore Hospital Annual Reports, 1923–32. OHA-L1/A2/1-10

[122] Medical Superintendent's Report, Annual Reports of the Littlemore Hospital, 1925/1926, 11.

[123] Medical Superintendent's Report, Annual Reports of the Littlemore Hospital, 10; and 1937/1938, 10.

[124] Ibid.

[125] Medical Superintendent's Report, Annual Reports of the Littlemore Hospital, 1939/1940, 11.

[126] Medical Superintendent's Report, Annual Reports of the Littlemore Hospital, 1938/1939, 10.

male and female patients.[127] Cricket and football teams of male patients were created and a schedule of both home and away matches was organised with patient teams from neighbouring hospitals.[128] Facilities for hockey and netball were provided for nurses and female patients, and badminton and tennis equipment was provided in some of the women's gardens. Indoor recreational facilities included table-tennis and darts for the men and table-tennis and card games for the women.[129] Nurse B, who joined Littlemore in 1936, remembered an annual sports day, as well as a fancy-dress ball at Christmas.[130] At the Littlemore, it was the Chaplain who managed the hospital library and was responsible for distributing books to patients each month. On hearing from the Chaplain that many books were being destroyed by patients, the Board suggested that such wastage could be reduced by starting a bookbinding workshop, which might also provide an interesting and useful occupation for patients who were not otherwise employed.[131]

Out of the three hospitals discussed, the financial implications of patient occupation appeared to be most pressing at the Littlemore. Here the medical superintendent chose not to invest in the services of an occupational therapist and the cost-savings generated by the sale and consumption of produce from the hospital farm continued to feature in the institution's accounts. Expenditure on entertainments was kept to a minimum through the organisation of "in house" events. Despite the legislation of 1913 and 1930, the Littlemore continued to care for a significant proportion of chronic and incurable patients, many of whom, had they been placed in a colony would have had to contribute to the costs of their care by working. Whilst this is not explicit, it could be that costs per capita were expected to be lower in institutions continuing to care for chronic and incurable patients. Occupation at the Maudsley was organised with therapy as the primary aim for all patients from the outset. By the time the Maudsley opened in 1923, the thinking around occupation had already started to move on from its pre-war pre-occupation with offsetting costs. Although some of the items produced by the occupational therapy department were used by the hospital, none of the products were evaluated for accounting

[127] Ibid.

[128] Ibid.

[129] Ibid.

[130] Interview with former nurse, Littlemore History Project, Oral History Interview by J.Goddard (1994/1995). OHA/LH/OT233.

[131] Board of Control Report, Annual Reports of the Littlemore Hospital, 1933/1934, 17.

purposes. Patients were engaged in both occupational therapy and work around the hospital. The recreational programme at the Maudsley was varied and the medical superintendent clearly felt it made an important contribution to treatment. Whilst some aspects of the programme, and the facilities required to deliver it, were paid for, the use of volunteers kept costs down. At Bethlem, where institutional finance came from charitable sources or the patients themselves, and where many patients were unaccustomed to manual labour, occupation was focused on recreation rather than labour, hence the high expenditure on sport and entertainments. That is not to say that Bethlem was oblivious to budgetary constraints. The Governors were reported as reluctant to make the initial outlay required to establish the occupational therapy department, which eventually opened in 1932, and questioned whether the patient magazine was really needed when sales failed to cover its production costs.

Conclusion

The different management structures of French and English asylums, coupled with the different levels of economic difficulty experienced after World War I, contributed significantly to the divergent attitudes towards patient work in France in England. They were not the only significant factors but added to the differences in opinion regarding the perceived usefulness of occupation as a means of therapy, particularly for acute patients, discussed in the previous chapter. In France, where the economy was more severely damaged by the war, the emphasis on keeping costs down increased the importance of patient work as a cost-saving device. The prolonged effects of the Great Depression (from which England recovered relatively quickly) also intensified reliance on patient work. Those psychiatrists who wished to change how patient work was organised by introducing occupational therapy, or simply by easing the production targets set by workshop managers, faced potential opposition from the asylum director and prefect for whom financial matters were a priority. The way to overcome such opposition, as Édouard Toulouse demonstrated, was to insist on the role of medical director which put a doctor in sole charge of the hospital, responsible for both medical services and administration. English medical superintendents had this dual responsibility and were able to establish therapeutic priorities for occupation within their institutions. An emphasis on therapy was encouraged by the Board of Control, who sought to downplay the financial contribution of patient work and advocated occupational therapy, despite the additional costs involved. This was in marked contrast to the Board's attitude towards the incurable patients placed in colonies, from whom productive work was expected. English mental hospital patients were not paid for their work, nor for what they produced in the occupational therapy department, which supported the notion that the purpose of this occupation (as opposed to the work carried out in colonies) was therapeutic. The French *pécule* system emphasised the transactional nature of patient work, making it appear more like work in the outside world, and, arguably, detracting from its role as therapy.

In terms of the entertainment provided for patients, it is interesting that the types of events in English and French institutions were quite similar, despite their being organised by a medical superintendent in England and an asylum director in France. In both countries, resources for entertainments were far more lavish in metropolitan institutions than in provincial ones. In both countries, efforts were made to secure at least some of the activities free of charge and to use the services of volunteers where possible, even at Bethlem where the programme was accorded such a high priority. The balance between work, occupational therapy and recreation, and the budgetary implications of each type of occupation, was a matter for negotiation between chief medical officer and asylum director in France. In England, the medical superintendent had overall control of patient occupation and the precise nature of the programme depended on his preferences and management style. The roles of these individuals are discussed in more depth in the next chapter.

Bibliography

Secondary Sources

Andrews, Jonathan, Asa Briggs, Roy Porter, Penny Tucker, and Keir Waddington. *The History of Bethlem*. Abingdon and New York: Routledge, 1997.

Beveridge, A. "A disquieting feeling of strangeness? The art of the mentally ill." *Journal of the Royal Society of Medicine* 94 (2001); 595–99.

Chaney, Sarah. "Useful members of society, or motiveless malingerers? Occupation and malingering in British asylum psychiatry, 1870–1914." In *Work, Psychiatry and Society, c.1750–2015* edited by Waltraud Ernst, 277–97. Manchester: Manchester University Press, 2016.

Eloranta, Jari and Mark Harrison. "War and Disintegration, 1914–1950." In *The Cambridge Economic History of Modern Europe*, edited by Stephen Broadberry and Kevin O'Rourke, 133–55. Cambridge: Cambridge University Press, 2010.

Freebody, Jane. "'The Root of All Evil is Inactivity': The response of French psychiatrists to new approaches to patient work and occupation, 1918–1939." In *Voices in the History of Madness: Personal and Professional Perspectives on Mental Health and Illness*, edited by Robert Ellis, Sarah Kendal and Steven J. Taylor, 71–94. Cham: Palgrave Macmillan, 2021.

Hall, John. "From Work and Occupation to Occupational Therapy: the Policies of Professionalisation in English Mental Hospitals from 1919 to 1959." In *Work, Psychiatry and Society, c.1750–2015*, edited by Waltraud Ernst, 314–33. Manchester: Manchester University Press, 2016.

Hide, Louise. *Class and Gender in English Asylums, 1890–1914*. Basingstoke: Palgrave Macmillan, 2014.

Jones, Kathleen. *Asylums and After. A Revised History of the Mental Health Services: From the Early 18th Century to the 1990s*. London and Atlantic Highlands, NJ: Athlone Press, 1993.

McMillan, James F. *Twentieth-Century France: Politics and Society 1898–1991*. London: Arnold, 1992.

Monk, Lee-Ann. "Exploiting patient labour at Kew Cottages, Australia, 1887–1950." *British Journal of Learning Disabilities* 38 (2010): 86–94.

Oswald, Ute. "'Distraction from Hurtful Thoughts': Recreational Activities as Agents of Healing in Nineteenth-Century British Asylums." *Medizinhistorisches Journal* 56 no. 1–2 (2021): 30–57.

Reaume, Geoffrey. "A Wall's Heritage: Making Mad People's History Public." *Public Disability History* 1 (2016): 20. Accessed 12/02/2022. https://www.public-disabilityhistory.org/2016/11/a-walls-heritage-making-mad-peoples.html.

Shepherd, Anne. "The Female Patient Experience in Two Late Nineteenth-Century Surrey Asylums." In *Sex and Seclusion, Class and Custody: Perspectives on Gender and Class in the History of British and Irish Psychiatry*, edited by Jonathan Andrews and Anne Digby, 223–48. Amsterdam and New York: Rodopi, 2004.

Thomson, Mathew. *The Problem of Mental Deficiency: Eugenics, Democracy and Social Policy in Britain, c.1870–1959*. Oxford: Oxford University Press, 1998.

Von Bueltzingsloewen, Isabelle. *L'Hécatombe des Fous: La Famine dans les Hôpitaux Psychiatriques Français sous l'Occupation*. Paris: Éditions Flammarion, 2007.

Warner, Richard. *Recovery from Schizophrenia: Psychiatry and Political Economy*. London and New York: Routledge and Kegan Paul, 1985.

Archival Sources

Fourth Annual Report of the Board of Control. London: HMSO, 1917.
Fifth Annual Report of the Board of Control. London: HMSO, 1919.

Tenth Annual Report of the Board of Control. London: HMSO, 1923.
Fifteenth Annual Report of the Board of Control. London: HMSO, 1928.
Sixteenth Annual Report of the Board of Control. London: HMSO, 1929.
Eighteenth Annual Report of the Board of Control. London: HMSO, 1931.
Nineteenth Annual Report of the Board of Control. London: HMSO, 1932.
Twenty-first Annual Report of the Board of Control. London: HMSO, 1934.
Author Unknown. "Royal Commission on Lunacy and Mental Disorder: Summary of Report." *The British Journal of Nursing* September (1926): 200-201.
Béquet, M. Rapport au nom de la 3e Commission sur l'Asile Clinique: budget additionnel de 1926 et budget de 1927. Paris : Conseil Général de la Seine, 1926. ADP/D.10K3/40/105.
Board of Control. *Memorandum on Occupation Therapy for Mental Patients.* London: HMSO, 1933. Accessed 04/05/2017. http://wellcomelibrary.org/item/b24957070.
Bulletin Officiel du Ministère de l'Intérieur. Circulaire No. 7 du 20 mars 1857, 42–79. Paris, 1857. Accessed 12/02/2018. https://gallica.bnf.fr/ark:/12148/bpt6k5539701n/f2.item.zoom
Chausse, M.E. Rapport au nom de la 3e Commission sur un projet de modification des cadres du personnel des asiles des aliénés. Paris : Conseil Général de la Seine, 1928. ADP/D.10K3/44/63.
Chausse, M.E. Rapport Général au nom de la 3e Commission sur le compte du Service des Aliénés pour 1932. Paris : Conseil Général de la Seine, 1933. ADP/D.10K3/50/110.
Collavéri, M. Rapport au nom de la 6e Commission sur les comptes du directeur de l'Asile Clinique en 1931. Paris : Conseil Général de la Seine, 1932. ADP/D.10K3/48/54.
Dausset, Louis. Rapport Général au nom de la 3e Commission sur les propositions budgétaires pour le Service des Aliénés (budget de 1919). Conseil Général de la Seine, 1918. ADP/D.10K3/27/20.
État des Produits en Nature. Compte Administratif. Le Mans : Asile de la Sarthe, 1923. ADS-1X964.
Fiancette, M. Rapport Général presenté au nom du Comité du budget, du compte et du contrôle 1933. Paris : Conseil Général de la Seine, 1933. ADP/D.10K3/49/64.
Goddard, J. Oral History Interview with Former Littlemore Nurse. Littlemore History Project. Oxford, 1994/1995. OHA/LH/OT233
Latour, François. Rapport au nom de la 6e Commission sur les comptes du directeur de l'Asile Clinique et du receveur des asiles relatifs à la gestion de cet établissement en 1924. Paris: Conseil Général de la Seine, 1925. ADP/D.10K3/38/75.
Latour, François. Rapport au nom de la 6e Commission sur les comptes du directeur de l'Asile Clinique et du receveur des asiles relatifs à la gestion de cet

établissement en 1927. Paris: Conseil Général de la Seine, 1928. ADP/D.10K3/44/64.

Latour, François. Rapport au nom de la 6e Commission sur les comptes du directeur de l'Asile Clinique et du receveur des asiles relatifs à la gestion de cet établissement en 1929. Paris: Conseil Général de la Seine, 1930. ADP/D.10K3/46/76.

Macdonald, Mary and Norah Haworth. *Theory of Occupational Therapy for Students and Nurses.* London: Bailliere, Tindall and Cox, 1940.

Marie, A. "L'Art Chez Les Aliénés." *Revue du Médecin* July (1930): 11–14.

Meyer, Adolf. "The Philosophy of Occupational Therapy." In *The Collected Papers of Adolf Meyer, Volume IV: Mental Hygiene,* edited by Eunice Winters, 86–92. Baltimore: Johns Hopkins Press, 1952.

Procès Verbal de la Séance du 28/11/1931, Commission de Surveillance des Asiles Publiques d'Aliénés de la Seine, Paris, 1931. ADP/D1X3-44/18.

Procès Verbal de la Visite de la Commission, 21-4 Avril 1923. Commission de Surveillance des Asiles Publiques d'Aliénés de la Seine, Paris, 1923. ADP/D1X3-44/6.

Rapport du Directeur. Asile de la Sarthe, Le Mans, 1924, 1925, 1936. ADS-1X964/5.

Rapports Moraux de l'Asile Clinique, 1932–1940. Paris : Préfecture de la Seine. ADP/3719-W68-W76.

Report of the Board of Control. Annual Reports of the Littlemore Hospital, Oxford, 1925/1926, 1933/1934, 1934/1935. OHA-L1/A2/3; OHA-L1/A2/11; OHA-L1/A2/12.

Report of the Medical Superintendent. Maudsley Hospital, London, 1924, 1925, 1926, 1927-31, 1935. BMM/MSR-01.

Report of the Medical Superintendent. Annual Reports of the Littlemore Hospital, Oxford, 1925/1926, 1927/1928, 1928/1929, 1934/1935, 1937/1938, 1938/1939, 1939/1940. OHA-L1/A2/3-17.

Report of the Physician Superintendent. Annual Reports of the Bethlem Royal Hospital, London, 1923, 1926, 1927, 1928, 1931, 1932, 1933, 1935, 1938. BMM/BAR-53.

Rousselle, Henri. Rapport au nom de la 3e Commission sur l'Asile Clinique et sur le Service libre de prophylaxie mentale. Paris : Conseil Général de la Seine, 1922. ADP/D.10K3/32/77.

Rousselle, Henri. Rapport au nom de la 3e Commission sur les budgets et comptes de l'Asile Clinique, Conseil Général de la Seine, 1915. ADP/D.10K3/25/20.

WMA. "Bethlem Royal Hospital Sports Club." *Orchard Leaves.* London: Bethlem Royal Hospital, Summer (1937).

WMA. "Sporting Activities." *Orchard Leaves.* London: Bethlem Royal Hospital, Winter (1935).

CHAPTER 6

The Medical Prescription of Patient Occupation

This chapter focuses on the role, authority and therapeutic preferences of the medical practitioners prescribing patient occupation. The attitude of individual medical superintendents and chief medical officers towards the therapeutic value of occupation for patients, and its suitability for certain types of patients, played an important role in whether the new methods of occupational therapy were adopted in French and English mental hospitals. They oversaw how patient work was organised and to whom it was allocated. This raises questions over whether French and English psychiatrists shared similar views regarding treatment and what other effective treatments for mental disorder existed during the interwar period. What influenced psychiatrists' responses to these different treatment methodologies? The matter of autonomy was also a significant issue; did both French and English psychiatrists enjoy sufficient authority within their institutions to impose their preferred treatment regimes, or were there others in the chain of command to whom they had to defer?

Management and Authority

Different asylum management structures, affecting the levels of authority enjoyed by English and French psychiatrists, originated in the legislative frameworks established to support the emerging national asylum systems in France and England. It is worth pausing to consider the legislation that

J. Freebody, *Work and Occupation in French and English Mental Hospitals, c.1918–1939*, Mental Health in Historical Perspective,
https://doi.org/10.1007/978-3-031-13105-9_6

developed. English local authorities were empowered to raise a county rate for the purpose of building an asylum for the county by the County Asylums Act of 1808; the Lunacy Act of 1845 made this a requirement. Although the asylums were to provide "constant" access to "medical assistance", there were no regulations regarding treatment.[1] Asylums were run by a medical superintendent, who was expected to be medically qualified and an able administrator. The standards of care that asylums should provide, including the application of moral therapy, and standards referring to food, cleanliness, exercise and occupation were indicated by the Report of the Select Committee on Pauper Lunatics in Middlesex and on Lunatic Asylums, published in 1827.[2]

The report included a questionnaire for asylum superintendents intended to assess their asylum's performance in these areas. The questionnaire included an inquiry into the provision of manual labour and activities for patients designed to "engage the attention to external objects" such as drawing, painting or gardening.[3] It asked whether "innocent amusement[s]", such as music, looking after poultry or domestic animals, or gardening were provided for patients whose conditions were severe, or whether intellectual pursuits were available for "patients of a superior description".[4] The questionnaire was sent out to the asylum authorities, but there was no law at that time requiring them to complete it, nor to comply with its implicit recommendations.[5] The principle of inspecting asylums, and of holding the medical superintendent to account, was introduced by the County Asylums Act of 1828. The Lunacy Act of 1845 made it a requirement for every county to provide an asylum for its pauper insane. The Act also established the Lunacy Commission, which was replaced by the Board of Control in 1913. These organisations were responsible for issuing directives concerning the internal regulation of institutions for the mentally disordered and the treatment of patients.[6] The Board of Control made it clear in 1932 that responsibility for the

[1] Kathleen Jones, *Asylums and After: A Revised History of the Mental Health Services: From the Early 18th Century to the 1990s* (London: Athlone Press, 1993), 37.

[2] Appendix 1, *Report of the Select Committee on Pauper Lunatics in Middlesex and on Lunatic Asylums* (London, 1827), 10–12.

[3] Ibid., 11.

[4] Ibid.

[5] Jones, *Asylums and After*, 63–4.

[6] Phil Fennell, *Treatment without Consent: Law, Psychiatry and the Treatment of Mentally Disordered People since 1845* (London: Routledge, 1996), 6.

direction of the medical and scientific work of each hospital lay with the medical superintendent.[7] The introduction of occupational therapy was therefore in his gift since all administrative and medical decisions rested with him. The English medical superintendent may not have been in control over admissions to a public institution, but he held the ultimate authority over the policies and procedures within it.

In France, legislation regarding care of the mentally disordered in asylums, initiated in the early 1800s, was delayed until 1838 due to successive political regime changes, a lack of finances and ongoing debates over whether asylums were to be considered as hospitals or as institutions of detention.[8] A requirement for each department to provide, or ensure access to, an asylum for the mentally disordered was established by the law of 30 June 1838.[9] Further legislation created on 18 December 1839 set out the principle of shared responsibility for asylum management between an asylum director and a chief medical officer.[10] The chief medical officer was responsible for all matters concerning the treatment of patients, while the asylum director took charge of finance and administration.[11] Detailed regulations regarding how a model institution should function were the subject of the ministerial order of 20 March 1857. The order included instructions concerning the medical service, administration, the daily regime, dietary, and the provision of work and occupation for patients. The Minister responsible for the order declared that it completed the law of 1838 and was based on 18 years' experience of asylum management in France and on observation of the workings of asylums overseas.[12] This order, together with the original legislation of 1838, provided the legal framework for the management of asylums for the first half of the twentieth century. A noticeable difference between the French and English legislation was the level of detail. The French Bulletin of 1857 specified very precisely how asylums should be managed and how patients should be

[7] Nineteenth Annual Report of the Board of Control, 1932, 9.

[8] Jacques Postel and Claude Quétel, *Nouvelle Histoire De La Psychiatrie* (Paris: Dunod, 2012), 180.

[9] Texte de la loi du 30 juin 1838, No. 7443: Loi sur les aliénés. https://www.ascodocpsy.org/wp-content/uploads/textes_officiels/Loi_30juin1838.pdf (accessed 03/10/2020).

[10] Ordonnance du 18 décembre 1839 portant règlements sur les établissements publics et privés consacrés aux aliénés. https://www.legifrance.gouv.fr/loda/id/LEGISCTA000006091614 (accessed 23/04/2019).

[11] Postel and Quétel, *Nouvelle Histoire*, 292–3.

[12] Ibid., 293.

treated. In England, the role of the medical superintendent, and the precise nature of patient occupation, was not codified in the same way.

The management structure of the French public asylum was quite different to that of an English institution. In England both medical and administrative decisions were taken by one person, the medical superintendent. Although the French chief medical officers had control over medical matters, patient work was both a medical and an administrative concern. Work was supposed to be prescribed by the chief medical officer as therapy for patients, according to the Bulletin of 1857,[13] but it also fulfilled an important financial role in controlling maintenance costs, which was the concern of the asylum director. The Bulletin specifically highlighted the fact that the "product" of patient work belonged to the establishment.[14] The departmental prefect was supposed to settle any disagreements between the asylum director and chief medical officer, such as those that might arise over the question of patient work. Since neither the asylum director nor the prefect was medically qualified, the medical perspective was in danger of being overruled, despite the regulations regarding the medical prescription of work. M. Reyrel, for example, appointed asylum director of the Asile Clinique in Paris in 1918, was a former a cabinet minister at the Ministry of the Interior and Prefect of the Seine. He was an experienced negotiator and financial manager, but not medically trained.[15]

Compared with their Dutch colleagues, as Seine psychiatrists Paul Courbon and A. Porot observed, French psychiatrists lacked authority. They were not respected, either by their staff or by the public and local authorities.[16] Dissatisfaction with services for the mentally ill had not abated since the anti-psychiatry movement of the late 1890s had branded French asylums as "modern Bastilles".[17] As Coffin has suggested,

[13] Article 150, Bulletin Officiel du Ministère de l'Intérieur, Circulaire No. 7, 20/03/1857, 75. www.gallica.bnf.fr (accessed 12/02/2018).

[14] Article 153, ibid., 77.

[15] Louis Dausset, Rapport Général au nom de la 3e Commission sur les propositions budgétaires pour le Service des Aliénés (Paris: Conseil Général de la Seine, 1918), 6. ADP/D.10K3/27/20.

[16] P. Courbon, "Un voyage d'étude dans les asiles de Hollande," *Annales Médico-psychologiques* no. 2 (1928):403; A. Porot, "L'Assistance par le travail dans les asiles Hollandais," *L'Hygiène mentale* no. 2 (1929): 50–1.

[17] Patricia Prestwich, "Family Strategies and Medical Power: 'voluntary' committal in a Parisian asylum, 1876–1914," *Journal of Social History,* 22 June (1994). *The Free Library:* https://www.thefreelibrary.com/Family+strategies+and+medical+power%3a+%27voluntary%27+committal+in+a…-a016108112 (accessed 02/04/2021).

psychiatrists in the 1920s felt "under attack" because asylums were so overcrowded that it was difficult to provide any proper treatment.[18] A lack of respect was also a consequence of French psychiatry's subordinate position in relation to neurology and the dearth of effective biological cures for mental disorder. The combination of a lack of authority and the combined administrative might of the asylum director and the prefect, may have rendered the chief medical officer powerless to instigate changes to the way patient work was organised.

It was noted by General Councillor Louis Dausset in 1918 that all the Seine asylums had workshops, but the number of patients working in them had diminished each year, despite the fact that many patients were artisans—mechanics, electricians, locksmiths, painters, stone-masons, and shoe-makers—whose skills had not been completely lost as a result of their illness.[19] The reason for the reduction in the numbers of patient workers was not addressed in this 1918 report, but pre-war reports had indicated that disagreements between doctors and technical staff, over how patients were managed, had deterred the medical teams from sending their patients to the workshops.[20] Such disagreements were exacerbated by the management structure of French asylums, since the workshop managers reported to the asylum director, whose agenda was financial, while the chief medical officers' priority was therapy. The situation did not appear to have improved by the 1930s. As doctors Maurice Legrain (chief medical officer of the Seine's Villejuif Asylum) and Georges Demay (chief medical officer of the Clermont Asylum in the Oise Department, northern France) wrote in a report commissioned in 1934 by the French government's Superior Council for Public Assistance (published in *L'Aliéniste français* in 1936), "it is clear that patient work is regarded differently by the economic and technical services and by the medical services".[21] They observed that for the Administration, patient workers were divided into two groups, good

[18] Jean-Christophe Coffin, "'Misery and Revolution': the organisation of French psychiatry, 1900–1980," in *Psychiatric Cutlures Compared: Psychiatry and Mental Health Care in the Twentieth Century. Comparisons and Approaches*, edited by Marijke Gijswijt-Hofstra, Harry Oosterhuis, Joost Vijselaar and Hugh Freeman (Amsterdam: Amsterdam University Press, 2005), 226.

[19] Dausset, Rapport Général, 1918, 45.

[20] Henri Rousselle, Rapport au nom de la 3e Commission sur les budgets et comptes de l'Asile Clinique (Paris: Conseil Général de la Seine, 1910), 17. ADP/D.10K3/20/63.

[21] M. Legrain et G. Demay, "Le Travail des Aliénés Convalescents", *L'Aliéniste français* 12 (1936), 283.

Fig. 6.1 Édouard Toulouse, Medical Director (front row, with the long scarf), and his staff, Henri Rousselle Hospital, Paris. (© *Collection Musée d'histoire de la psychiatrie et des neurosciences, GHU Paris, photographie Direction de la communication du GHU*)

workers and the rest, while for the medical team, productivity was not the main aim of the work.[22]

Dr Édouard Toulouse (1865–1947) was keen to avoid such conflicts and insisted on the role of "medical director" when appointed to manage the Henri Rousselle Hospital in 1922, combining the functions of asylum director and chief medical officer.[23] Toulouse is pictured with his team at the Henri Rousselle in Fig. 6.1 (above). As medical director, he had control over both administrative and medical matters. Toulouse, like his English counterparts, could focus on the therapeutic aspects of work and occupation when prescribing them to patients. General Councillor Henri Rousselle noted in his 1923 report that patients "were not to be subjected to any work, but were recommended to occupy themselves".[24] While work was available for patients (for which they would be paid), it was stated in the Henri Rousselle regulations that the *sole* aim of this work was

[22] Ibid.

[23] Édouard Toulouse, Rapport du médecin-directeur sur le Centre de prophylaxie mentale et l'Hôpital psychiatrique Henri-Rousselle (Paris: Conseil Général de la Seine, 1930), 65. ADP/D.10K3/46/78 Annexe.

[24] Henri Rousselle, Rapport au nom de la 3e Commission sur le Service départemental de prophylaxie mentale et les service annexes (Paris: Conseil Général de la Seine, 1923), 6. ADP/D.10K3/34/87.

therapy.[25] In other words, there was no obligation for patients to contribute to institutional running costs, but keeping busy was advised. While Toulouse, and the English medical superintendents, still had to deliver a balanced hospital budget, their decisions were not constrained by the agenda of an asylum director who was not medically qualified and whose priority was management efficiency rather than treatment. Decisions made by the chief medical officers of the Asile Clinique and the Asile de la Sarthe regarding patient work were subject to negotiation with the asylum administration.

As Demay highlighted in 1929, it was perfectly legitimate for the work carried out by patients to benefit the asylum and help reduce running costs, but the interests of the patients were always to take priority. The concepts of work-as-therapy ("*le travail-traitement*") and work-for-profit ("*le travail-rendement*") should not be mutually exclusive, but problems of interpretation could arise. A medical director, he argued, invested with both medical and administrative authority, would be able to balance the interests of patients and asylum management.[26] Hermann Simon (1847–1945), the German psychiatrist who developed "more active therapy", was also adamant that asylums should be medically directed to ensure that work was oriented towards a therapeutic rather than an economic goal.[27] The matter of medical control of the asylum was also raised by Dr Ferrio who criticised the French law of 1838 in an article appearing in the *Annales Médico-psychologiques.* Ferrio insisted that "the director of the psychiatric hospital must be a doctor, exclusively a doctor".[28] This, he believed, was the only way of achieving harmony between all the various hospital services. Ferrio supported Toulouse's assumption of the role of both asylum director and chief medical officer at the Henri Rousselle Hospital, insisting that this was the only way forward.[29] Only a medical director had the authority to see through reforms and ensure the

[25] Article 28, Règlement du service départemental de prophylaxie mentale, dans M. Grangier, Rapport au nom de la 3e Commission sur le Service départemental de prophylaxie mentale et les services annexes (Paris: Conseil Général de la Seine, 1925), 75. ADP/D.10K3/38/79.

[26] G. Demay, "Les conditions de la thérapeutique par le travail dans les asiles," *L'Hygiène Mentale*, no. 2 (1929): 40.

[27] J. Vié, "Le traitement des malades mentaux par le travail: les idées et les réalisation de H. Simon," *L'Aliéniste francais* (1934): 597.

[28] Carlo Ferrio, "Le centenaire de la loi française du 30 juin 1838. Ce qu'en pense un étranger," *Annales Médico-psychologiques* no. 1 (1939): 751.

[29] Ibid., 752.

primacy of a therapeutic agenda. In England, the medical superintendent, who was in sole charge of the hospital, had this authority.

Although the English medical superintendent did not have to contend with the aims and objectives of an asylum director, Peter Bartlett argues that his authority was compromised by the close association between the poor law and county asylums. Bartlett maintains that the asylum system was created out of the poor law administrative infrastructure; the county asylums and the other poor law institutions were thus parts of the same system, administered by the same people. [30] Asylum doctors, according to Bartlett, had little say in how asylums would be constructed, nor in the admission or discharge of pauper patients. The medical professionals, far from being autonomous, were simply part of an administrative network comprising local justices of the peace and poor law officials overseeing provision for the poor. The latter might receive "outdoor" relief or be admitted to the workhouse or county asylum.[31] Whilst it is true that the medical superintendent lacked control over the confinement of pauper patients, once admitted, patients were subject to his medical regimen. The level of care and the nature of the treatment they received were decided upon by the medical superintendent. It was up to the medical superintendent to "balance the books" by deciding how much to spend on personnel, treatment and pharmaceuticals, entertainments and recreation, and on the basic necessities such as food, fuel and clothing. His remit included the decision on whether to employ an occupational therapist and whether patients should be engaged in occupational therapy or work around the hospital.

The importance of the medical superintendent's role was recognised by the Board of Control, who remarked that how a hospital was managed "depends more on the personality, the outlook and the experience of the Superintendent than on anything else".[32] The medical superintendent was "the inspiration" behind the medical and scientific work of the hospital and needed to be the "physician in chief" and not just a good administrator.[33] In other words, if the medical superintendent or chief medical officer believed that patient occupation was important, this would be prioritised

[30] Peter Bartlett, *The Poor Law of Lunacy: The Administration of Pauper Lunatics in Mid-nineteenth Century England* (London and New York: University of Leicester Press, 1999), 50.

[31] Ibid., 2–4.

[32] Nineteenth Annual Report of the Board of Control, 1932, 9.

[33] Ibid.

(although the converse was equally true). It was up to the medical superintendent to make the necessary staffing, financial and practical arrangements to facilitate occupational therapy. The Board of Control insisted that occupation should be supervised by doctors and used throughout the hospital as "a socialising factor or as a method of cure".[34] The Board were keen that medical officers "direct the course of occupational therapy in every phase", watching the patients in their classes, the occupation centre, the gardens and recreational hall, or wherever they may be. They should study the effect of the treatment on different types of patients, using their observations to prepare prescriptions to guide staff and "ensur[e] that every occupation and recreation is used therapeutically".[35]

The Views of Psychiatrists in Paris and London

The priority accorded to occupation by doctors depended on the availability of, and the doctors' faith in, alternative treatments, such as biological remedies. Modern treatment, as identified by the French Asylums Medical Society in 1918, comprised hydrotherapy, isolation, UV-ray treatment, electricity, radiography, psychotherapy, work and distractions. Of these, the only really modern mode of treatment was psychotherapy, but this was not accepted by psychiatrists outside Toulouse's circle in Paris.[36] Experiments with a variety of biological remedies were carried out during the 1920s, including treatment with organ extracts, hypnotics, barbiturates and anti-syphilitic drugs, but these did not provide the breakthrough that psychiatrists had hoped for and were regarded with scepticism by many English psychiatrists. French psychiatrists, desperate to find a "scientific" treatment for mental disorder that would elevate their status amongst the rest of the medical profession, viewed them more positively. Great claims were made for malaria therapy, found to be an effective treatment for some sufferers of neurosyphilis or general paralysis of the insane (GPI) in 1917. Malaria therapy came into general use in the mid 1920s. Its safety and effectiveness were questioned by some, particularly in England, and it was only effective for one condition (despite attempts to use it for other conditions). GPI became the first treatable brain disease causing

[34] Board of Control, *Memorandum on Occupation Therapy for Mental Patients*. London: HMSO, 1933, 15.

[35] Ibid., 16.

[36] Dausset, "Rapport Général," 23–29.

serious psychiatric symptoms and malaria therapy marked the beginning of physical therapies for mental disorders.[37] It was not until the late 1930s, however, that shock treatments (such as insulin coma therapy and electro-convulsive therapy) and surgical interventions (focal sepsis and leucotomy or lobotomy) emerged on the scene. In other words, during the 1920s, as Jones *et al.* emphasise, psychiatrists had "little to offer in the way of treatment".[38]

In Paris, Toulouse's interest in the use of therapeutic occupation, like his understanding of heredity, was atypical of his French colleagues. He believed that social; psychological; moral; physical and pathological factors could stimulate an inherited disposition towards mental illness. Crucially, however, he did not believe that mental illness itself could be transmitted through heredity. He was critical of such theories, held by French alienists since they led to a fatalistic tendency to regard all mental illness as incurable. Toulouse was clear that in his view mental illness was both curable and preventable.[39] Influenced by contact with American psychiatry through the Mental Hygiene Movement, Toulouse adopted a more psychobiological stance after World War I. He advocated psychotherapy for both in- and out-patients, as well as biological and physical treatments, all of which were available at the Henri Rousselle Hospital. Toulouse's use of drugs is indicated by the steep rise in expenditure by the Asile Clinique pharmacy (the Henri Rousselle Hospital was administratively dependent on the Asile Clinique until 1926). Between 1921 and 1922 (when the Henri Rousselle Hospital was founded) expenditure rose from 43,000F to 58,000F, despite the Henri Rousselle having just one fifth of the number of patients as the Asile Clinique.[40] Although clearly not averse to prescribing biological remedies, Toulouse also expected his hospital patients to keep themselves busy and ensured that the facilities were available for them to do so.

Psychiatrists at the Asile Clinique did not consider patient work a suitable treatment for patients at the acute stage of their illness. When the

[37] Edward Shorter, *A Historical Dictionary of Psychiatry* (Oxford: Oxford University Press, 2005), 299.

[38] Edgar Jones, Shahina Rahman and Brian Everitt, "Psychiatric Case Notes: Symptoms of Mental Illness and Their Attribution at the Maudsley Hospital, 1924–1935," *History of Psychiatry* 23 (2012): 163.

[39] Michel Huteau, *Psychologie, Psychiatrie et Société Sous La Troisième Républisque, La biocratie d'Édouard Toulouse (1865–1947)* (Paris: L'Harmattan, 2002), 149–50.

[40] Dr Thabuis, Rapports du Pharmacien en Chef, Asile Clinique (Paris: Préfecture de la Seine, 1921), 71; and 1922, 101. ADP/PER-566-5.

Asile Clinique became an acute service, only convalescent patients in the treatment sections worked. Doctors preferred "active" biological treatments for acute-stage patients, resulting in another significant rise in expenditure by the pharmacy in 1927.[41] In 1928, the pharmacist remarked that not only was the cost of medicines rising, but so was their use by the various services within Ste Anne's.[42] Although work was not considered suitable for patients at the acute stage of their illness, once passed this stage patients were able to work. Treatment of GPI with malaria therapy and Stovarsol, an arsenic-based drug, at the Asile Clinique, following the transformation of the Asile Clinique's services in 1927, was reported to have enabled several patients to take up work once more, either within the hospital as convalescent patients, or after discharge. In 1928, for example, Dr Leroy reported that eight of his GPI patients had been able to leave the asylum in a condition that allowed them to re-enter *la vie sociale* and take up their professions once more.[43] In the women's first section, which became the Malaria Therapy Centre in 1930, 20 patients were discharged "cured" and able to resume their previous professions.[44] In 1935, Dr Marchand discharged four more GPI patients from the Men's First Division, following treatment with Stovarsol, who also returned to productive work.[45] After discharge, GPI patients who had been treated with malaria therapy or Stovarsol as in-patients of the Asile Clinique continued to receive treatment (twice or three times per week) as out-patients at the Henri Rousselle Hospital, under the therapeutic direction of Dr Barbé.[46]

Insulin shock treatment and Cardiazol were used in all the services of the Asile Clinique in 1938.[47] The judicious application of shock treatments was responsible for a "considerable reduction" in the length of a patient's stay in hospital, which was good for the patient (who could

[41] Dr Thabuis, Rapport du Pharmacien en Chef, Asile Clinique, 1927; 77. ADP/PER-566-5.

[42] M. Lévêque, Rapport du Pharmacien en Chef, Asile Clinique, 1928; 102. ADP/PER-566-5.

[43] Ibid.

[44] Dr Leroy, Rapport de la 1re section des femmes, Asile Clinique, 1932; 76. ADP/PER-566-6.

[45] Dr Marchand, Rapport de la 1re section des hommes, Asile Clinique, 1935; 90. ADP/PER-566-6.

[46] Ibid.

[47] Rapport Moral de l'Asile Clinique (Paris: Préfecture de la Seine, 1938), 17. ADP/3719-W74.

convalesce at home) and for departmental finances.[48] Treatment of GPI with malaria therapy and arsenic had continued, apparently with good results. Dr Guiraud had noted, by following patients' progress since 1934, that the results were not transitory, but definitive in most cases provided the treatment was continued for a long period after discharge.[49] UV-ray and electrical treatment had also been continued.[50] The report was optimistic, suggesting that the increasingly varied treatment offered at the Asile Clinique emphasised its role as a "proper hospital", where all acute patients were given appropriate medical care.[51] In the same report, the patient workforce was described as "practically non-existent" (*à peu près nulle*). The work of the general services and technical departments had been performed by the permanent and auxiliary personnel, rather than by patients.[52] This underscores the reluctance of doctors to prescribe work to acute patients. The apparent lack of interest in psychological methods of treatment, such as psychotherapy or occupational therapy, that were not even mentioned by doctors at the Asile Clinique, contrasted with the approaches taken by Toulouse at the Henri Rousselle Hospital (which was literally next door) and Edward Mapother (1881–1940), medical superintendent of the Maudsley Hospital in London. Mapother, in particular, was far more cautious with regard to biological and physical treatments than his French counterparts.

Both Toulouse's and Mapother's understanding of mental illness evolved during their careers. Mapother's early papers (1911–1914) were strongly neurological in tenor, although he worked with some of the leading names in psychology (including Bernard Hart) at the Long Grove Asylum before World War I. During the conflict, Mapother served in the Royal Army Medical Corps, and completed the three-month course in military psychiatry at the Maghull Hospital in 1917. Immediately after the war he was appointed to the Maudsley Hospital whilst it was operating as a Ministry of Pensions specialist treatment centre for soldiers suffering from severe war neuroses. [53] These wartime and post-war experiences led Mapother to become interested in psychopathology. He maintained that

[48] Ibid.
[49] Ibid.
[50] Ibid., 10.
[51] Ibid., 17.
[52] Ibid., 18.
[53] Edgar Jones, "Aubrey Lewis, Edward Mapother and the Maudsley," *Medical History* 47 Supplement no. 22 (2003): 7.

"in all properly investigated cases of insanity, it is found that it is the result of the summation of multiple causes, effective in combination, though inadequate singly".[54] He declared futile the controversy between those who believed mental disorder had a psychogenic origin and those for whom its origin was physiological.[55] Mapother stressed the "continuity of all forms of mental disorder and for the compatibility of treatment … of all grades of it".[56] His policy at the Maudsley was "to encourage an unprejudiced trial of every form of treatment offering a reasonable prospect of benefit rather than harm".[57] Wary of the latest biological and physical treatments, Mapother maintained in 1925 that "certain long-established measures still form the foundation of any successful treatment of neuroses and psychoses". These he identified as "suitable feeding, fresh air and sun, the regulation of rest, exercise and occupation, [and] the procuring of sleep", reminiscent of the balanced "regimen" of the six non-naturals.[58]

This preference for "long-established measures" did not preclude the Maudsley's experimentation with organotherapy, as Bonnie Evans and Edgar Jones have shown, but this, and other biological treatments, were blended with efforts to help a patient to adapt more effectively to their environment through psychotherapy and occupation.[59] The Maudsley's approach was "interdisciplinary" and "pragmatic" and drew on Adolf Meyer's methodology. It incorporated "psychology, biology, evolutionary theory, and even the moral and social sciences". Each patient's personal circumstances and family histories were integrated with their symptoms to develop a coherent understanding of their problems.[60] Mapother (pictured below in Fig. 6.2) recognised the need to "elucidate both recent and remote sources of mental trouble" through psychotherapy but felt this

[54] Thomas Bewley, "Online Archive 13: Edward Mapother (1881–1940)," in Thomas Bewley, *Madness to Mental Illness: A History of the Royal College of Psychiatrists* (London: Royal College of Psychiatrists, 2008), 9. https://www.rcpsych.ac.uk/docs/default-source/about-us/library-archives/archives/madness-to-mental-illness-online-archive/people/edward-mapother-1881-1940.pdf?sfvrsn=909358a2_6 (accessed 23/05/2020).

[55] Ibid, 10.

[56] Ibid., 11.

[57] Ibid., 5.

[58] Report of the Medical Superintendent, Annual Reports of the Maudsley Hospital, London, 1924, 8. BMM/MSR-01.

[59] Bonnie Evans and Edgar Jones, "Organ Extracts and the Development of Psychiatry: Hormonal Treatments at the Maudsley Hospital 1923–1938," *Journal of the History of the Behavioural Sciences* 48, no. 3 (2012), 258.

[60] Ibid., 263.

Fig. 6.2 Edward Mapother, Medical superintendent of the Maudsley Hospital, London. (© *By permission of Bethlem Museum of the Mind, EM-01*)

was better achieved after the patient's emotional state had been improved by established measures outlined above, and by the removal of the patient from home, which was often at the heart of a patient's anxieties.[61] He acknowledged that psychotherapy was the most effective means of combating mental stress in cases of "functional disorder", following his experiences in World War I. In this therapeutically tolerant atmosphere, occupational therapy, which had the advantage of doing no harm to patients (unlike many of the treatments of the 1930s) was able to flourish. Mapother first reported employing an occupational therapist in 1925.[62] A

[61] Report of the Medical Superintendent, Annual Reports of the Maudsley Hospital, London, 1925–6, 10. BMM/MSR-01.

[62] Report of the Medical Superintendent, Annual Reports of the Maudsley Hospital, London, 1925, 3. BMM/MSR-01.

school of occupational therapy was established at the Maudsley in 1932, underlining its perceived importance by the superintendent.

The discovery of insulin shock treatment for schizophrenia in Vienna in 1936 generated international excitement at a time when the Board of Control were starting to despair of ever seeing "any noticeable improvement in the discharge rate".[63] The procedure was not without risks, however, and the following year, an alternative, safer method of shock treatment, by injections of the drug Cardiazol, was introduced. Cardiazol became the most widely used somatic treatment for schizophrenia and affective disorders in British public mental hospitals.[64] Mapother was wary of the new shock treatments.[65] Until their safety and efficacy could be proven, he opposed their use. In line with Mapother's commitment to avoiding treatment that could cause his patients harm, clinical trials of Cardiazol were banned at the Maudsley because the convulsions generated could cause extreme anxiety and fear. Similarly, Mapother delayed the introduction of insulin coma therapy until November 1938 owing to the medical risks posed to patients.[66] While these shock treatments were being greeted with caution by Mapother, the provision of occupational therapy was expanding at the hospital, following the department's relocation to larger quarters after the hospital was extended in 1936. That year, a male occupation officer was appointed to provide carpentry classes for male patients; a dedicated carpentry workshop was planned for the hospital's second extension in 1939.[67]

In contrast with the holistic approach of Mapother, John Porter-Phillips, physician superintendent of the Bethlem Royal Hospital from 1914 until 1944, maintained a predominantly physicalist approach to treating acute-stage patients for most of his tenure at Bethlem. When the hospital was amalgamated with the Maudsley in 1948, his treatment methods were denounced as "old-fashioned" by Aubrey Lewis, who became the Maudsley's superintendent in 1939.[68] The records indicate that sedation, bed-rest, a milk diet and electrotherapy were the preferred methods

[63] Twenty-third Annual Report of the Board of Control, 1936, 2–4.

[64] Niall McCrae, "A Violent Thunderstorm: Cardiazol Treatment in British Mental Hospitals," *History of Psychiatry* 17 no. 1 (2006): 67.

[65] Jones, "Aubrey Lewis, Edward Mapother and the Maudsley Hospital," 13–14.

[66] Ibid.

[67] Twenty-third Annual Report of the Board of Control, 1936, 490.

[68] Jonathan Andrews *et al.*, *The History of Bethlem* (Abingdon and New York: Routledge, 1997), 680–1.

of treating acute patients during the interwar period at Bethlem.[69] All patients, whatever their condition, were sedated on admission (sulphonal and paraldehyde were the sedatives most commonly used), a custom established in the late nineteenth century.[70] The use of mechanical restraint was regularly reported by the Board of Control's inspectors. For example, in 1929 they noted that two women had been mechanically restrained on 11 occasions for a total of 62 hours "to prevent self-injury".[71]

Porter-Phillips' initial focus on the "physical side of mental illness" was indicated by the Medical School's neurological orientation and by his particular interest in focal sepsis. The latter became a significant method of treatment at Bethlem after the appointment of dental surgeon William Bulleid to Bethlem in 1922.[72] In his annual report of 1926, Porter-Phillips stressed the importance of treating "the large number of patients … admitted with marked dental disease" given the "possible role this focal sepsis plays in the causation of mental disorder".[73] Although more concerned with focal sepsis and other physical or biological remedies, Porter-Phillips appointed a psychologist in 1923 as he felt that psychology should be included in the medical school's syllabus. Psychology at Bethlem, however, was limited to the psychometric testing of patients and there was little cooperation between the psychology department and other departments of the hospital.[74] Psychotherapy was not widely used at Bethlem until after World War II.

A lack of enthusiasm from Porter-Phillips, coupled with the Bethlem Governors' unwillingness to set aside the requisite funds, delayed the introduction of occupational therapy until 1932. That said, Porter-Phillips was committed to the provision of a comprehensive programme of entertainment and recreation. The programme of plays, operas, dances, whist drives, and fancy dress ball at Christmas, and the carriage drives into London, visits to see the Boat Race and to Epsom Races, sporting activities, walks and picnics in summer, formed, in Porter-Phillips' opinion,

[69] Ibid.

[70] Ibid., 679.

[71] Sixteenth Annual Report of the Board of Control, 1929, 322.

[72] Andrews *et al., History of Bethlem*, 682.

[73] Ibid., 683.

[74] Ibid., 686.

"one of our most potent factors in treatment".[75] He claimed that the programme helped patients "focus their attention in a healthy manner and thus readjust themselves to our so-called normal social life".[76] The Chaplain gave lantern-slide lectures; patients were provided with a piano and a library; contributions to the hospital magazine *Under The Dome* (renamed *Orchard Leaves* after the hospital's move to Monks Orchard in 1930) were encouraged. Classes in Swedish Drill, Morris and Country dancing were introduced in the mid-1920s, from which patients were said to derive "enormous benefit in mind and body".[77] These activities were well suited to the Bethlem's increasingly middle-class clientele.

The early nineteenth-century moral therapists had long since recognised the value of entertainment and recreational activities that diverted the patient's attention and encouraged social interaction (just as they had advocated the therapeutic potential of work). The programme at Bethlem was not novel, although it was far more lavish than the entertainment offered in public asylums. Porter-Phillips also realised that "employment in the open air", which had been highly recommended by Pinel and Tuke, could provide a "great auxiliary to the medical treatment" particularly amongst patients in "late adolescence or early manhood" (despite that fact that manual work was unsuitable for many of his middle-class patients).[78] Porter-Phillips' evaluation of entertainment and work appeared to be rooted in the nineteenth-century concept of moral therapy rather than in modern notions of occupational therapy. Patient complaints of boredom eventually persuaded the Governors that occupational therapy should be introduced, and an occupational therapist was duly appointed in 1932.[79] A photograph of the new occupational therapy department was included in the hospital prospectus in 1932 (see Fig. 6.3), which boasted of the range of arts and crafts on offer.[80] Although supportive of occupational therapy once it had been introduced, no doubt encouraged by its popularity amongst staff and patients, Porter-Phillips remained committed to

[75] Report of the Physician Superintendent, Annual Reports of the Bethlem Royal Hospital, London, 1926, 20.

[76] Ibid., 21.

[77] Ibid., 1927, 19.

[78] Ibid., 1924, 20.

[79] Andrews et al., *History of Bethlem*, 689.

[80] J. Walke, *Repute and Remedy: Psychiatric Patients and their Treatment at Bethlem Royal Hospital 1930–1983,* Thesis (London School of Hygiene and Tropical Medicine, 2015), 228.

Arts and Crafts Department.

Fig. 6.3 The arts and crafts department at Bethlem Royal Hospital, Monks Orchard. (© *By permission of Bethlem Museum of the Mind, ECB-07*)

searching for the physical causes of, and biological treatment for, mental illness.[81]

Only a few GPI patients were admitted to Bethlem, owing to its policy of restricting admission to curable patients, and Porter-Phillips was not enthusiastic about malaria therapy, believing it only offered "a slight hope of complete cure".[82] At Bethlem in the 1930s, patients continued to be sedated, with hyoscine prescribed to deal with head-banging and other destructive habits. Barbiturates, such as the hypnotics veronal and medinal, were increasingly used at this time to calm patients and promote sleep. Depressed patients were prescribed stimulants, including Benzedrine, that produced a sense of euphoria.[83] Staff often ignored the complaints of patients that the drugs made them feel "drunk" and relied on sedation to

[81] Andrews *et al.*, *The History of Bethlem*, 683.

[82] Ibid., 684.

[83] Ibid., 679.

keep patients calm and manageable.[84] Porter-Phillips remained unconvinced by the "experimental" shock treatments emerging in the late 1930s.[85] He observed that "each method after trial appears to be withdrawn in favour of some other therapeutic agent more suitable and promising".[86] The results of insulin shock treatments delivered at Bethlem were inconclusive, but Porter-Phillips conceded that they led to a definite alteration in the mental state or attitude of patients which made their management by the nurses much easier.[87] This was also noted by the Board of Control inspectors who recorded in 1938 that no seclusion or mechanical restraint had been used in the previous 12 months.[88] Mechanical restraint was not a method associated with modern psychiatric practice.

Bethlem's physician superintendent, Porter-Phillips, was conservative in his approach, continuing to rely on sedation, isolation and restraint (until the advent of shock treatments) as the main means of calming turbulent patients, rather than prescribing occupational therapy or work around the hospital. The lavish programme of sport and entertainment, designed to appeal to a middle-class clientele, had changed little since before the war. Occupational therapy was only introduced after patients complained of boredom. Porter-Philips remained committed to finding a physical cause of mental disorder and never embraced the possibility of a psycho-social cause. At the Maudsley, on the other hand, Mapother introduced occupational therapy from the outset. He believed that mental illness might have a psycho-social origin, and that psychotherapy and occupational therapy could help patients re-adjust to their environment. In this, Mapother was similar in outlook to Toulouse at the Henri Rousselle Hospital. Toulouse also adopted a holistic stance, using psychotherapy and advocating some form of occupation for his patients, as well as biological treatments. The latter were the preferred methods of treatment for the psychiatrists of the Asile Clinique. Here, the active treatment of acute patients was just as much a priority as it was at the Maudsley and Henri Rousselle hospitals, but the emphasis was on delivering physical and biological treatments. Occupation was not deemed appropriate for acute cases.

[84] Ibid.

[85] Report of the Physician Superintendent, Annual Reports of the Bethlem Royal Hospital, London, 1938, 6. BMM/BAR-53.

[86] Ibid.

[87] Ibid.

[88] Report of the Board of Control, Annual Reports of the Bethlem Royal Hospital, London, 1938, 26. BMM/BAR-53.

The Views of Psychiatrists in the Provinces

The long-standing medical superintendent of the Littlemore and the successive chief medical officers of the Asile de la Sarthe had quite different therapeutic approaches, which were related to the different pathways of professional development taken by French and English psychiatry discussed in Chap. 3.[89] The different approaches were reflected in their prescription of patient work. Dr Thomas Saxty Good, (1870–1945), medical superintendent of the Littlemore from 1906 to 1936, put great faith in psychotherapy, following his experience of treating shell shock during World War I.[90] The chief medical officers of the Asile de la Sarthe took a more "biological" approach. Psychotherapy became the main method of treatment at the Littlemore after 1922, when the hospital re-opened for civilian use after the war, along with hydrotherapy, "work therapy" and amusements.[91] The Board of Control commented on Good's "judicious exercise of psychotherapy in conjunction with close study of physical conditions".[92] While drug treatments were not eschewed, the Board of Control noted with approval the Littlemore's "sparing use of sedatives".[93]

Good confirmed in 1930 that sedatives, narcotics and hypnotics were only used for extreme cases. He claimed that 15 female patients (or 2% of the average daily number resident), but no males, were given paraldehyde (a hypnotic drug used to treat insomnia) continuously throughout the year, while bromides (used to control the seizures associated with epilepsy, a neurological condition) were given in 48% of male epileptic cases and 38% of female epileptics. Two dozen tablets of "dial" and an equal number of "didial" were administered during the year, but no other barbiturates. Hydrotherapy was used for cases of confusional excitement.[94] Good maintained that as a "matter of practice" all methods of treatment were tried at the Littlemore, but he had found that "mental analysis" was the "most certain".[95] Occupation complemented this psychological approach.

[89] The chief medical officers of the Asile de la Sarthe during the interwar period were Dr Victor Bourdin (1904–1932); Dr Schutzenberger (1932–1934); Dr Henry Christy (1935–1939) and Dr Hédouin (1939–1945).

[90] See: John Stewart, "Tackling Shell Shock in Great War Oxford: Thomas Saxty Good, William McDougall, and James Arthur Hadfield," *Canadian Bulletin of the History of Medicine* 33 no. 1 (2016): 205–227.

[91] Thomas Saxty Good, "History and Progress of Littlemore Hospital," *Journal of Mental Science* October (1930): 614.

[92] Report of the Board of Control, Annual Reports of the Littlemore Hospital, Oxford, 1925/1926, 14.

[93] Ibid., 1929/1930, 14.

[94] Good, "History and Progress," 618.

[95] Ibid., 621.

Good used psychotherapy not only in his treatment of patients, but also in his dealings with members of staff.[96] Psychotherapeutic principles permeated Good's management style. Because he believed that all departments of the hospital, including the kitchen, laundry, stores and offices, contributed to the well-being of patients, Good insisted on close communication between them. While units operated independently, there was "frequent discussion amongst the various head of units". In this way, "every person in the hospital staff has an opportunity of hearing the ideas of others, and therefore perceives not only the effect of his own unit, but of others".[97] This was an important factor when patients were working all over the hospital, assisting in various departments where they might not be medically supervised. As for the nursing staff, Good ensured that nurses were freed from administrative duties in the laundry and kitchen so that they could spend more time with patients.[98] He made it a priority that nurses were not only able to deal with physical aspects of illness, but also to "sympathise with the psychological disposition of the individuals under their care".[99]

Described by the *British Medical Journal* as "one of the pioneers of modern psychiatry", Good's approach was endorsed by his appointments as lecturer in psychiatry and nervous disorders at the University of Oxford in 1928, as President of the Royal Medico-Psychological Association in 1930 and of the British Medical Association's Section on Neurology and Psychological Medicine in 1936.[100] Good's approach was continued by Dr Robert William Armstrong after the former's retirement in 1936. During the late 1930s, many more biological treatments were beginning to be explored, raising the question of whether psychotherapy, together with occupation, remained an important means of therapy. From the Littlemore's annual reports, it appears that Dr Armstrong was keen to trial new treatments, but psychotherapy and occupational therapy remained a priority at the Littlemore. Armstrong reported in 1938 that malaria therapy was being continued and had produced very satisfactory results in several cases.[101] Good had mentioned in 1926 that the effects of malaria

[96] Ibid., 614.

[97] Ibid., 613.

[98] Ibid., 614.

[99] Ibid., 615.

[100] E.M., "Obituary: T. Saxty Good, OBE, MRCS," *British Medical Journal*, 24 Nov (1945): 747.

[101] Report of the Medical Superintendent, Annual Reports of the Littlemore Hospital, 1938/1939, 15. OHA-L1/A2/16.

therapy on cases of GPI and encephalitis had been trialled by his medical team,[102] but he did not mention it in subsequent annual reports, nor in his history of the Littlemore. Armstrong's comment, however, suggests that its use had been on-going. The 1938/1939 report also referred to the use of shock therapy in the treatment of schizophrenia. Cardiazol had been used at the Littlemore and Armstrong hoped to try other forms of convulsive drugs such as Triazol.[103] He had not attempted insulin therapy due to a shortage of medical staff, but this was planned following the appointment of a third assistant medical officer.[104] This appointment, the Board of Control believed, was justified in the light of the number of acute patients being admitted on a voluntary basis; the need for thorough and sometimes prolonged investigation; and the time taken up with psychotherapy at the Littlemore.[105] The latter comment suggests that psychotherapy remained a significant method means of therapy.

The year 1939 was reported as "exceedingly active" from the point of view of the medical treatment of patients. A "large number" of patients were treated with various forms of shock therapy and "nearly a dozen" with insulin therapy resulting in "a few striking successes and a great many disappointments".[106] More success had been achieved using continuous narcosis, which Armstrong believed was "definitely a useful addition to the therapeutic armamentarium".[107] On balance, however, he considered that these "heroic measures of treatment" were "unlikely to realise the high hopes with which they were introduced".[108] His ongoing commitment to occupational therapy, on the other hand, was indicated in the reports from 1936/1937 to 1939/1940. These revealed that provision of occupational therapy had expanded both in terms of the numbers of patients involved and the range of items produced. He confirmed that this "important form of treatment" was being provided for "all grades of

[102] Ibid., 1926/1927, 11. OHA-L1/A2/4.

[103] Triazol was another form of fit-inducing drug that had fewer unpleasant side effects than Cardiazol.

[104] Report of the Medical Superintendent, Annual Reports of the Littlemore Hospital, Oxford, 1938/1939, 15. OHA-L1/A2/16.

[105] Ibid., 16.

[106] Report of the Board of Control, Annual Reports of the Littlemore Hospital, Oxford, 1939/1940, 22. OHA-L1/A2/17.

[107] Report of the Medical Superintendent, Annual Reports of the Littlemore Hospital, Oxford, 1939/1940, 11. OHA-L1/A2/17.

[108] Ibid., 11.

patients" including those at the agitated, acute stages of their illness.[109] Occupation was therefore being used to treat curable cases as well as to distract the more turbulent, incurable patients. Armstrong appeared equally supportive of occupational therapy as his predecessor.

The Littlemore had just two superintendents during the interwar period; Good and Armstrong, both of whom valued psychotherapy and occupation. Although Armstrong was more open to modern biological interventions, he remained committed to occupational therapy. In contrast, while the Asile de la Sarthe enjoyed continuity of medical leadership until the retirement of Dr Victor Bourdin in 1931, during the 1930s three chief medical officers were appointed in fairly rapid succession. None of these chief medical officers mentioned psychotherapy and all appeared to favour biological interventions for curable cases, in line with their organicist principles. Bourdin's reports were dominated by his trials of new medicines, such as Borosodine to treat epilepsy, and the hypnotic somnifène, but the results were not promising.[110] The families of some GPI patients had asked Bourdin to experiment on their relatives with anti-syphilitic treatments based on bismuth or arsenic, but he felt that, apart from being very expensive, these drugs were risky to the patient and only gave temporary relief. Bourdin prescribed them if relatives insisted, however.[111] The fact that patient work was not mentioned in Bourdin's reports, while his use of various drugs was paid significant attention, suggests that Bourdin preferred biological to psychological methods of treatment, in keeping with French psychiatric tradition. The records appended to the Asile de la Sarthe accounts, however, indicated that patient work was occurring on a similar scale to before World War I. It can be assumed that this patient work was carried out by calm, chronic and incurable patients and the few convalescents who had made a recovery (as was the practice before the war). There is no indication that occupation was being used to treat acute patients.

When Dr. Schutzenberger arrived in Le Mans from the Asile Clinique in Paris, becoming chief medical officer of the Asile de la Sarthe in 1932, he was keen to offer "complete, effective and modern treatment" using the

[109] Ibid., 1937/1938, 7–8. OHA-L1/A2/15.

[110] Rapport du Directeur et Rapport Médical, Asile de la Sarthe, Le Mans, 1922, 35–6. ADS-1X964.

[111] Rapport du Directeur et Rapport Médical, Asile de la Sarthe, 1923, Le Mans, 35. ADS-1X964.

full range of therapeutic methods available.[112] His comments suggest that he found the treatment deployed at the Asile de la Sarthe old-fashioned and inadequate. He was able to obtain additional funds from the asylum director that enabled him to purchase the equipment required to perform blood tests, lumbar punctures and the examination of fluids that would allow biological tests, such as the Bordet-Wassermann test for syphilis, to be undertaken.[113] Schutzenberger, taking advice from his former senior colleagues at the Asile Clinique, Drs Guiraud and Truelle, introduced malaria therapy to the Asile de la Sarthe in 1933.[114] Henceforth, he maintained, this method would be used to treat all GPI sufferers. Rates of cure and improvement for 1933–1935 did not increase significantly, however. Schutzenberger's medical report of 1933 was the first interwar medical report to mention patient work, itemising the numbers of days worked in each category, and the breakdown of work between male and female patients.[115] It is interesting that he chose to mention work, given his background at the Asile Clinique where work was not undertaken by patients in the acute, treatment wards. Schutzenberger's report did not discuss the work in therapeutic terms, however, so it may be that its prescription was limited to incurable, chronic and convalescent patients. Schutzenberger's strong views on the extent of overcrowding at the Asile de la Sarthe brought him into conflict with the prefect and he did not remain long in post.

When Dr Henry Christy took over as chief medical officer in 1934, his first report was designed to impress upon the prefecture the "scientific trends" that guided his treatment methodology.[116] His emphasis on the scientific nature of his methods is indicative of French psychiatrists' desire to be taken seriously by the rest of the medical profession. Christy divided patients into two categories, the intellectually impaired, who were incurable, and the mentally disordered, who responded to treatment. The two categories required different therapeutic techniques. The intellectually impaired required treatment for any physical ailments; support; comfort; and the moral discipline provided by work. The mentally disordered, on the other hand, required "aggressive biological treatment".[117] The two types of treatment required different skill sets of the doctors. When treat-

[112] Rapport du Médecin en Chef, Asile de la Sarthe, Le Mans, 1933 [no page nos.].
[113] Ibid.
[114] Ibid.
[115] Ibid.
[116] Rapport du Médecin en Chef, Asile de la Sarthe, Le Mans, 1935, 7.
[117] Ibid.

ing the intellectually impaired, the doctor had to be a psychologist, while the skills of a neuro-psychiatrist were required to treat the mentally disordered.[118] Christy's preference for a neuro-psychiatric approach towards treatment of acute cases underscores French psychiatry's lack of independence from neurology. Christy made it clear in 1936 that he was more interested in neuro-psychiatry than psychological medicine.[119] He wanted to focus on treating acute cases, but the demands of modern (biological) medicine made this difficult with so many patients under the care of just one doctor.

Although Christy emphasised biological treatments for curable patients in his reports, he also made several references to patient work in a rare outline of specific cases presented in 1935. A 19-year-old woman, Berthe, was diagnosed with dementia praecox. She recovered after shock treatment and was reported to have been "docile and working". Berthe then contracted pleurisy, but on recovering from this illness she returned to regular work within the institution and was soon able to leave the asylum. Patient P.J., aged 21, also suffering from dementia praecox, improved considerably after shock treatment and was currently working at the asylum. His family were going to take him home in a short while. Mme D., 42, suffered from acute depression and had attempted suicide. She was malnourished and deficient in vitamins. She recovered, and after receiving the appropriate food, she returned home and was able to take up her profession as a market trader once more.[120] The first two cases indicate that Christy prescribed work for patients who were convalescing, following successful shock treatment. The third case indicates that Christy regarded a patient's ability to return to work as evidence of recovery. An ability to work was considered a requirement for discharge, as indicated by the annual examination records of patients throughout the interwar period. These revealed that if a patient was unable to work and had no-one at home to support them, the patient should remain in the asylum. The phrase "*à maintenir*" appeared next to their record.[121]

Dr Hédouin, who took over from Christy in 1939, agreed that one chief medical officer could only deal with the most pressing cases and that

[118] Ibid.

[119] Rapport du Médecin en Chef, Asile de la Sarthe, Le Mans, 1937, 12. ADS-1X965.

[120] Rapport du Médecin en Chef, Asile de la Sarthe, Le Mans, 1935, 10. ADS-1X965.

[121] Hervé Guillemain, *Chronique De La Psychiatrie Ordinaire: Patients, Soignants et Institutions en Sarthe du 19e au 21e siècle* (Tours: Éditions de la Reinette, 2010), 45.

it was impossible to follow the progress of each patient.[122] Following the outbreak of World War II, patient numbers at the Asile de la Sarthe swelled to over 1,000. Le Mans had been invaded by the Germans and the Asile de la Sarthe was obliged to admit patients evacuated from a neighbouring hospital, Dury-les-Amiens. Hédouin found it impossible to categorise or separate patients. He was obliged to lodge acute psychopaths next to the intellectually impaired and impressionable adolescents with the medico-legal cases.[123] The accounts show that patient work continued throughout 1939. It can be assumed that its contribution to asylum maintenance and food production was essential during the Occupation when conditions were harsh and most French mental hospitals were deprived of food and other essential resources.[124]

The focus of the interwar chief medical officers of the Asile de la Sarthe was biological or physical treatment rather than psychotherapy or occupational therapy for acute patients, while patient work was prescribed to the chronic, incurable and convalescent patients. Psychotherapy, as the English Board of Control observed, was time-consuming, and a single chief medical officer in charge of between 600 and 1,000 patients did not have time to deliver it. But neither was psychotherapy an area of interest for the succession of chief medical officers who presided over the Asile de la Sarthe. Christy actually stated his preference for neuro-psychiatry. This preference for biological interventions appears to have influenced the chief medical officers' views on patient occupation. Christy felt that work was appropriate for incurable patients, while the curable should receive "aggressive biological treatment", followed by work during their convalescence. There was no question of developing occupations for those who were confined to bed, or for other patients whose conditions precluded work around the hospital. However, the ability of the Asile de la Sarthe chief medical officer to recruit interns and medical officers compromised the provision of biological treatment for patients. By 1939, when Hédouin joined the Asile de la Sarthe, overcrowding and understaffing, as a result of the war, meant that little active treatment of any kind could be delivered.

Psychiatrists at both the Littlemore, and the Asile de la Sarthe after 1932, were striving to provide active treatment for their patients, unlike many

[122] Rapport du Médecin en Chef, Asile de la Sarthe, Le Mans, 1939 [no page nos]. ADS-1X965.

[123] Ibid.

[124] Compte Administratif, Asile de la Sarthe, Le Mans, 1939. ADS-1X964.

provincial establishments (particularly in France) where the model of custodial care remained in place. But each institution's ways of providing that active treatment were quite different, and in the case of the Asile de la Sarthe, such attempts were thwarted by issues of resources and overcrowding. At the Littlemore, Good's war experiences had led him to prioritise psychotherapy and occupational therapy, treatments that he would not have considered before the war. Good was also an advocate of treating patients at the early stage of their illness, before their symptoms became entrenched, at the outpatients' facility he established at the Radcliffe Infirmary in 1918. Christy of the Asile de la Sarthe was quite clear that his preference, as a neurologist, was for the "aggressive biological" treatment of acute cases, and that occupation was the preserve of incurable and chronic patients. The stances of Good and Christy were indicative of the different approaches taken by English and French psychiatrists towards the active treatment of acute cases.

Conclusion

This chapter has shown that the introduction of occupational therapy as an active treatment was intrinsically linked to the outlook and treatment preferences of individual psychiatrists. Those psychiatrists who supported a psychosocial or psychobiological view of mental disorder, as espoused by Adolf Meyer, valued treatments that acted on the psyche of the patient, helping them to re-adapt to their environment and re-establish behaviours that allowed them to function in society. This type of treatment was favoured by those in charge of the Maudsley and Littlemore Hospitals in England, and the Henri Rousselle Hospital in Paris. These psychiatrists all enjoyed therapeutic autonomy. At Bethlem, the physician superintendent retained an organic, pre-war approach to treatment, focused on sedation followed by attempts at re-socialisation through entertainment and sport. Occupational therapy was added almost as an afterthought at Bethlem, in response to the Board of Control's insistence that occupational therapy was the hallmark of a truly modern hospital, and to patient complaints of boredom. The introduction or absence of occupational therapy highlights the different approaches of the psychiatrists in charge of institutions in the same country, and even the same city.

The Asile de la Sarthe was run along custodial lines between 1918 and 1932, heavily reliant on sedation to calm the turbulent. The arrival of the progressive Dr Schultzenberger, who was keen to introduce the modern therapies for acute patients with which he had become familiar at the Asile

Clinique, witnessed the introduction of a more active approach to treatment. Schultzenberger's successor, Christy, was equally keen to deploy "aggressive biological" treatment methods for acute, curable patients. While innovative treatments such as malaria therapy and the shock treatments were attempted, occupational therapy was dismissed as inappropriate. Occupation was not an option for acute patients, either at the Asile de la Sarthe or at the Asile Clinique where doctors failed to recognise its curative value. Patient work remained restricted to the calm, incurable and chronic patients, just as it had been before the war. "Modern" treatments therefore meant different things to different psychiatrists, depending on their training, professional networks and outlook. The delivery of these modern treatments, whether inclusive of psychotherapy and occupational therapy or not, depended on the quality and quantity of nursing staff and others in supporting roles to provide adequate assistance to the psychiatrists.

Bibliography

Secondary Sources

Andrews, Jonathan, Asa Briggs, Roy Porter, Penny Tucker, and Keir Waddington. *The History of Bethlem*. Abingdon and New York: Routledge, 1997.

Bartlett, Peter. *The Poor Law of Lunacy: The Administration of Pauper Lunatics in Mid-nineteenth Century England*. London and New York: University of Leicester Press, 1999.

Bewley, Thomas. "Online Archive 13: Edward Mapother (1881–1940)." In *Madness to Mental Illness: A History of the Royal College of Psychiatrists*, edited by Thomas Bewley. London: Royal College of Psychiatrists, 2008. Accessed 23/05/2020. https://www.rcpsych.ac.uk/docs/default-source/about-us/library-archives/archives/madness-to-mental-illness-online-archive/people/edward-mapother-1881-1940.pdf?sfvrsn=909358a2_6.

Coffin, Jean-Christophe. "'Misery and Revolution': The Organisation of French Psychiatry, 1900–1980." In *Psychiatric Cultures Compared: Psychiatry and Mental Health Care in the Twentieth Century: Comparisons and Approaches*, edited by Marijke Gijswijt-Hofstra, Harry Oosterhuis, Joost Vijselaar and Hugh Freeman, 225–48. Amsterdam: Amsterdam University Press, 2005.

E.M., "Obituary: T. Saxty Good, OBE, MRCS," *British Medical Journal*, 24 November (1945):747.

Evans, Bonnie and Edgar Jones. "Organ Extracts and the Development of Psychiatry: Hormonal Treatments at the Maudsley Hospital 1923–1938." *Journal of the History of the Behavioural Sciences* 48 no. 3 (2012): 251–76.

Fennell, Phil. *Treatment without Consent: Law, Psychiatry and the Treatment of Mentally Disordered People since 1845*. London: Routledge, 1996.

Guillemain, Hervé. *Chronique de la Psychiatrie Ordinaire: Patients, Soignants et Institutions en Sarthe du 19e au 21e Siècle*. Tours: Éditions de la Reinette, 2010.

Huteau, Michel. *Psychologie, Psychiatrie et Société Sous La Troisième République, La biocratie d'Édouard Toulouse (1865–1947)*. Paris: L'Harmattan, 2002.

Jones, Edgar, Shahina Rahman and Brian Everitt. "Psychiatric Case Notes: Symptoms of Mental Illness and Their Attribution at the Maudsley Hospital, 1924–1935." *History of Psychiatry*, 23 no. 2 (2012): 156–68.

Jones, Edgar. "Aubrey Lewis, Edward Mapother and the Maudsley." *Medical History* 47 Supplement no. 22 (2003): 3–38.

Jones, Kathleen. *Asylums and After. A Revised History of the Mental Health Services: From the Early 18th Century to the 1990s*. London: Athlone Press, 1993.

McCrae, Niall. "A Violent Thunderstorm: Cardiazol Treatment in British Mental Hospitals." *History of Psychiatry* 17 no. 1 (2006): 67–90.

Postel, Jacques and Claude Quétel. *Nouvelle Histoire de la Psychiatrie*. Paris: Dunod, 2012.

Prestwich, Patricia. "Family Strategies and Medical Power: 'voluntary' committal in a Parisian asylum, 1876–1914." *Journal of Social History,* 22 June (1994). Accessed 2 April 2021 from *The Free Library*: https://www.thefreelibrary.com/Family+strategies+and+medical+power%3a+%27voluntary%27+committal+in+a...-a016108112.

Shorter, Edward. *A Historical Dictionary of Psychiatry*. Oxford: Oxford University Press, 2005.

Stewart, John. "Tackling Shell Shock in Great War Oxford: Thomas Saxty Good, William McDougall, and James Arthur Hadfield." *Canadian Bulletin of the History of Medicine* 33 no. 1 (2016): 205–227.

Walke, J. *Repute and Remedy: Psychiatric Patients and their Treatment at Bethlem Royal Hospital 1930–1983*. Thesis. London: London School of Hygiene and Tropical Medicine, 2015.

Archival Sources

Appendix 1. *Report from the Select Committee on Pauper Lunatics in the County of Middlesex and on Lunatic Asylums,* 9–12. HMSO: London, 1827. Accessed 17/02/2022. [web site—ch1]

Board of Control. *Memorandum on Occupation Therapy for Mental Patients*. London: HMSO, 1933.

Bulletin Officiel du Ministère de l'Intérieur. Circulaire No. 7 du 20 mars 1857, 42–79. Paris, 1857. Accessed 12/02/2018. https://gallica.bnf.fr/ark:/12148/bpt6k5539701n/f2.item.zoom

Compte Administratif. Asile de la Sarthe, Le Mans, 1939. ADS-1X964.

Courbon, P. "Un voyage d'étude dans les asiles de Hollande." *Annales Médico-psychologiques* no. 2 (1928): 289–306 and 385–405.

Dausset, Louis. Rapport Général au nom de la 3e Commission sur les propositions budgétaires pour le service des Aliénés (budget de 1919). Paris: Conseil Général de la Seine, 1918. ADP/D.10K3/27/20.

Demay, G. "Les conditions de la thérapeutique par le travail dans les asiles." *L'Hygiène Mentale* no. 1 (1929): 33–40.

Ferrio, Carlo. "Le centenaire de la loi française du 30 juin 1838. Ce qu'en pense un étranger," *Annales Médico-psychologiques* no. 1 (1939): 725–555.

Good, Thomas Saxty. "The History and Progress of Littlemore Hospital." *Journal of Mental Science* October (1930): 602–21.

Grangier, M. Rapport au nom de la 3e Commission sur le Service Libre de prophylaxie mentale et les services annexes. Paris: Conseil Général, 1925. ADP/D.10K3/38/79.

Legrain, D-M. and G. Demay, "Le travail des aliénés convalescents: rapport présenté à la 4e Section du Conseil supérieur de l'Assistance publique en 1934." *L'Aliéniste français,* no. 2 (1936): 281–97.

Leroy, Dr. Rapport de la 1re Section des femmes, Asile Clinique. Paris: Préfecture de la Seine, 1932. PDS. ADP/PER-566.

Lévêque, M. Rapport du Pharmacien en Chef, Asile Clinique. Paris: Préfecture de la Seine, 1928. PDS. ADP/PER-566-5.

Marchand, Dr. Rapport de la 1re Section (Hommes), Asile Clinique. Paris: Préfecture de la Seine, 1935. PDS. ADP/PER-566-6.

Nineteenth Annual Report of the Board of Control. London: HMSO, 1932.

Ordonnance du 18 décembre 1839 portant règlements sur les établissements publics et privés consacrés aux aliénés. Accessed 23/05/2020. https://archive.org/details/b30459291/page/12/mode/2up?view=theater&q=Appendix+1, https://www.legifrance.gouv.fr/loda/id/LEGISCTA000006091614

Porot, A. "L'Assistance par le travail dans les Asiles Hollandais," *L'Hygiène mentale* no. 2 (1929): 41–54.

Rapport du Directeur et Rapport Médical. Asile de la Sarthe, Le Mans, 1922 and 1923. ADS-1X964.

Rapport du Médecin en Chef. Asile de la Sarthe, Le Mans, 1933, 1935, 1937, 1939. ADS-1X965.

Rapport Moral de l'Asile Clinique, 1938. Paris: Préfecture de la Seine. ADP/3719-W68-W74.

Report of the Board of Control. Annual Reports of the Bethlem Royal Hospital, London, 1938. BMM/BAR-53.

Report of the Board of Control. Annual Reports of the Littlemore Hospital. Oxford, 1925/6; 1938/9 and 1939/40. OHA-L1/A2/4; OHA-L1/A2/16 and OHA-L1/A2/17.

Report of the Medical Superintendent. Annual Reports of the Littlemore Hospital. Oxford, 1926/7 and 1938/9. OHA-L1/A2/4 and OHA-L1/A2/16.

Report of the Medical Superintendent. Maudsley Hospital, London, 1924; 1925; 1926. BMM/MSR-01.

Report of the Physician Superintendent. Annual Reports of the Bethlem Royal Hospital. London, 1924, 1926, 1927, 1938. BMM/BAR-53.

Rousselle, Henri. Rapport au nom de la 3e Commission sur le Service départemental de prophylaxie mentale et les services annexes. Paris: Conseil Général de la Seine, 1923. ADP/D.10K3/34/87.

Rousselle, Henri. Rapport au nom de la 3e Commission sur les budgets et comptes de l'Asile Clinique, 1910. Paris: Conseil Général de la Seine, 1911. ADP/D.10K3/20/63.

Sixteenth Annual Report of the Board of Control. London: HMSO, 1929.

Texte de la loi du 30 juin 1838. No. 7443: Loi sur les Aliénés. Accessed 03/10/2020. https://www.ascodocpsy.org/wp-content/uploads/textes_officiels/Loi_30juin1838.pdf

Thabuis, Dr. Rapports du Pharmacien en Chef, Asile Clinique. Paris: Préfecture de la Seine, 1921, 1922, 1927. PDS. ADP/PER-566-5.

Toulouse, Édouard. Rapport du médecin-directeur sur le Centre de prophylaxie mentale et l'Hôpital psychiatrique Henri-Rousselle. Paris: Conseil Général de la Seine, 1930. ADP/D.10K3/46/78 Annexe.

Twenty-third Annual Report of the Board of Control. London: HMSO, 1936.

Vié, J. "Le Traitement des malades mentaux par le travail." *L'Aliéniste français* no. 1(1934): 589–98.

CHAPTER 7

The Supervision of Patient Occupation

The experience and level of professionalisation of the staff who oversaw patient work and occupation influenced the type of occupation that could be prescribed by psychiatrists and the success of its application. Those who spent the most time with patients on a daily basis were the mental nurses. They had a significant impact on the management and treatment of patients in mental hospitals, and thus on patient occupation. Whether patients received adequate supervision, or were encouraged in their work depended on whether nurses had sufficient time, in addition to their regular duties, had received the requisite training and were temperamentally suited to the role. As John Crammer has emphasised, mentally ill patients could be "unpredictable in behaviour, restless, impossible to reason with, uninhibited in aggression when thwarted or frightened [and] incomprehensible in feelings and reactions".[1] The role of socialising and re-educating patients, encouraging them to fit in and co-operate with others on the wards took considerable patience and skill.[2] Psychiatrist Charles Mercier commented in 1894 that "the happiness and welfare of the patients … depend far more on the character and conduct of the attendants [male

[1] John Crammer, *Buckinghamshire County Pauper Lunatic Asylum—St John's* (London: Royal College of Psychiatrists, 1990), 89–90.
[2] Ibid., 89.

J. Freebody, *Work and Occupation in French and English Mental Hospitals, c.1918–1939*, Mental Health in Historical Perspective,
https://doi.org/10.1007/978-3-031-13105-9_7

nurses] than on those of all the rest of the asylum put together".[3] So, who was attracted to a career in mental nursing and how similar was the profession in France and England? The calibre of the mental nurses supervising occupation had a significant impact on a patient's willingness to engage with, and derive benefit from, a task. This chapter examines the roles of those who, in addition to the mental nurses, were involved in the supervision of patient occupation, including the new group of professionals, the occupational therapists.

Mental Nursing: Professionalisation and Training

By 1918, the professionalisation of mental nursing had developed further in England than in France in terms of the availability of training, the existence of a standardised manual, and the establishment of nationally recognised qualifications. That said, professional training and qualifications were not obligatory in either country and the quality of mental nursing attracted criticism in both France and England during the 1920s. Montagu Lomax, for example, was highly critical about the standard of English mental nursing at the Prestwich Asylum. He accused attendants of lacking "patience, tact, sympathy, and an understanding of the insane mind".[4] Whilst an "attendant's character and disposition" were paramount, he also believed that "certification or registration should be compulsory upon all attendants, male and female".[5] His remarks led to an enquiry by the Board of Control's Committee on Nursing in County and Borough Mental Hospitals. Published in the *Journal of Mental Science* in 1925, the report resulting from the enquiry recommended a "national nursing service … a service in which the same qualifications are recognised and required at all institutions for the same positions".[6] The Committee emphasised that quality nursing depended upon "adequate training" based on theoretical and practical instruction by qualified teachers.[7] This was already available; the MPA had produced a handbook for mental nurses (first published in 1885) and established a standardised, national training scheme. From 1891, trainees could

[3] Charles Mercier, *Lunatic Asylums: Their Organisation and Management* (London: Charles Griffin & Co., 1894), 284.

[4] Montagu Lomax, *Experiences of an Asylum Doctor: With suggestions for asylum and lunacy law reform* (London: G. Allen and Unwin Ltd, 1921), 190.

[5] Ibid., 189.

[6] AU, "Board of Control Report from Committee on Nursing in County and Borough Mental Hospitals," *Journal of Mental Science* April (1925): 289–90.

[7] Ibid.

sit an examination resulting in a Certificate of Proficiency in Nursing the Insane. By 1899, 100 asylums were participating in the training scheme, the duration of which was extended to three years in 1908.[8] In 1919, the General Nursing Council established its own training programme for mental nurses, resulting in parallel training schemes based on almost identical curricula.[9] The availability of such instruction, however, did not obligate nurses to undertake it, nor hospitals to provide or insist upon it.

Deficiencies in the English system were also underlined by the Macmillan Report of 1926 which emphasised that nurses should be carefully recruited and properly trained.[10] There appeared to be a considerable difference between the quality of nurses working at the Maudsley, Bethlem and Littlemore hospitals, where nurses were given proper training and where the pay and conditions were relatively good, and the other public mental hospitals at which the criticisms were directed. At the Maudsley Hospital, when it opened in 1923, the Board noted that all senior nurses were required to hold a certificate representing at least three years' training at a general hospital or a diploma from a recognised nursing school. All the sisters and most of the staff nurses also possessed experience at either mental or neurological hospitals.[11] Pay and conditions, considered crucial to attracting a high calibre of nurse and maintaining high standards, were considerably more favourable in city institutions, such as the Maudsley and Bethlem Hospitals.[12] At the Maudsley, nurses' accommodation was described as "excellent" in 1923, while at Bethlem, the pay for probationer nurses was described in an article in *The Hospital* in 1921 as "far in excess of that usually accorded to learners".[13] Nursing care at the Maudsley was praised in the Board of Control's annual inspection reports. In 1926, the Board remarked that "everything is being done not only for

[8] Peter Nolan, "Mental Health Nursing in Great Britain," in *150 Years of British Psychiatry, Vol II: The Aftermath*, eds Hugh Freeman and German E. Berrios (London: Athlone Press, 1996), 178.

[9] Ibid., 180. Nurses who passed their final examinations were "certificated" by the MPA or became "registered" with the General Nursing Council (GNC).

[10] John Crammer, "Training and Education in British Psychiatry 1770–1970," in *150 Years of British Psychiatry, Vol II: The Aftermath*, eds Hugh Freeman and German E. Berrios (London: Athlone Press, 1996), 228.

[11] Tenth Annual Report of the Board of Control (London: HMSO, 1923), 201.

[12] AU, "Board of Control Report from Committee on Nursing in County and Borough Mental Hospitals," *Journal of Mental Science* April (1925): 301.

[13] AU, "Nursing in the Bethlem Royal Hospital," *The Hospital* 8 Jan (1921): 346.

the restoration to health but for the comfort of the patients".[14] The nursing care was described as "excellent" in 1935.[15] A nine-month course in handicrafts was established for nurses in 1935; nurses attended voluntarily in their free time. Thirty nurses had joined the two classes, for which craft teachers were supplied by the London County Council.[16] Maudsley nurses who underwent such training could therefore assist in the provision of occupational therapy for patients.

In contrast to the Maudsley, where all senior nursing staff were qualified, at Bethlem, during the mid-1920s, approximately 65% of male nurses and 30% of female nurses were registered or certificated. Despite the fact that training was incentivised at Bethlem by the offer of an additional two shillings per week for those obtaining the MPA Nursing Certificate, the numbers of trainees fell.[17] By 1934, the percentage of certificated nurses had fallen to 48% of male nurses, and 19% of female nurses.[18] Although nurses at Bethlem were not involved in instructing patients in occupational therapy (introduced in 1932), they were expected to encourage patients to take exercise and "to enter into amusements".[19] Bethlem's "Rules and Orders", prepared in 1932, stated that "the Nurses shall devote the whole of their time during the day to the Patients, and shall execute with diligence all the direction they shall receive respecting their treatment, medicine, food, dress, occupation, exercise and amusement".[20] There was no mention of work around the hospital in the Rules, but one nurse remembers that "we tried to keep everyone occupied…they [the patients] played cards, table tennis, did knitting [and] sewing".[21]

Although Bethlem's Rules and Orders did not emphasise the benefits of manual labour, in other respects they appeared to be closely aligned with the principles of moral therapy, and with the practices recommended in the mental nursing handbook produced by the MPA. The first edition

[14] Thirteenth Annual Report of the Board of Control, 1926, 145–6.

[15] Sixteenth Annual Report of the Board of Control (London: HMSO, 1929), 222; Twenty-second Annual Report of the Board of Control (London: HMSO, 1935), 484.

[16] Twenty-second Annual Report of the Board of Control, 1935, 484.

[17] AU, "Nursing in the Bethlem Royal Hospital," 346.

[18] Report of the Physician Superintendent Report, Bethlem Royal Hospital, London 1934, 20. BMM/BAR-53.

[19] Bethlem Royal Hospital Rules and Orders, 1932; 6. BMM/BRH/SRO-24.

[20] Ibid., 10.

[21] Cited in Jonathan Andrews et al., *The History of Bethlem* (Abingdon and New York: Routledge, 1997), 688.

of the handbook was published in 1885 and remained in circulation until 1902.[22] Numerous subsequent editions (the seventh appearing in 1923) were produced between 1902 and 1978. Early editions of *The Handbook* encouraged staff to train patients to adopt "proper healthy habits", by adhering to the daily routine of the asylum with its "regular hours for rising, taking food, work, exercise, amusement and retiring to bed".[23] Occupation in the form of work was reported as having "a most salutary effect on both the body and the mind".[24] Housework, work in the gardens or workshops, on the farm or in the laundry, needlework, drawing or writing, or whatever work was "congenial" to the individual, diverted the attention away from "morbid fancies" and helped patients to focus on healthier matters. It was the attendants' or nurses' duty to ensure that patients were properly engaged with the work and not allowed to "lounge about idly" to ensure the maximum benefit was derived from the activity. It was also made clear that "the willing must not be over-tasked".[25] The value of providing amusements for patients in the form of dancing and games was also highlighted, and nurses were to encourage patients to participate.[26] Although the first edition of the handbook was produced some forty years after the heyday of moral therapy, the passages relating to occupation bore a marked resemblance to the teaching of the moral therapists. The wording did not change significantly in subsequent editions produced during the interwar period.

As in England, standards of nursing care in French asylums attracted widespread criticism. Formal training in mental nursing was rare in France in the immediate aftermath of World War I, despite the efforts of psychiatrists Théodore Simon and Georges Daumézon to professionalise mental nursing and improve instruction.[27] Training was provided in some departments, including the Seine, where a mental nursing school had been established at Ste Anne's in 1882 and at other Seine asylums in 1907, but

[22] Medico-Psychological Association, *Handbook for the Instruction of Attendants on the Insane* (London: Ballière, Tindall and Cox, 1885).

[23] Ibid., 50.

[24] Ibid.

[25] Ibid.

[26] Ibid., 51.

[27] Alexander Klein, "Théodore Simon (1873–1961): Itinéraire d'un Psychiatre Engagé pour la Professionnalisation des Infirmiers et Infirmières d'Asile," *Recherche en soins infirmiers* no. 135 (2018): 91–106.

provision outside the capital was patchy.[28] A State Diploma, requiring a year's training (increased to two years in 1924), was introduced in 1922, but it was not compulsory. The general education of those nurses who took the examination compromised their chances of success. In the Seine, 28 out of 40 student nurses failed the diploma in 1922, as a result of poor communication skills.[29] Thirteen years later, Daumézon found that the level of education achieved by mental nurses was still extremely poor. Only 5% of mental nurses held a certificate of primary education in 1935, and some were illiterate.[30] Manuals were available, including Roger Mignot and Ludovic Marchand's *Manuel technique de l'infirmier des établissements psychiatriques* (first published in 1912, with a second expanded edition appearing in 1930), and Antony Rodiet's *Manuel des infirmiers et infirmières des hôpitaux et des asiles* (1928) which were based on "modern republican medicine".[31] But the manuals' usefulness depended on the nurses' ability and inclination to study them. The Mignot and Marchand manual indicated that work for patients was a distraction and helped to maintain a level of intellectual activity. It also highlighted that in many asylums, patient work generated important cost savings.[32] The influence of moral therapy can be detected in the notion of work as a distraction and a stimulus to the intellect, while its benefit as a means of making cost savings can be linked to a concern for economy that grew in importance after *c*.1850. Neither of the later editions of the manuals advocated the more sophisticated techniques of occupational therapy.

[28] Elisabeth Jean-Louis and Aline Valentin, "Un Institut Au Coeur de L'Histoire de La Profession Infirmière," in *L'Hôpital Sainte-Anne: Pionnier De La Psychiatrie Et Des Neurosciences Au Coeur De Paris*, ed. Stéphane Henry, Catherine Lavielle, and Florence Patenotte (Paris: Somogy Éditions d'Art, 2016), 170.

[29] Henri Rousselle, Rapport au nom de la 3e Commission sur l'Asile Clinique et sur le Service libre de prophylaxie mentale (Paris: Conseil Général de la Seine, 1922), 3. ADP/D.10K3/32/77.

[30] Patrice Krzyzaniak, *Georges Daumézon (1912–1979): Un Camisard Psychiatre et Pédagogue: Une contribution singulière aux sciences de l'éducation*, Thèse (Paris: Université Charles de Gaulles, 2018), 89.

[31] Hervé Guillemain, "Le Soin En Psychiatrie Dans La France Des Années 1930," *Histoire, Médecine et Santé* Spring, no. 7 (2015): 81–2.

[32] Roger Mignot and L. Marchand, *Manuel Technique de l'Infirmier des Établissements d'Aliénés* (Paris: Octave Doin et Fils, 1912), 244.

In 1925, General Councillor Chausse referred to the need to eliminate "useless" staff from Seine asylums.[33] Nurses' salaries had been raised in an attempt to recruit competent, devoted staff who could give their patients the best care, but this had not had the desired effect.[34] French nurses were compared unfavourably with their Dutch counterparts by Seine psychiatrist, Paul Courbon. After visiting the Dutch asylum at Santpoort in 1929, Courbon observed that Dutch nurses regarded the profession as a vocation, rather than just a job, and they were recruited from a more highly educated and cultivated class than in France.[35] Similar observations were made by Dauмézon in 1935. He noted that nurses tended to be recruited from peasant or labouring stock and lacked a vocational calling to their profession.[36] From 1930, a new five-year training scheme was offered to nurses, but this was optional and did little to improve overall standards.[37] The need for nurses to be highly trained and committed to their role was essential for the successful introduction of Simon's "more active therapy", since, as Legrain and Demay highlighted in their 1934 report, the application of therapeutic work rested with the nurses.[38] Nurses needed to have the skills to motivate patients, to direct the activity of distracted or confused patients, to intervene if a patient became agitated and to modify the work according to how a patient was coping. They needed to be familiar with their patients' conditions, interests and capabilities, and able to handle them with patience and tact.[39] French mental nurses, including those in the Seine department, lacked the training and experience to direct patient work effectively, particularly when dealing with more challenging patients.[40]

Édouard Toulouse was keen to address these shortcomings. He drew up a proposal for a school of mental health, to be established within the Henri Rousselle Hospital. The aim was to provide technical, theoretical

[33] M.E. Chausse, Rapport Général au nom de la 3e Commission sur le compte du Service des aliénés pour 1924, le budget rectificatif de 1925 et le projet de budget pour 1926 (Paris: Conseil Général de la Seine, 1925). ADP/D.10K3/38/83.

[34] Chausse, Rapport Général, 1925; 8–10.

[35] Paul Courbon, "Voyage d'études dans les asiles de Hollande," *Annales Médico-Psychologiques* no. 2 (1928): 399.

[36] Ibid.

[37] Krzyzaniak, "Georges Dauмézon," 91.

[38] D-M. Legrain and G. Demay, "Le travail des aliénés convalescents: rapport présenté à la 4e Section du Conseil supérieur de l'Assistance publique en 1934," *L'Aliéniste français* no. 2 (1936): 283.

[39] Ibid.

[40] Demay, "Les conditions," 33.

and practical training to mental nursing staff, laboratory assistants and social workers. The course was aimed at those who already had a basic knowledge of anatomy, physiology and hygiene; it was designed to complement, rather than replace, that which was taught at nursing school. Instruction in psychiatry and mental health would be taken by everyone, followed by specialist classes in nursing, social work and laboratory work.[41] Although the school was described as in the "process of being created" in 1931,[42] Toulouse's plans do not appear to have been realised. His intentions, however, are indicative of the general dissatisfaction with mental nursing standards at that time.

Standards of nurse training and skills differed quite markedly at the rural Littlemore and the Asile de la Sarthe. They were of a much higher standard at the Littlemore (although standards here were not typical of all English provincial mental hospitals). Dr Thomas Saxty Good expected every Littlemore nurse to learn a craft that they could teach to patients. By developing a rapport with patients, nurses were able to encourage them to take up the craft in which they specialised. As Good put it, a patient's attachment to a nurse "will often induce them to start that nurse's particular craft".[43] Nurses in both the male and female divisions were expected to learn and teach handcrafts to patients. The Board of Control was impressed that this interaction with patients enabled them to produce "detailed and helpful notes … on the behaviour and conversation of patients" which aided the doctors in their treatment decisions.[44] Doctors discussed each case fully with the senior nurses and involved them in decisions regarding treatment.[45] The hospital did not suffer from overcrowding (attributed to the existence of the outpatient clinic) so the nursing staff were able to maximise the amount of time spent with patients, engaging them in conversation and helping them with their craft activities.

At the Asile de la Sarthe, on the other hand, it was difficult to find reliable, experienced male nursing staff who could supervise patient work at all but the most rudimentary level. Staff shortages were compounded by

[41] Édouard Toulouse, Rapport sur le Centre de Prophylaxie Mentale et L'hopital Henri Rousselle (Paris: Conseil Général de la Seine, 1928–1929), 129. ADP/D.10K3/46/83 Annexe.

[42] Édouard Toulouse, Rapport sur le Centre de Prophylaxie Mentale et L'hopital Henri Rousselle (Paris: Conseil Général de la Seine, 1930–1931), 147. ADP/D.10K3/46/86 Annexe.

[43] Thomas Saxty Good, "The History and Progress of Littlemore Hospital," *Journal of Mental Science* Oct (1930): 614.

[44] Report of the Board of Control, Annual Reports of the Littlemore Hospital, Oxford, 1933/1934, 19. OHA-L1/A2/11.

[45] Good, "History and Progress," 615.

overcrowding, which meant that nurses had less time to spend with patients. In 1920 the Asile de la Sarthe had barely enough staff to give nurses a rest day every week.[46] Despite a salary rise in 1920 which put the Asile de la Sarthe on a level with other, similar establishments, few male applicants were able to satisfy the Asile de la Sarthe's requirements for nursing experience gained at other local hospitals or asylums.[47] Those who were recruited did not remain long in post, which was unsettling for patients.[48] It had not been possible to institute the eight-hour day for nurses in 1919 (although it had been applied to administrative staff) because recruitment was so difficult. Splitting the nursing teams into three would have been impossible, since the existing two teams were rarely complete. The asylum director, who was in charge of recruitment, rather disparagingly observed that "a good nurse, let us say, is the exception" and that it was probably unwise to allow nurses 16 hours of liberty because they would probably abuse it![49] These comments lend weight to the criticisms of nursing staff made by Paul Courbon in 1929.[50]

The nursing situation was rendered more complex at the Asile de la Sarthe, because as was the case in many French provincial asylums, nursing on the female side was provided by nuns from the *Charité d'Evron*. This was the result of a long-standing arrangement between the Prefect of La Sarthe and the Mother Superior of the religious order dating back to 1870.[51] In Paris and other large cities, where republican values predominated, the secularisation of asylum personnel took place in (or before) 1905,[52] when all state institutions officially became secular.[53] In the

[46] Rapport du Directeur, ADLS, 1920 [no page nos.]. ADS-1X965.

[47] Ibid.

[48] Rapport du Directeur, ADLS, 1923, 10. ADS-1X965.

[49] Rapport du Directeur, ADLS, 1919 [no page nos]. ADS-1X965.

[50] P. Courbon, "Un voyage d'étude dans les asiles de Hollande," *Annales Médico-psychologiques* no. 2 (1928): 399.

[51] Hervé Guillemain, *Chronique De La Psychiatrie Ordinaire: Patients, Soignants et Institutions en Sarthe du 19e au 21e Siècle* (Tours: Éditions de la Reinette, 2010), 39.

[52] The Bill of Separation passed in 1905 ended the Napoleonic Concordat that had protected the relationship between the Catholic Church and the French state; henceforth all state institutions, including schools and hospitals, were to be run on a secular basis. See: Roger Price, *A Concise History of France*, second edition (Cambridge: Cambridge University Press, 2005), 234–5.

[53] Évelyne Salem and Stéphane Henry, "Laicisation et Formation du Corps Soignant à l'Asile Sainte-Anne," in *L'Hôpital Sainte-Anne: Pionnier de la Psychiatrie et des Neurosciences au Coeur de Paris*, ed. Stépane Henry, Catherine Lavielle, and Florence Patenotte (Paris: Somogy Éditions d'Art, 2016), 167.

provinces, however, particularly in areas where support for the Church remained strong, institutional secularisation was piecemeal and occurred much later.[54] Secularisation of the nursing staff did not occur at the Asile de la Sarthe until the 1960s.[55] Fears were expressed by the secular psychiatrists of the Seine that employing nuns as nurses created a "state within a state" and impeded the introduction of modern methods.[56] Although Dr Christy had come to Le Mans from the republican city of Lyon, such criticisms did not appear to trouble him. He regularly praised the diligence and devotion to duty of his nursing "sisters" and did not regard them as an impediment to his pursuit of modern treatment methods. [57]

The nursing situation on the male side of the Asile de la Sarthe contrasted with that at the Littlemore, where nurse training based on the RMPA syllabus was given by Dr Good, his medical officers, the matron and head male nurse. Approximately 40–45% of Littlemore nurses (pictured taking a break from their duties in Fig. 7.1, below) were registered or certificated. Examination results were highlighted each year by the medical superintendent in the hospital's annual report. For example, in 1936, nine male nurses and 16 female nurses were reported to have passed their preliminary examinations, while six women and one man passed their finals, one woman gaining a distinction.[58] General training was supplemented at the Littlemore by a system of nurse exchange between the Littlemore and the Radcliffe Infirmary, Oxford's general hospital, enabling Littlemore nurses to gain experience of physical illness and Radcliffe nurses to learn something of mental nursing.[59] The Board of Control remarked that nurse training at the Littlemore was "calculated" to ensure that nurses could provide "valuable co-operation" in the treatment of patients.[60]

An important aspect of developing a rapport with patients was continuity of care. Frequent staff changes were unsettling for patients. At the

[54] Guillemain, *Chronique De La Psychiatrie Ordinaire,* 41.

[55] Ibid., 42.

[56] A. Porot, "L'assistance par le travail dans les asiles Hollandais," *L'Hygiène mentale* no. 2 (1929): 51.

[57] Several of the Asile de la Sarthe Medical Reports ended with a reference to the invaluable services of the nursing "sisters", such as that of Dr Christy in 1935 (see above).

[58] Report of the Medical Superintendent, Annual Reports of the Littlemore Hospital, Oxford, 1936/1937, 10. OHA-L1/A2/14.

[59] Report of the Medical Superintendent, Annual Reports of the Littlemore Hospital, Oxford, 1929/1930, 12. OHA-L1/A2/7; Good, "History and Progress," 615.

[60] Report of the Board of Control Report, Annual Reports of the Littlemore Hospital, Oxford, 1929/1930, 14.

Fig. 7.1 Nurses relaxing at the Littlemore Hospital, Oxford, 1930s. (© *Oxfordshire County Council, Oxfordshire History Centre, POX016605*)

Littlemore, patients were able to establish stable, trusting relationships with their nurses because so many of the latter remained in post for a long time. In 1929, the medical superintendent announced that the Deputy Head Male Nurse, Henry Shattock, was retiring after 41 years of service. Four other male nurses also retired that year after working for between 30 and 35 years at the Littlemore.[61] Good praised the "general excellence and almost universal good behaviour of the Staff" which he attributed to the "keenness and efficiency" of the Matron and Head Male Nurse.[62] The Board's inspectors noted in 1934 that "the [senior] nurses with whom we spoke showed both knowledge and interest in their patients and the general standard of nursing appeared to us to be high."[63] Bethlem also appeared to inspire devotion to duty and long-service amongst its senior

[61] Report of the Medical Superintendent, Annual Reports of the Littlemore Hospital, Oxford, 1929/1930, 12.

[62] Ibid.

[63] Report of the Board of Control Report, Annual Reports of the Littlemore Hospital, Oxford, 1934/1935, 16. OHA-L1/A2/12.

nursing staff. In 1925, Nurse Eva Scott retired after 21 years of service and in 1927 Male Nurse William Redaway retired after 41 years.[64] In 1935, the physician superintendent, John Porter-Phillips, noted that two of the nursing sisters had achieved 25 years of service, and in 1938 he congratulated Sister Neave on her retirement after 33 years at Bethlem.[65] He remarked in 1929 that "…it is gratifying to note that the senior members of the Nursing Staff continue to give us loyal and faithful service".[66] He also commented on the "appreciation and gratitude" expressed by patients and their relatives for the "sympathetic care, kindness and devotion" of the nursing staff.[67] Continuity of care was not something that the Asile de la Sarthe could offer. On the male side, it was difficult to recruit sufficient nurses, and many did not stay long. On the female side, the requirements of their religious order meant that the nuns were continually being replaced.

As well as providing care, nursing staff in English mental institutions had been expected to supervise and participate in the programme of leisure activities and entertainments for patients ever since the advent of moral therapy in the early nineteenth century. As Jocelyn Goddard highlights, the ability to play a musical instrument or demonstrate proficiency in one or more sports continued to be highly valued attributes amongst English mental nursing staff until the 1950s. One former nurse at the Littlemore remembered, "when the male staff were enlisted, they had to either be good at some sport or a musical instrument". Such attributes were even stipulated in the advertisements for nursing positions.[68] The Littlemore's cricket, hockey, football, badminton, and table tennis teams regularly played other hospital teams, travelling as far as Birmingham and Portsmouth for matches.[69] Bethlem also boasted hospital teams in all sports; match schedules and results were included in the hospital magazine, *Orchard Leaves*. For example, in the summer of 1935, it was reported that 28 cricket matches were played, of which Bethlem won 12, lost eight and drew five, while three had to be abandoned due to bad weather. Eleven tennis clubs sent teams to play at Bethlem, and the contests were

[64] Report of the Physician Superintendent, Annual Reports of the Bethlem Royal Hospital, 1925 and 1927. BMM/BAR-53.

[65] Ibid., 1935, 11–12; Ibid., 1938, 21.

[66] Ibid., 1929, 12.

[67] Ibid., 1938, 20.

[68] J. Goddard, *Mixed Feelings: Littlemore Hospital—An Oral History Project* (Oxford: Oxfordshire County Council, 1996), 39.

[69] Ibid., 35.

described as "very keen".[70] A Sports Club was established in 1933.[71] The matches provided spectator sport for those patients who were not included in the teams. Hospital orchestras played at the weekly dances for patients, until music was provided by phonograph in the 1930s.[72] The involvement of staff in the provision of entertainment was not such a feature of French institutions. This may have been related to a lack of accomplishments amongst the nursing staff, or lack of staff time, or to the fact that entertainments were organised by the asylum director, rather than the medical staff.

How much time staff had to supervise patient occupation, and thus the extent to which occupation could be used as a curative agent, depended to a large extent on the ratio of nurses to patients. This was particularly important in institutions where there were no occupation officers or occupational therapists, as at the Asile Clinique. Ladame and Demay observed that in most French asylums there was usually one nurse to every ten patients.[73] This ratio was similar, according to the Board of Control to that of most English mental hospitals where the ratio was one nurse to nine or ten patients.[74] Ratios at the metropolitan hospitals of Bethlem, the Maudsley and the Henri Rousselle hospitals, which were all dedicated to the care of acute patients, were far more favourable. At Bethlem and the Maudsley the ratio was two nurses to every five patients, and at the Henri Rousselle Hospital it was one to seven. At the Faculty Clinic, however, Professor Claude complained in 1922 that nurse numbers were so inadequate, particularly during the holiday season, that he was unable to use his isolation rooms for lack of nurses to supervise them.[75] The Asile Clinique's ratio improved following transformation into an acute hospital and the "doubling" of the services in 1927. This resulted in one chief medical officer becoming responsible for *c.*200 cases instead of 400, and the addition of 34 medical and nursing staff who were recruited to provide the more intensive care required for acute patients undergoing biological

[70] WMA, "Sporting Activities," *Orchard Leaves*, Bethlem Royal Hospital, London, Winter (1935): 8.

[71] Andrews *et al.*, *History of Bethlem*, 688.

[72] Goddard, *Mixed Feelings*, 39; Report of the Physician Superintendent, Annual Reports of the Bethlem Royal Hospital, London, 1932, 14.

[73] Legrain and Demay, "Le travail des aliénés convalescents," 283.

[74] AU, "Board of Control Report from the Committee on Nursing," 292.

[75] Henri Claude, Rapport de la Clinique des maladies mentales, Asile Clinique (Paris: Préfecture de la Seine, 1922); 95–6. ADP/PER-566-5.

treatment, such as malaria therapy.[76] However, even in the 1930s there were still complaints regarding staff shortages at the Asile Clinique. Dr Marchand, for example, noted in 1931 that nurses had to take patients from the First Quarter, where there was no bathroom, over to the general baths, leaving acute patients unattended on the ward.[77] At the Seine's Villejuif Asylum, the "sister" institution to Ste Anne's, one chief medical officer claimed that "moral treatment" [*sic*] had become almost impossible since patient surveillance and routine tasks took up all the nurses' time and they had no time to talk to patients.[78] If nurses had no time to talk to patients, they were unlikely to have time to supervise patient occupation on anything other than a rudimentary level.

At the Asile de la Sarthe, an insufficient quantity of nurses and junior medical staff compromised treatment. In 1933, the new chief medical officer Dr Schutzenberger highlighted that although "on paper" there might be an appropriate number of patients per nurse, in reality, the number of nurses was insufficient. Nurses were often required to be somewhere other than the patients' quarters, such as supervising bathing, and days off and holidays had to be taken into consideration. In the Men's Division, there were 335 patients and 42 nurses, but only 26 of these were on the wards at any one time, giving a ratio of one nurse to 13 patients. On the women's side, there were 496 patients and 43 sisters, but one worked in the pharmacy and two more supervised bathing, giving a ratio of one nurse to 12 patients. Lack of continuity of service amongst the sisters added to the problems caused by the patient to staff ratio. The introduction of the 40-hour week in 1937 forced the Asile de la Sarthe to increase the number of male nurses from 46 to 125. Many of the new recruits had no experience of nursing and none were qualified.[79] The chief medical officer, Dr Christy, was compelled to organise courses for all male nurses, including elementary instruction in anatomy, physiology and hygiene, since he could achieve little if his staff lacked any understanding of patient care and had scant interest in their work.[80] Until then, there was no formal mental nurse

[76] The administration of malaria therapy, for example, was particularly labour-intensive.

[77] Dr Marchand, Rapport de la 1ere Section des hommes, Asile Clinique (Paris: Préfecture de la Seine, 1931), 93. ADP/PER-566-6.

[78] Maurice Quentin et Alphonse Loyau, Rapport Général au nom de la 3e Commission sur le compte du Service des Aliénés (Paris: Conseil Général de la Seine, 1936), 2. ADP/D.10K3/54/114.

[79] Rapport du Médecin en Chef, Asile de la Sarthe, 1937, (Le Mans, 1938), 13. ADS-1X964/5.

[80] Ibid.

training at the Asile de la Sarthe. On the female side, however, the majority of the nuns were qualified. In 1928, a third of the nuns had gained nursing diplomas from the general hospital in Le Mans, and half held a "certificate of professional capability" recognised by the Ministry of Hygiene.[81] There was therefore a significant imbalance in the qualifications held by nurses at the Asile de la Sarthe. Male nurses were unlikely to have the time, knowledge or inclination to become involved with patient occupation, other than delivering the patients for whom it had been prescribed to their places of work, while the nuns may not have had the time between their nursing and religious duties.

Apart from the workshop managers, there were no other members of staff who could have supervised occupational therapy (or administered biological treatments, for that matter) at the Asile de la Sarthe. Dr Schutzenberger highlighted that it was impossible for a single chief medical officer to treat *c.*850 patients without either a medical assistant or interns.[82] Christy admitted that he had only achieved modest results in the treatment of GPI with malaria therapy because there were insufficient staff to carry out all the necessary procedures.[83] Until the mid-1930s, the chief medical officer had been the only medically qualified member of staff. Le Mans was a small town without a university (at that time), which made the recruitment of interns difficult, although one had been engaged for a short period.[84] In this, the situation of the Asile de la Sarthe was quite different to that of the Littlemore, which was located in the university town of Oxford, where there existed an abundance of medical students. Dr Christy was delighted to have a medical assistant in 1937 but having been without an intern for several months he had been forced to abandon some of the more labour-intensive therapies that he had begun. He concluded that without interns, he could not offer the modern treatments that he would like to use, particularly as he was also running courses for the male nursing staff.[85]

[81] Guillemain, *Chronique de la Psychiatrie Ordinaire*, 41.

[82] Rapport du Médecin en Chef, Asile de la Sarthe, Le Mans, 1933 [hand-written, no page nos.]. ADS-1X261.

[83] Christy also noted that there far fewer cases of GPI in Le Mans than Lyons, where he had practised before, leading him to conclude that the disease spread more slowly in rural areas.

[84] Rapport du Médecin en Chef, Asile de la Sarthe, 1937 (Le Mans, 1938), 22. ADS-1X964/5.

[85] Ibid, 24.

Unlike Le Mans, the existence of the university in Oxford, made the recruitment of medical students for the Littlemore much easier. As the Board of Control highlighted in 1928, "Dr Good's recent appointment as lecturer of psychiatry ... will undoubtedly benefit this Hospital, as the periodical introduction of keen medical students has always been found to be a stimulus to high class work".[86] In addition to medical students, the medical superintendent was supported by two assistant medical officers throughout the interwar period. The ratio of nurses to patients was also more favourable at the Littlemore, with one nurse to between seven and nine patients, while at the Asile de la Sarthe the ratio was one nurse to 12 patients. This meant that both medical and nursing staff had more time to spend with patients, and more time to supervise patient occupation. The medical superintendent highlighted the fact that Littlemore nurses were very dedicated and keen to further the work of the medical staff.[87]

Nurses at the Bethlem and Maudsley hospitals had more to offer their patients in terms of time, training and aptitudes than those at the Asile Clinique, and the English hospitals also benefited from the existence of trained occupational therapists. The Henri Rousselle Hospital enjoyed a more generous nurse to patient ratio than other French hospitals, but there was no occupational therapist here either. French nurses, even in the capital, where training was more readily available than in the provinces, lacked the requisite skills to deliver occupational therapy. In the English provincial hospital at Littlemore, most of the nursing staff were dedicated and genuinely interested in their patients according to Dr Good's article in the *Journal of Mental Science* and the hospital's annual reports. Both male and female nurses were willing to learn a craft that they could teach to patients, and in this they were encouraged by the medical staff who recognised the benefits of keeping patients "of all grades" occupied. At the Asile de la Sarthe, male nurses lacked the time and the knowledge to become involved with patient occupation, and this was not an area that appeared to be prioritised by the asylum's chief medical offers. The latter were more concerned with focusing their limited resources on biological treatments for acute cases.

[86] Report of the Board of Control, Annual Reports of the Littlemore Hospital, Oxford, 1928/1929, 16. OHA-L1/A2/6.

[87] Ibid., 1926/1927, 15–16. OHA-L1/A2/4.

The New Profession: Occupational Therapy

One of the fundamental differences between French and English mental hospitals was the employment of an occupational therapist. In France, because the new approaches to occupational therapy had not been championed by chief medical officers, there was no requirement for occupational therapists and the profession had not developed. Patients working in the hospital workshops, as they had done in the nineteenth century, were supervised by the workshop managers. Nurses supervised patients working on the wards. There were no opportunities to train as an occupational therapist during the interwar period and the position of occupations officer or occupational therapist did not exist in France at that time. The first schools of occupational therapy (*ergothérapie*) opened in Paris and Nancy in 1954. Craft activities were not taught to patients at the Asile Clinique, although at the Henri Rousselle Hospital local artists gave art classes to patients.

In England, during the early 1920s, the Board of Control encouraged every large hospital to employ an occupations officer, to put the organisation of patient occupation "on a better footing".[88] From 1933, they recommended the employment of a trained occupational therapist, someone with "considerable knowledge of the theory and practice of occupation therapy and who has the education and mentality to interpret the doctors' instructions in the widest therapeutic sense".[89] David Henderson, who had introduced occupational therapy to Scotland in the 1920s, recommended in an article appearing in the *Lancet* in 1924 that occupational therapy would suit "a well-educated and intelligent, refined type of girl".[90] The "refinements" to which Henderson alluded may have been the arts and crafts taught to many middle-class girls and young women in the late nineteenth and early twentieth centuries. Mary Macdonald, principal of the Dorset House School of Occupational Therapy, wrote that the success of occupational therapy depended on the "tact, sympathy and power of understanding the patient's mental state and individual needs" of the

[88] Tenth Annual Report of the Board of Control (London: HMSO, 1923), 10.

[89] Board of Control, *Memorandum on Occupation Therapy for Mental Patients* (London: HMSO, 1933), 16.

[90] Cited in John Hall, "From Work and Occupation to Occupational Therapy: the Policies of Professionalisation in English Mental Hospitals from 1919 to 1959," in *Work, Psychiatry and Society, c.1750–2015*, ed. Waltraud Ernst (Manchester: Manchester University Press, 2016), 319.

therapist, on "a thorough knowledge of the crafts and occupations she employs", and finally on her "common sense".[91] A rather different set of skills was therefore expected of occupational therapists to those held by regular nursing staff, although at an RMPA conference in 1934, it was observed that "a good matron was, *ipso facto*, an occupational therapist".[92] The Association of Occupational Therapists (formed in 1936) defined an occupational therapist as:

> Any person who is appointed as responsible for the treatment of patients by occupation, and who is qualified by training and experience to administer the prescription of a Physician or Surgeon in the treatment of any patient by occupation.[93]

When the Maudsley first opened, patients confined to bed were given activities by the assistant matron, but she soon became "too busy".[94] An occupations officer was therefore engaged in 1924. Her working hours were extended from three afternoons per week to five in 1925, and she became full-time in 1929.[95] She was assisted by seven volunteers during 1925, one of whom had trained as an occupational therapist in the USA.[96] In 1924, the medical superintendent claimed that patients were engaged in household duties, needlework, clerical work, gardening, carpentry and upholstery.[97] The occupational therapist also supervised the new occupational therapy centre that opened in 1931, where patients who were fit enough could carry out a wider range of crafts than those practised on the wards. The occupational therapist was assisted by occupational therapy students and nurses who had received training in occupational therapy.[98] Although there was only one occupational therapist employed by the hospital by 1936, patient occupation was also supervised by trained volunteers, nurses and students.

[91] Mary Macdonald and Norah Haworth, *The Theory of Occupational Therapy for Students and Nurses* (London: Bailliere, Tindall and Cox, 1940), 1.

[92] Hall, "From Work and Occupation," 320.

[93] Ibid., 326.

[94] Report of the Medical Superintendent, Maudsley Hospital, London, 1924, 9. BMM/MSR-01.

[95] Report of the Medical Superintendent, Maudsley Hospital, London, 1926, 6; Sixteenth Report of the Board of Control (London: HMSO, 1929), 222.

[96] Report of the Medical Superintendent, Maudsley Hospital, London, 1926, 6. BMM/MSR-01.

[97] Ibid., 8–9.

[98] Report of the Medical Superintendent, Maudsley Hospital, London, 1936, 40. BMM/MSR-01.

The Maudsley Hospital was one of three English institutions offering training in occupational therapy during the interwar period.[99] Initially, it was offered on an informal basis. Those who wished to take up a career in supervising occupations for the mentally disordered could gain practical experience through working with the Maudsley's occupations officer, appointed in 1924.[100] From 1932, the Maudsley offered a regular six-month course in occupational therapy in conjunction with the Royal College of Nursing; the course was extended to 12 months in 1935. Camberwell School of Arts and Crafts provided the technical training in weaving, rug-making, basketry, bookbinding, cardboard construction, leather-work, sewing, embroidery and design.[101] Nurses were encouraged to join some of the practical training. In this way, the nurses could become a "useful supplement to any possible provision for the instruction of patients by more highly skilled occupational therapists".[102] At Bethlem, the physician superintendent's report of 1936 indicated that "the training of new students in Occupational Therapy has become a regular feature of the Department" enabling "work to be done on a much wider field".[103] Bethlem did not become a recognised school of occupational therapy, suggesting that this training was carried out on an informal basis.

Workshop Staff: Managers or Therapists?

In England, occupational therapists were not involved with the work performed by patients around the hospital; their time was spent supervising ward activities and arts and crafts in the occupational therapy department. The Board of Control recommended the employment of one occupational therapist on "each side" (that is, the male and female sides) of a hospital for *c*.1,000 patients.[104] A therapist responsible for 500 patients would not have time to supervise work as well as arts and crafts. The supervision of patients undertaking office work, gardening or work in the kitchen at the Maudsley was not mentioned; it may have been the responsibility of

[99] The others were Dorset House (founded in 1930) and the London School (founded in 1935).

[100] Report of the Medical Superintendent, Maudsley Hospital, London, 1935, 41. BMM/MSR-01.

[101] Ibid.

[102] Ibid.

[103] Report of the Physician Superintendent, Annual Reports of the Bethlem Royal Hospital, 1936, 10. BMM/BAR-53.

[104] Board of Control, *Memorandum,* 23.

nurses, or the regular hospital staff. There is no indication that the regular staff received any training, but all ultimately reported to the medical superintendent. It was good practice for doctors to complete an occupational therapy prescription, either for the occupational therapist or for whoever was supervising the patient, so this may have helped workshop staff in their management of patients. The prescription used by the American Occupational Therapy Association was recommended by John Ivison Russell, medical superintendent of the North Riding Mental Hospital, in his book on occupational therapy. The form included details of the patient's diagnosis, mental traits or characteristics, previous employment, special aptitudes or interests, the results desired, and the duration and frequency of treatment.[105]

In France, Demay observed in 1929 that workshop managers, who reported to the Bursar and ultimately the asylum director, appeared unaware of the nature of mental illness and many showed neither kindness nor patience towards the patients working in their workshops.[106] They were overly concerned with productivity and failed to prioritise the therapeutic, re-educative purpose of patient work. Patients were sometimes excluded from workshops for being disruptive, or too placid, or for showing a "lack of respect" for an employee. Demay recommended (as Dr Marie had done in 1913) that each workshop manager should undergo a period of training in medical services. At the very least, the doctor needed to give the workshop staff precise instructions regarding the patients conferred to their care.[107] Between 25% and 45% of the Asile Clinique's patient workforce were employed in the workshops, so this daily interaction with workshop staff affected a significant proportion of working patients.

The Role of the Chaplain

Most English mental hospitals, including the three studied here, had their own chaplain and chapel within the hospital grounds. Patient attendance at church services was encouraged and the numbers recorded by the Board of Control inspectors before World War I. Although attendance details

[105] John Ivison Russell, *The Occupational Treatment of Mental Illness* (London: Bailliere, Tindall and Cox, 1938), 33.

[106] G. Demay, "Les conditions de la thérapeutique par le travail dans les asiles," *L'Hygiène mentale* no. 1 (1929) 37.

[107] Ibid.

ceased to be recorded after the war, religious observances continued and provided a social activity for patients on a Sunday. The chaplain also played an important role outside the provision of religious services and spiritual guidance, often taking responsibility for the hospital library, as was the case at the Littlemore, accompanying patients on outings, and giving talks and lantern lectures. When, in 1935, the Littlemore chaplain was moved to another post after 26 years, he remarked, "I hope that my ministrations have been the means of brightening the lives of many of the patients and making them happier".[108] He gave services every Sunday, took communion to patients on the wards who were unable to come to chapel, and distributed books from the library every month. The new chaplain began confirmation classes and baptised an infant born in the hospital during his first year.[109] Roman Catholic patients (of whom there were 62 at the Littlemore in 1935) received ministrations from a Catholic priest, as they did at the Maudsley, where patients also had access to a Jewish Rabbi and a Nonconformist minister.[110] At Bethlem, where running the library was entrusted to a nurse, the chaplain edited the hospital magazine, *Under the Dome*, and his lantern lectures were very popular with patients.[111]

In France, following the secularisation of public services in 1905, provision for religious observances had been phased out in most mental institutions, particularly those in the republican large cities, and the role of chaplain no longer existed. The chapel at Ste Anne's had been converted into a Salle des Fêtes to provide space for the entertainments programme following the chapel's decommissioning in 1908, an event described by Henri Rousselle (who had campaigned for the chapel's conversion for years) as a "victory for republican values".[112] There was, however, an "*aumônier*" (or chaplain) at the Asile de la Sarthe, where secularisation did not take place until the 1960s. Religious events organised by the Catholic chaplain, such as the Fête-Dieu, a procession that took place 60 days after Easter, punctuated the year at the Asile de la Sarthe and provided a diversion for patients. There was a clear divide

[108] Report of the Medical Superintendent, Annual Reports of the Littlemore Hospital, Oxford, 1935/1936, 14. OHA-L1/A2/13.

[109] Ibid., 1936/1937, 13. OHA-L1/A2/14.

[110] Ibid., 1935/1936, 15.

[111] *Under The Dome* was renamed *Orchard Leaves* after Bethlem's move to Monks Orchard in 1930.

[112] Henri Rousselle, Rapport au nom de la 3e Commission sur les budgets et comptes de l'Asile Clinique our 1908 (Paris: Conseil Général de la Seine, 1909), 1. ADP/D.10K3/19/50.

between the secular metropolitan and the more religiously orientated provincial French asylums. The English institutions fell somewhere in the middle, with a lower profile of religious observances, administered by chaplains who also fulfilled secular functions.

Conclusion

The successful deployment of occupation as a means of therapy in French and English mental hospitals depended in large measure on the training, skills and number of non-medical staff, including nurses, workshop managers, chaplains and, in England, occupational therapists. The authority and therapeutic preferences of the psychiatrist were crucial, but so too were the means at his disposal to implement those preferences. In England, the medical superintendent, who had sole responsibility for management of the hospital, had the authority to drive through the introduction of the news methods of occupational therapy, but he had to have staff willing and able, and preferably trained, to supervise patients in these new methods. Mental nurse training was more widely available in England than in France, particularly in London, but while mental nurse training emphasised the importance of occupation, it did not necessarily follow that a trained mental nurse would be able to deliver occupational therapy. For the successful introduction of the latter, it was desirable to employ an occupational therapist. The new profession of occupational therapy emerged in England during the interwar period. Formal training in occupational therapy was available in England from 1930, while in France, the profession did not exist until after World War II. Here occupation was supervised by the nursing staff, whose lack of training and competence attracted considerable criticism, and by the workshop managers, who had no training in the supervision of mental patients. While English workshop managers reported to the medical superintendent, in France (other than at the Henri Rousselle Hospital) they reported to the asylum director whose priority was the contribution that patient labour could make to the hospital rather than its therapeutic benefit to the patient.

At the Maudsley Hospital, occupational therapists were supported by well-qualified nurses, trained in the basics of occupational therapy. Workshop managers at the Maudsley and Bethlem were ultimately responsible to the medical superintendent who believed in the curative properties of occupation and could insist on its deployment as therapy. At the Asile Clinique, the workshop managers who supervised *c.*30% of the patient workforce, were not always sympathetic to patient needs, and most of the acute patients in the treatment divisions were unoccupied. This was as much a matter of the psychiatrist's choice as it was to do with the competence and availability of staff. The situation at the Asile de la Sarthe was similar, with acute patients remaining unoccupied and the calm, chronic and incurable patient workers supervised by nursing or workshop staff. At the Henri Rousselle, all staff reported to Toulouse, who as medical director, had the authority to prioritise the therapeutic benefits of occupation. There were no occupational therapists here, nor at the Littlemore Hospital. At the latter, occupational therapy was delivered by nurses, who were each expected to know a craft that they could teach to patients. The nurses at the Littlemore, who were of a different calibre to those at the Asile de la Sarthe, were able to pass on useful information to the medical staff as a result of the time they spent with patients undertaking occupational therapy. The fact that the Littlemore did not suffer from overcrowding, and the nurses' exemption from administrative duties (something the medical superintendent insisted upon) meant that nurses were able to devote time to their patients, a commodity that was impossible to find in most French asylums. The types of activities that patients undertook, supervised by mental nurses, occupational therapists or workshop managers, are examined in the following chapter.

Bibliography

Secondary Sources

Andrews, Jonathan, Asa Briggs, Roy Porter, Penny Tucker, and Keir Waddington. *The History of Bethlem*. Abingdon and New York: Routledge, 1997.

Crammer, John. "Training and Education in British Psychiatry 1770–1970." In *150 Years of British Psychiatry, Volume II: The Aftermath*, edited by Hugh Freeman and German E. Berrios, 209–42. London: Athlone Press, 1996.

Crammer, John. *Buckinghamshire County Pauper Lunatic Asylum—St John's*. London: Royal College of Psychiatrists, 1990.

Goddard, J. *Mixed Feelings: Littlemore Hospital—An Oral History Project*. Oxford: Oxfordshire County Council, 1996.

Guillemain, Hervé. "Le Soin En Psychiatrie Dans La France Des Années 1930." *Histoire, Médecine et Santé* Spring no. 7 (2015): 77–99.

Guillemain, Hervé. *Chronique de la Psychiatrie Ordinaire: Patients, Soignants et Institutions en Sarthe du 19e au 21e Siècle*. Tours: Éditions de la Reinette, 2010.

Hall, John. "From Work and Occupation to Occupational Therapy: the Policies of Professionalisation in English Mental Hospitals from 1919 to 1959." In *Work, Psychiatry and Society, c.1750–2015*, edited by Waltraud Ernst, 314–33. Manchester: Manchester University Press, 2016.

Jean-Louis, Elisabeth and Aline Valentin. "Un Institut au Coeur de l'Histoire de la Profession Infirmière." In *L'Hôpital Sainte-Anne: Pionnier de la Psychiatrie et des Neurosciences au Coeur de Paris*, edited by Stéphane Henry, Catherine Lavielle, and Florence Patenotte, 170–71. Paris: Somogy Éditions d'Art, 2016.

Klein, Alexander. "Théodore Simon (1872–1961): Itinéraire d'un Psychiatre Engagé pour la Professionalisation des Infirmiers et Infirmières d'Asile." *Recherche en soins infirmiers* no. 135 (2018): 91–106.

Krzyzaniak, Patrice. *Georges Daumézon (1912–1979): Un Camisard Psychiatre et Pédagogue: Une contribution singulière aux Sciences de l'éducation*. Thèse. Paris: Université Charles de Gaulles, 2018.

Nolan, Peter. "Mental Health Nursing in Great Britain." In *150 Years of British Psychiatry, Volume II: The Aftermath*, edited by Hugh Freeman and German E. Berrios, 171–92. London: Athlone Press, 1996.

Price, Roger. *A Concise History of France*. Second edition. Cambridge: Cambridge University Press, 2005.

Salem, Évelyne and Stéphane Henry. "Laicisation et Formation du Corps Soignant à l'Asile Sainte-Anne." In *L'Hôpital Sainte-Anne: Pionnier de la Psychiatrie et des Neurosciences au Coeur de Paris*, edited by Stépane Henry, Catherine Lavielle, and Florence Patenotte, 172–3. Paris: Somogy Éditions d'Art, 2016.

Archival Sources

Author Unknown. "Board of Control Report from the Committee on Nursing in County and Borough Mental Hospitals." *Journal of Mental Science* April (1925): 289–305.

Author Unknown. "Nursing in the Bethlem Royal Hospital." *The Hospital* 8 January (1921): 346.

Bethlem Royal Hospital Rules and Orders. Bethlem Royal Hospital, London, 1932. BMM/BRH/SRO-24.

Board of Control Report. Annual Reports of the Littlemore Hospital, Oxford, 1933/4. OHA-L1/A2/12.

Board of Control. *Memorandum on Occupation Therapy for Mental Patients*. London: HMSO, 1933.

Chausse, M.E. "Rapport Général au nom de la 3e Commission sur le compte du Service des aliénés pour 1924, le budget rectificatif de 1925 et le projet de budget pour 1926." Paris: Conseil Général de la Seine, 1925. ADP/D.10K3/38/83.

Claude, Henri. Rapport de la Clinique des maladies mentales, Asile Clinique. Paris: Préfecture de la Seine, 1922. ADP/PER-566-5.

Courbon, P. "Un voyage d'étude dans les asiles de Hollande." *Annales Médico-psychologiques* no. 2 (1928): 289–306 and 385–405.

Demay, G. "Les conditions de la thérapeutique par le travail dans les asiles." *L'Hygiène mentale* no. 1 (1929): 33–40.

Good, Thomas Saxty. "The History and Progress of Littlemore Hospital." *Journal of Mental Science* October (1930): 602–21.

Legrain, M. and G. Demay. "Le travail des aliénés convalescents: rapport présenté à la 4e Section du Conseil supérieur de l'Assistance publique en 1934." *L'Aliéniste français* no. 2 (1936): 281–97.

Lomax, Montagu. *Experiences of an Asylum Doctor: With suggestions for asylum and lunacy law reform.* London: G. Allen and Unwin Ltd, 1921.

Macdonald, Mary and Norah Haworth. *The Theory of Occupational Therapy for Students and Nurses.* London: Baillière, Tindall and Cox, 1940.

Marchand, Dr. Rapport de la 1e Section des hommes, Asile Clinique. Paris: Préfecture de la Seine, 1931. ADP/PER-566-6.

Medico-Psychological Association. *Handbook for the Instruction of Attendants on the Insane.* London: Ballière, Tindall and Cox, 1885.

Mercier, Charles. *Lunatic Asylums: Their Organisation and Management.* London: Charles Griffin & Co., 1894.

Mignot, Roger and L. Marchand. *Manuel Technique de l'Infirmier des Établissements d'Aliénés.* Paris: Octave Doin et Fils, 1912.

Porot, A. "L'Assistance par le Travail dans les Asiles Hollandais." *L'Hygiène mentale* no. 2 (1929): 41–54.

Quentin, Maurice et Alphonse Loyau. Rapport Général au nom de la 3e Commission sur le compte du Service des Aliénés. Paris: Conseil Général de la Seine, 1936. ADP/D.10K3/54/114.

Rapport du Directeur. Asile de la Sarthe, 1919. Le Mans, 1920. ADS-1X965.

Rapport du Directeur. Asile de la Sarthe, 1920. Le Mans, 1921. ADS-1X965.

Rapport du Directeur. Asile de la Sarthe, 1923. Le Mans, 1924. ADS-1X965.

Rapport du Médecin en Chef. Asile de la Sarthe, 1937. Le Mans, 1938. ADS-1X965.

Rapport du Médecin en Chef. Asile de la Sarthe, 1933. Le Mans, 1934. ADS-1X965.

Report of the Board of Control. Annual Reports of the Littlemore Hospital, Oxford, 1926/7; 1928/9 and 1929/30. OHA-L1/A2/4; OHA-L1/A2/6 and OHA-L1/A2/7.

Report of the Medical Superintendent. Annual Reports of the Littlemore Hospital, Oxford, 1929/30; 1934/5; 1935/6 and 1936/7. OHA-L1/A2/7; OHA-L1/A2/12; OHA-L1/A2/13 and OHA-L1/A2/14.

Report of the Medical Superintendent. Reports of the Maudsley Hospital, London, 1924; 1926; 1935; 1936. BMM/MSR-01.

Report of the Physician Superintendent. Annual Reports of Bethlem Royal Hospital, London, 1925; 1927; 1929; 1934; 1935; 1936 and 1938. BMM/BAR-53.

Rousselle, Henri. "Rapport au nom de la 3e Commission sur l'Asile Clinique et sur le Service libre de prophylaxie mentale." Paris: Conseil Général de la Seine, 1922. ADP/D.10K3/32/77.

Rousselle, Henri. Rapport au nom de la 3e Commission sur les budgets et comptes de l'Asile Clinique pour 1908. Paris: Conseil Général de la Seine, 1909. ADP/D.10K3/19/50.

Russell, John Ivison. *The Occupational Treatment of Mental Illness.* London: Bailliere, Tindall and Cox, 1938.

Sixteenth Annual Report of the Board of Control. London: HMSO, 1929.

Tenth Annual Report of the Board of Control. London: HMSO, 1923.

Thirteenth Annual Report of the Board of Control. London: HMSO, 1926.

Toulouse, E. Rapports sur le Centre de prophylaxie mentale et l'hôpital Henri-Rousselle. Paris: Conseil Générale de la Seine, 1928–29 and 1930–31. ADP/D.10K3/46/83 Annexe and ADP/D.10K3/47/86 Annexe.

Twenty-second Annual Report of the Board of Control. London: HMSO, 1935.

WMA. "Sporting Activities." *Orchard Leaves*, Bethlem Royal Hospital, London, Winter (1935): 8.

CHAPTER 8

The Patient Workers Inside Hospital

This chapter focuses on the patients who were prescribed (or not prescribed) occupation in French and English mental hospitals. The extent to which class, gender, age, physical health and mental condition influenced whether patients were allocated some form of work or occupational therapy, is examined. The matter of patient choice—whether or not they wanted to be occupied—is difficult to ascertain, but the existence of incentives to work indicates that some encouragement was necessary. This raises the question of what happened to patients who were unable or unwilling to work, such as the elderly, who may have been too frail, and the middle classes, for whom manual work was something of an anathema in both France and England. Were these patients presented with alternatives to work? The influence of the psychiatrists prescribing occupation, discussed in chap. 6, is also relevant here since psychiatrists' views on the appropriateness of occupation for acute cases is fundamental to the allocation of work or the prescription of occupational therapy. The combination of these influences contributed to the experience of patient work, and to the overall experience of life in each of the mental hospitals examined.

Patient Class

Before the emergence of new methods of occupational therapy, a patient's class affected whether they were expected to carry out work around the hospital, irrespective of their physical or mental condition. Establishments

J. Freebody, *Work and Occupation in French and English Mental Hospitals, c.1918–1939*, Mental Health in Historical Perspective,
https://doi.org/10.1007/978-3-031-13105-9_8

catering for a middle-class clientele, such as Bethlem and the private section of the Asile de la Sarthe, had fewer working patients than public institutions, because manual work was not considered appropriate for the middle-classes. Bethlem, as a registered hospital, catered for the "deserving poor", who were "received gratuitously" (that is, they paid no fees) and the poorer middle classes who could not afford the high fees of a private institution, but could still afford to pay for all or part of their treatment at Bethlem.[1] Manual work was unlikely to be considered appropriate for these paying patients. Physician superintendent John Porter-Phillips noted that admissions in 1921 included "many professional men, officers of Her Majesty's Forces and civil servants".[2] In the early 1920s, *c.*70% of patients paid no fees; the group of professional men referred to by Porter-Phillips were probably among the remaining 30%. By the late 1930s, the proportion of fee-paying patients to those paying no fees had reversed, with only around 20% of patients "received gratuitously", suggesting that the profile of Bethlem patients became progressively middle class during the interwar period. It is plausible to assume that the numbers of patients engaged in work decreased as a result, while those of patients occupied with occupational therapy (introduced at Bethlem in 1932), leisure activities and entertainments increased.

Between 25% and 31% of Bethlem's patients were recorded as "usefully employed" by the Board of Control's inspectors between 1920 and 1928 (after 1928 figures for "useful employment" ceased to be recorded).[3] These working patients were most likely to have been selected from the non-paying cohort, but even when this is taken into account, the figures are considerably lower than the average of 57% of public mental hospital patients who worked, as identified by the Cobb Report of 1922.[4] This can be explained by the fact that Bethlem's admissions policy was to accept only curable cases. Those at the acute stage of their illness, who were most likely to respond to treatment, were sedated at Bethlem and would not have been prescribed work, which was allocated to convalescent patients. The fact that

[1] The costs of care at Bethlem were paid for out of charitable funds; patients who were unable to make a contribution towards these costs were "received gratuitously" if they met Bethlem's criteria (for example, the hospital did not accept paupers); those who were able to pay fees, did so.

[2] Report of the Physician Superintendent, Annual Reports of Bethlem Royal Hospital, London, 1921, 7. BMM/BAR-53.

[3] Report of the Board of Control, Annual Reports of Bethlem Royal Hospital, London, 1920–1928. BMM/BAR-53.

[4] Report of the Committee on Administration of Public Mental Hospitals, Ministry of Health (London: HMSO, 1922), 52.

some of his patients worked, was not something that Porter-Phillips wished to publicise, since, as Andrews *et al.* have suggested, "work did not agree with the Hospital's bourgeois character".[5] Porter-Phillips preferred to emphasise Bethlem's provision of sport, leisure and entertainment activities (in which between 34% and 49% of patients participated), and occupational therapy after 1932, which were designed to appeal to a middle-class clientele. Class was an issue at Bethlem. In 1911, the physician superintendent noted that although the "social status of patients has been rather better in recent years", he had been obliged to admit out of "common humanity", several patients on a voluntary basis who were "not of the educated classes". He was relieved to note that once "the new hospital on Denmark Hill" (the Maudsley) had opened, he would be able to refuse such patients.[6]

The Maudsley was a public mental hospital, but it was unique among public mental institutions in that it specialised in patients at the early, acute stage of their disease when their symptoms were not so severe as to warrant certification; and admission was voluntary. Other English public mental hospitals accepted a mix of curable, acute, incurable and chronic patients but only patients who were certified were admitted before the Mental Treatment Act of 1930. At the Maudsley *c.*75% of patients were "rate-aided" (a term that carried less stigma than pauper) and *c.*25% paid fees.[7] There was no indication that work or occupational therapy was allocated differently between the private patients and those whose care was paid for by the London County Council. All patients (with the exception of children under 14 years, who comprised 3–7% of Maudsley patients) were obliged to undertake some form of ward work as soon as they were able to get out of bed. Patients were occupied with craft activities whilst still in bed.[8] Although admission was determined solely on medical grounds, and not by "social or financial considerations", Maudsley patients included a high proportion of middle-class individuals, including educated professionals, such as doctors, clergymen and school teachers.[9] In 1925, the largest professional group were artisans (26%), while the next most highly

[5] Jonathan Andrews, Asa Briggs, Roy Porter, Penny Tucker, and Keir Waddington, *History of Bethlem* (Abingdon and New York: Routledge, 1997), 689.

[6] Report of the Physician Superintendent, Annual Reports of Bethlem Royal Hospital, London, 1911, 5. BMM/BAR-53.

[7] According to the statistics featured in the Reports of the Medical Superintendent, Maudsley Hospital, London, 1923–1931. BMM/MSR-01.

[8] Report of the Medical Superintendent, Maudsley Hospital, London, 1926, 5.

[9] Ibid.

Table 8.1 Table to show the occupation of patients at the Maudsley Hospital, London, prior to admission in 1925. [Source: Report of the Superintendent, Maudsley Hospital, 1925. BMM/MSR-01]

Occupation of Patients Prior to Admission	
Professionals*	123 = 13%
Clerks	99 = 10%
Hospital nurses	16
Farmers	6
Shopkeepers	93 =10%
Policemen, postmen and transport workers	57 = 6%
Artisans	243 = 26%
Domestic servants	78 = 8%
Labourers	91 = 10%
Unemployed	138 = 15%
TOTAL	944
*Professionals comprised:	
Doctors	9
Engineers	9
Teachers	6
Clergymen	5
Naval officers	3
Army officers	2
Artists	2

Source: Report of the Superintendent, Maudsley Hospital, 1925. BMM/MSR-01

represented group were the unemployed (15%) [see Table 8.1 above].[10] Medical superintendent Edward Mapother was aware of the wide range of social classes and educational attainment amongst patients, although he observed that such "social differences have never given rise to difficulties".[11] Class therefore appeared to have less relevance to the type of activities allocated to patients at the Maudsley than it did at Bethlem. The high proportion of middle-class patients at the Maudsley might be explained in economic terms—patients who were insecure economically were less likely to seek voluntary treatment than those who had more of a financial "cushion" and could afford to spend time in hospital without working.

The Asile Clinique in Paris was also a public mental hospital where the majority of patients were paupers.[12] Details regarding the previous occupations of patients passing through the Admissions Service (15% of whom were admitted to the Asile Clinique) reveal that between 1919 and 1926,

[10] Ibid., 2.

[11] Report of the Medical Superintendent, Maudsley Hospital, London, 1924, 6.

[12] Patients who could afford to pay fees went to private asylums, or *maisons de santé*.

Table 8.2 Table to show the sectors in which male (M) and female (F) Seine patients were employed prior to admission to one of the Seine's six asylums (of which the Asile Clinique was one) during the period 1918–1926. [Source: Reports to the General Council, 1919-26. ADP-D.10K3 26-34]

	1919 M	*1919 F*	*1923 M*	*1923 F*	*1926 M*	*1926 F*
Liberal professions	27	19	24	–	32	10
Industrial workers	226 (19%)	188 (9%)	285 (17%)	194 (10%)	195 (12%)	126 (7%)
Farmers	18	4	15	–	13	–
Property owners	9	6	7	14	37	16
Beveridge trade	6	3	13	–	73	21
Other trades	26	9	18	4	67	15
Travelling salespeople	–	1	5	7	9	4
Employees of business and administration = clerical	107 (9%)	62	115 (7%)	40	84 (5%)	35
Military/nurses/wardens	10	10	4	4	13	6
Domestic servants	28	95 (5%)	25	143 (7%)	31	127 (7%)
Day labourers	59	73	45	60	93	75
Religious/clergy	–	1	2	–	–	2
Without profession	671 (56%)	1548 (77%)	1145 (67%)	1527 (76%)	1043 (62%)	1448 (77%)
Total	1187	2019	1709	2001	1690	1885

Source: Reports to the General Council, 1919–26. ADP-D.10K3 26-34

56%-77% of patients had no previous profession [see Table 8.2 above]. Patients of public asylums were expected to work to contribute to the costs of their care if they were sufficiently fit, whether or not they had worked outside the institutions prior to admission. However, the acute nature of patients admitted to the Asile Clinique after 1927, meant that very few patients in the treatment divisions were prescribed occupation, despite their pauper status, because doctors did not consider work a suitable treatment for patients at the acute stage of their illness. The ability of patients to work, as discussed in chap. 5, was one of the factors that delayed the transformation of the Asile Clinique to an acute hospital. The authorities were concerned about the budgetary implications of focusing on acute cases, since this would result in a significant reduction in the number of patient workers, upon whose economic contribution they relied. Before the transformation of the hospital to an acute service in 1927, an average of just 25% of patients worked. The figure of 25% was already considerably lower than the national average for French public asylums serving pauper patients, which stood at between 50% and 55%.[13] Patient health, discussed below, may have been influential in lowering the proportion of patient workers at the Asile Clinique prior to 1927. The asylum's city location was also a factor, since this limited the availability of land for cultivation (although vegetable and flowers were grown in the gardens). After 1927, it was the acute nature of patients' mental condition that kept numbers of workers numbers from the treatment divisions very low.

The Asile de la Sarthe provided care for both private patients who were able to pay for their care, and pauper patients whose fees were paid by the local authority. In 1919, 27% of patients paid for their care; this figure decreased to 25% by 1928 and to 19% by 1939. There was a marked difference between the classes in terms of the proportions of working patients. Only 18%–33% of male private patients worked, compared with 46%–63% of male pauper patients. Private patients were divided into four classes. The top three classes of patients, some of whom had their own personal servants in the asylum, would not have been expected to work, due to their middle-class status. Patients in class four, who comprised an average of 50% of the private patients [see Table 8.3, below], shared the same regime (*le régime en commun*) as the publicly funded patients. Relatives of private patients belonging to this fourth tier were charged roughly the same rates as the local authorities for the maintenance of pauper patients, who were expected to work.

The Littlemore Hospital was a public asylum, catering for predominantly pauper patients. Patients were referred by one of the Oxfordshire Poor Law Unions, or by local authorities from outside Oxfordshire [see Appendix A.14]. The hospital was under contract with the London County Council to receive 20 patients per year, and with the Middlesex local authority (*c.*125 patients). These patients, as well as those coming to the Littlemore from Croydon, Nottingham, Buckinghamshire and East Ham, would have had quite different backgrounds to patients coming from rural Oxfordshire, but all were pauper patients. The only private patients were Service, or ex-Service, patients who were classed as "private" in order to avoid the stigma of certification. There were 20–23 such patients between 1923 and 1935. As at the Asile Clinique, the suitability of work for middle-class patients therefore was not an issue at the Littlemore. Here, as in all the asylums, mental condition, gender, age and physical health played important roles in the type of occupation prescribed.

[13] Dr. Lautier, "Les Exagérations de la Thérapie par le Travail," *L'Hygiène Mentale* no. 2 (1929): 191.

Table 8.3 Table to show the class divisions, based on average numbers of patients resident per year, of paying patients at the Asile de la Sarthe Sarthe [Source: Report of the Asylum Director, Asile de la Sarthe, 1919-39. ADS-1X964/5]

	1919	*1922*	*1925*	*1928*	*1931*	*1934*	*1936*	*1939*
All paying patients (Male: Female)	167	127	173	188	198	219	215	175
1st class (Male: Female)	20 3:17	14 2:12	23 6:17	24 (4: 20)	27 3:24	28 5:23	33 9:24	31 12:19
2nd class (Male: Female)	27 5:22	14 5:9	18 7:11	35 (19: 16)	37 25:12	43 30:13	34 25:9	26 12:14
3rd class (Male: Female)	38 12:26	24 6:18	33 10:23	38 (15: 23)	31 6:25	34 9:25	31 9:22	33 17:16
4th class (Male: Female)	82 29:53	75 20:55	99 37:62	91 (31: 60)	103 31:72	114 35:79	117 34:83	85 32:53
All pauper patients (Male: Female)	442 159:283	491 168:323	519 187:332	573 (220:353)	598 237:361	624 269:355	639 261:378	743 316:427
Average total	609	618	692	761	796	843	854	918

Source: Report of the Asylum Director, Asile de la Sarthe, 1919-39. ADS-1X964/5

Patient Gender

The patient populations at Bethlem, the Maudsley and Littlemore hospitals showed a consistent female bias during the interwar period, with an average of *c*.45% men and *c*.55% women.[14] This bias was typical of English mental hospitals in the early twentieth century. In 1915, of the 137,188 cases of notified mental illness in England and Wales, 54% were women. This proportion remained fairly stable until 1946.[15] As the Board of Control noted in 1934, following the introduction of voluntary treatment in 1930, female voluntary admissions were increasing faster than male. The Board felt this was to be expected since "the man is generally the breadwinner of the family, and therefore is compelled by economic considerations to defer applying for treatment".[16] Just 32% of British women in 1921 and 34% in 1931 were in paid employment.[17] Women comprised

[14] The gender split for all the institutions can be seen in the Appendices (A.1-A.13).

[15] Andrews *et al.*, *The History of Bethlem*, 657.

[16] Twenty-first Annual Report of the Board of Control (London: HMSO, 1934), 2.

[17] Ian Gazeley, "Manual Work and Pay, 1900–1970," in *Work and Pay in 20th Century Britain*, eds Nicholas Crafts, Ian Gazeley, and Andrew Newell (Oxford: Oxford University Press, 2007), 56.

c.30% of the British labour force in 1921. Paid employment for women was concentrated in five sectors: domestic service; commercial, financial and insurance occupations; clerical work; textiles and clothing.[18]

There are no details regarding the gender split of occupations at the English mental hospitals during this period, as the recording of the numbers of patients who were "usefully employed", and at what tasks, ceased at public mental hospitals when war broke out in 1914. It was noted, however, that at the Maudsley in 1925 the new carpentry and upholstery workshops "greatly increased opportunities for the employment of suitable male patients", while "both sexes of patient" had been engaged in gardening and clerical work.[19] At the Littlemore, in 1913 (just before such details ceased to be recorded and prior to the introduction of occupational therapy), 38% of male patient workers cleaned the wards; 34% worked in the gardens; 12% were engaged in hair picking and another 12% in bed-making. Amongst the female patient workforce, cleaning occupied 43% of women; needlework 37%; laundry 13% and work in the kitchen 4%, revealing a clear split between the genders [see Table 2.1]. In 1937 the Board of Control inspectors noted in that although occupational therapy on the wards was well organised for the female patients, very little was being done for the men. Dr. Good explained that many of the male patients came from rural areas where they were "not accustomed to hand-crafts", but this explanation was not accepted by the inspectors who felt that with some encouragement, many of the male patients from both urban and rural backgrounds, would benefit from "this form of employment".[20] These remarks indicate that occupational therapy, as well as work around the hospital, was gendered.

In France, numerous laws passed between 1874 and 1919 were aimed at encouraging women to remain at home and become mothers, rather than seeking paid employment outside the home.[21] The promotion of motherhood was explicit in the maternity laws of 1909–1913 amid concerns regarding France's low birth-rate and high level of infant mortality.[22]

[18] Ibid., 59.

[19] Report of the Medical Superintendent, Maudsley Hospital, London, 1925, 3.

[20] Report of the Board of Control, Annual Reports of the Littlemore Hospital, Oxford, 1936/1937, 21. OHA-L1/A2/14.

[21] Elinor Accampo, Rachel Fuchs, and Mary Lynn Stewart, *Gender and the Politics of Social Reform in France, 1870–1914* (Baltimore and London: Johns Hopkins University Press, 1995), 9.

[22] Ibid.

That said, women comprised nearly 40% of the French labour force in 1921. Even though this figure had decreased to 36% by 1936, the female proportion of the workforce was higher in France than in Britain, where, paradoxically, there was less emphasis on women's role as mothers.[23] At the Asile Clinique, the proportion of male and female patients was more evenly balanced than in the English institutions, with a slight male bias.[24] Work for the female patients was associated with their domestic roles, as it was at the Littlemore, and included mending and making clothes, cleaning, ironing, working in the laundry and kitchen. The majority of working female patients were employed as cleaners on the wards or in the sewing (*couture*) workshops. Opportunities for men were much more diverse, and included work in the gardens, plumbing, decorating, electrical or mechanical work, shoemaking, carpentry, building, road mending, working in the wine cellar, library, pharmacy or the surgical wing. Some roles were divided between men and women, such as housework, scrubbing floors, kitchen and laundry work because these tasks were performed in their respective quarters. Male patients were not allowed in the female quarters and vice versa. At the Henri Rousselle Hospital, the proportion of male and female patients fluctuated, with a higher proportion of female admissions in the 1920s and becoming more evenly balanced in the 1930s as the Depression took its toll on the mental health of men unable to find employment. Patients of both sexes were encouraged to keep busy, but they were not obliged to work in the hospital. A sewing room was available for use by the female patients. A piano and gramophone were also provided, and patients could play tennis or boules in the hospital grounds. Artists visited the hospital free of charge to give art classes to the patients.[25]

In the provincial Asile de la Sarthe, the proportion of female patients was even higher than in the English institutions. The proportion of male patients (both pauper and private) gradually increased from 32% in 1921 to 41% in 1939. Amongst the paying patients, a higher percentage of men than women worked (an average of 26% of men, and 18% of women,

[23] Susan Pedersen, *Family, Dependence, and the Origins of the Welfare State: Britain and France 1914–1945* (Cambridge: Cambridge University Press, 1993), 124–5.

[24] The proportion of male to female patients varied from 51% male in 1919 to 54% male in 1937. See Table A.1 in the Appendices.

[25] Michel Huteau, *Psychologie, Psychiatrie et Société sous la Troisième République: La biocratie d'Édouard Toulouse (1865–1947)* (Paris: L'Harmattan, 2002), 199.

Table 8.4 Table to show the number and percentage of male and female pauper patients working at the Asile de la Sarthe, 1923–1937. [Source: Reports of the Asylum Director, Asile de la Sarthe, 1923-37. ADS-1X964/5]

	All Pauper men	*Pauper male workers*	*% Pauper men who worked*	*All Pauper women*	*Pauper female workers*	*% Pauper women who worked*
1923	176	104	59%	334	193	58%
1925	187	119	63%	332	193	58%
1927	293	111	39%	345	182	53%
1929	225	103	46%	357	178	50%
1930	234	113	48%	359	179	50%
1932	242	117	48%	360	191	53%
1934	269	140	52%	355	199	56%
1936	261	140	54%	378	204	54%
1937	272	133	49%	389	199	51%

Source: Reports of the Asylum Director, Asile de la Sarthe, 1923–37. ADS-1X964/5

worked, as indicated in Table 8.5). This might be explained by attitudes towards middle-class women, around whom a "cult of domesticity" had been created that emphasised middle-class women's reproductive, maternal and homemaking activities.[26] In contrast, an average of 54% of pauper women, and 51% of pauper men worked within the asylum, which was close to the average for French asylums observed by Dr. Lautier in 1929 [see Table 8.4]. Thus, three times as many female pauper patients than their private counterparts worked, and twice as many male pauper patients than male private patients, worked. Class and gender were therefore significant influences on the prescription of patient work in provincial France, despite its alleged therapeutic benefits for *all* classes and both genders. As in England, manual work was considered unsuitable for middle-class men, and the fact that patients were paying for their care meant that they were not expected to contribute to asylum maintenance costs with their labour. This exemption from work for middle-class patients was made explicit in the regulations of 1857.[27]

[26] Accampo, *et al.*, *Gender and the Politics of Social Reform*, 13.

[27] Article 151, Bulletin Officiel du Ministère de l'Intérieur, Circulaire No. 7, 20/03/1857, 57. https://gallica.bnf.fr/ark:/12148/bpt6k5539701n/f2.item.zoom. (accessed 12/02/2018).

Table 8.5 Table to show the number and percentage of male and female paying patients working at the Asile de la Sarthe, 1923–1937. [Source: Reports of the Asylum Director, Asile de la Sarthe, 1923-37. ADS-1X964/5]

	All paying male patients	*Paying male patients who worked*	*% paying men who worked*	*All paying female patients*	*Paying female patients who worked*	*% paying women who worked*
1923	35	7	20%	96	22	23%
1925	60	19	32%	113	25	22%
1927	72	29	40%	116	19	16%
1929	66	22	33%	129	20	16%
1930	67	17	25%	131	21	16%
1932	57	17	30%	143	20	14%
1934	79	15	19%	140	28	20%
1936	77	16	21%	138	23	17%
1937	79	14	18%	123	21	17%

Source: Reports of the Asylum Director, Asile de la Sarthe, 1923–37. ADS-1X964/5

The Mental Condition of Patients

Whether a patient's condition was perceived as curable or incurable, and whether the condition was at its early, acute stage (when it was most likely to respond to treatment) or well established (and more likely to become chronic), influenced the likelihood of the patient being prescribed occupation, and of what type. What was not clear from the records, was the influence of different types of mental disorder on the prescription of occupation. Hermann Simon, the originator of MAT, believed that patients with all types of condition, from mania to melancholia, could be employed in some capacity if they were physically fit. This view was shared by the Swiss psychiatrist Eugen Bleuler (1857–1939), who maintained that "every mental institution" should be able "to offer every patient some kind of work at all times".[28] Mary Macdonald's textbook of occupational therapy distinguishes between the therapeutic occupation "definitely prescribed as treatment with a view to improving the patient's condition" and the non-therapeutic work assigned to chronic cases. Work for the chronic and incurable, she maintained, was valuable because it kept patients happier, healthier and able to live as "normal" a life as possible.[29] Macdonald's book outlined the

[28] Richard Warner, *Recovery from Schizophrenia: Psychiatry and Political Economy*, Third ed. (Philadelphia: Routledge and Kegan Paul, 2004), 16.

[29] Mary Macdonald and Norah Haworth, *Theory of Occupational Therapy for Student and Nurses* (London: Baillière, Tindall and Cox, 1940), 8.

different types of *therapeutic* occupation that might be suitable for patients suffering from various types of mental disorder, such as depression (for example, a past hobby), schizophrenia (rug-making, weaving, basketry) or confusional insanity (winding yarn, polishing brass).[30] She advised that "the work chosen should demand the patient's whole attention and the standard must be raised as the patient improves" which resonated with the different grades of occupations outlined by Simon.[31] Macdonald recommended that the calm chronic and incurable patients were occupied in supervised work around the hospital, and the more turbulent in simple tasks on the wards that kept them busy and "out of mischief".[32] But it was not clear from the institutional records how specific diagnoses influenced the types of occupation prescribed in each of the hospitals examined. This issue was further complicated by changing nosology during the interwar period. Psychiatrists of the same nationality used different classification systems, which changed over time, and a comparison of French and English disease categories was impeded for the same reasons. Much clearer was the influence on occupation of the perceived curability of the patient.

Metropolitan Hospitals

All four metropolitan hospitals discussed in this chapter specialised in acute cases, but they treated these acute cases differently. Before World War I patient work had been considered unsuitable for acute patients and was reserved for chronic, incurable and convalescent patients. Bed-rest (*l'alitement*), sometimes for several weeks, had been the main treatment for acute patients at the Asile Clinique since 1896 when it was first advocated by the influential chief medical officer of the Ste. Anne's Admissions Service, Valentin Magnan.[33] His ideas were widely supported in Francophone psychiatric circles. Reporting at the First Belgian Congress of Neurology and Psychiatry in 1908, Dr. Cuylitz, speaking on the alleged therapeutic benefits of work for the acutely mentally ill, claimed that "the best exercise was rest". He rejected on scientific grounds any possibility that work might be

[30] Ibid., 8–17.
[31] Ibid., 20.
[32] Ibid., 18–19.
[33] Henri Rousselle, Rapport au nom de la 3e Commission sur les budgets et comptes de l'Asile Clinique (Paris: Conseil Général, 1903), 5–7; ADP/D.10 K3/15/38; Ibid., 1911; 4–6, ADP/D.10 K3/21/42.

therapeutic.[34] Although not all Magnan's colleagues in Paris agreed with him,[35] his views set the tone for the treatment of acute patients at the Asile Clinique, none of whom were prescribed occupation at the acute stage of their illness. In this, the Asile Clinique patients were treated similarly to those at Bethlem. Most patients, on admission to Bethlem, were sedated, often for several months, and therefore unoccupied.[36] Doctors at the Asile Clinique and Bethlem did not respond positively to the new theories concerning patient occupation, discussed in chap. 4, that emerged after World War I. The new theories put forward by Hermann Simon and advocated by Adolf Meyer challenged the view that prolonged, complete rest was the most effective treatment for acute cases. The continued use of bed-rest at the Asile Clinique was indicated in 1929 by chief medical officer Dr. Capgras' observation of the difficulties of supervising patients undergoing bed-rest in the Asile Clinique's overcrowded wards.[37]

The medical superintendent of the Maudsley, Edward Mapother, embraced the new approach to occupation for acute patients. While he believed that bed-rest (preferably in the open air) for "a considerable time after admission" was necessary for most acute cases, he recognised "the need of industries for those confined to bed, or … to the wards".[38] Once patients were up, all were given some sort of work, either "household duties", needlework, clerical work, gardening, upholstery or carpentry.[39] Patients made bed tables, letter boxes for the wards, wardrobe lockers, washstands and mortuary trolleys, among other items, and maintained the hospital furniture.[40] An occupations officer was employed in 1925 who taught "a large variety of handicrafts" including rug- and basket-making, leather, pewter, embroidery and raffia work to patients on the wards.[41] The results were very successful and had an "undoubted beneficial effect".[42]

[34] A. Marie, Rapport sur la service des aliénés du département de la Seine pendant l'année (Paris: Le Préfecture de la Seine, 1908), 20. ADP/PER566.

[35] Both Édourard Toulouse and August Marie of the Villejuif asylum were keen advocates of the therapeutic value of occupation before World War I. Marie insisted that work was the "touchstone of therapy" and "can and must be employed". (Ibid.)

[36] Andrews *et al.*, *The History of Bethlem*, 677.

[37] Dr. Capgras, Rapport de la 1re section des hommes, Asile Clinique (Paris: Préfecture de la Seine, 1929), 78. ADP/PER-566-5.

[38] Report of the Medical Superintendent, Maudsley Hospital, London, 1924, 9.

[39] Ibid., 8–9.

[40] Report of the Medical Superintendent, Maudsley Hospital, London, 1926, 5.

[41] Report of the Medical Superintendent, Maudsley Hospital, London, 1925, 3.

[42] Ibid., 3.

Patients were less inclined to "morbid preoccupation with their troubles" and less vulnerable to "boredom and deterioration".[43] Mapother was aware that because the average length of stay in hospital for Maudsley patients was short (usually around three months), occupational therapy had to be organised differently to that provided in an asylum or orthopaedic hospital where patients remained for long periods. The arts and crafts activities taught by the occupational therapy department had to be relatively simple, so the techniques could be grasped quickly to produce rapid but pleasing results.[44]

All Maudsley patients were at the acute stage of their illness, but their symptoms were not certifiable at the time of admission.[45] The fact that patients' symptoms were relatively mild, and their prognosis generally good, suggests that they were likely to be capable of work or occupational therapy as soon as they had undergone the initial rest period recommended by Mapother. The records do not indicate the numbers of patients occupied at the Maudsley, but Mapother suggests that everyone who was "up" was given work and those in bed or on the wards were occupied with occupational therapy except at the very beginning of their stay.[46] At Bethlem, on the other hand, a significant (if declining) proportion of patients were certified, indicating that their symptoms were more severe.[47] When occupational therapy was introduced in 1932, it was initially only provided in the occupational therapy workroom, suggesting that patients had to have recovered sufficiently to leave the wards.[48] It was only two years later that occupational therapy was provided to patients confined to the wards.[49]

At the Asile Clinique, after its transformation to an acute service in 1927, the bulk of the work around the hospital was performed by chronic and incurable patients specially drafted in for the purpose, and accommodated in a separate "workers" block. Only a small percentage of patients from the treatment divisions worked. These were most likely to be

[43] Ibid., 3, 26.

[44] Report of the Medical Superintendent, Maudsley Hospital, London, 1926, 5.

[45] Report of the Medical Superintendent, Maudsley Hospital, London, 1924, 12.

[46] Report of the Medical Superintendent, Maudsley Hospital, London, 1925–1926, 5.

[47] In 1920, 75% of admissions to Bethlem were certified; this figure fell to *c*.50% in 1927 and to *c*.25% by 1939. The Maudsley did not accept certified patients; Maudsley patients' symptoms were sufficiently mild not to warrant certification.

[48] As indicated earlier, nearly all patients were sedated for several weeks on admission to Bethlem.

[49] Report of the Physician Superintendent, Annual Reports of Bethlem Royal Hospital, London, 1934, 11. BMM/BAR-53.

convalescent patients, such as those in remission from GPI following malaria therapy or recovering alcoholics. Alcoholism and syphilis were rife in inter-war Paris; alcoholics and those suffering from GPI featured prominently amongst male admissions to the Asile Clinique. Between 18% and 22% of male patients, and between 8% and 16% of female patients, were admitted with GPI between 1919 and 1925. The Women's First Section became a centre for malaria therapy in 1931. Women diagnosed as suffering from GPI at other asylums were transferred to the Asile Clinique causing the number of female GPI patients to increase significantly after 1931 (GPI accounted for 83% of admissions to the Women's First Section in 1933).[50]

Alcoholism was acknowledged as a curable condition, as indicated by the fact that in 1933, alcoholics accounted for 70% of patients discharged cured.[51] Work had long been recognised as an effective treatment for mental disorders associated with alcoholism; the Asile Clinique's patient workforce in 1901 comprised "almost exclusively alcoholics and chronics".[52] In the past, work had not been considered suitable for patients suffering from GPI; they were too turbulent in the early stages of the disease, and likely to cause accidents, and too weak at the end.[53] But since the introduction of malaria therapy, which resulted in the recovery or partial recovery of some patients, many were able to work after receiving the treatment. Following its introduction in 1927, the prognosis for GPI patients improved. A convalescent quarter providing work for recovering GPI patients was established in the Women's First Section, after it became a centre for malaria therapy. The number of women engaged in sewing rose from 33 in 1928 to 49 in 1934.

Provincial Asylums

Unlike the Asile Clinique in Paris, the provincial Asile de la Sarthe admitted patients with both curable and incurable conditions. Because incurable patients, and those with chronic conditions, tended to remain in the

[50] Dr. Leroy, Rapport de la 1re section des femmes, Asile Clinique (Paris: Préfecture de la Seine, 1931), 77. ADP/PER-566-6.

[51] Dr. Marchand, Rapport de la 1re section des hommes, Asile Clinique (Paris: Préfecture de la Seine, 1933), 75. ADP/PER-566-6.

[52] Henri Rousselle, Rapport au nom de la 3e Commission sur les budgets et comptes de l'Asile Clinique (Paris: Conseil Général de la Seine, 1901), 6. ADP/D.10 K3/14/52.

[53] M. Navarre, Rapport au nom de la 3e Commission sur les propositions budgétaires pour le service des aliénés (Paris: Conseil Général de la Seine, 1908), 6. ADP/D.10 K3/18/35.

asylum for many years (and often for life), movement of the patient population was slow and opportunities to admit new, curable cases were limited. In the department of La Sarthe, there were no alternative institutions for incurable patients, as there were in the Seine department. As highlighted in the previous chapter, there was only one medical officer for the whole asylum; one individual could only do so much when treating *c.*800 patients. When Dr Henry Christy arrived at the Asile de la Sarthe in 1935, he made it clear where his priorities lay. He identified two main categories of patients: those who were intellectually impaired (*les infirmes du cerveau*) and had been so since birth, or on account of illness or injury, and were regarded as incurable; and those who had an "evolving cerebral illness" which was considered curable, either completely or partially.[54] Dr Christy prioritised the second group, whom he sought to treat biologically, and not with occupational therapy. He had been critical of the "exaggerated claims" of certain "Germanic methods" of work therapy (clearly those of Hermann Simon) which had been greeted with scepticism by many French psychiatrists, himself included, who sought "realistic, concrete objectives and not just words", as he put it.[55]

Christy, like most of his French colleagues, believed that work was for chronic, incurable and convalescent patients. He reported in 1937 that a significant number of intellectually impaired patients had been admitted, for whom the only treatment was re-education through the discipline of work.[56] It was not always clear from the annual medical reports, however, which of the conditions ascribed to patients were considered curable or incurable, as in the case of "degeneracy" and "chronic alcoholism". Christy highlighted in 1937 that although few cases of "degeneracy" were being discharged, not all were incurable and that many more would be curable if identified and treated earlier, and if more staff were at his disposal to treat them.[57] The categories of "mental debility"," idiocy", "cretinism" and "imbecility" were evidently incurable conditions associated with intellectual impairment. Intellectual impairment was attributed to 16% of male residents in 1921 and 24% in 1931. The admissions records suggest that significantly more pauper than private patients suffered from

[54] Rapport du Médecin en Chef, Asile de la Sarthe, 1935 (Le Mans, 1936), 7. ADS-1X964/5.
[55] Ibid.
[56] Rapport du Médecin en Chef, Asile de la Sarthe, 1937 (Le Mans, 1938), 11. ADS-1X964/5.
[57] Ibid., 12–13.

intellectual impairment (between 11% and 15% of pauper patients, compared to between 2% and 9% of private patients admitted in 1921–1931) and were deemed incurable.

At the Asile de la Sarthe, those diagnosed with mania, melancholia and delusions of persecution or paranoia, accounted for 57% of all female residents in 1921 and 54% in 1931. From the admissions data, it appears that such diagnoses were particularly common amongst the female private patients. The condition was ascribed to 54%–71% of female private patients between 1921 and 1931, while to just 37%–38% of female pauper patients. Hermann Simon, who developed MAT, advocated work for such conditions, including mania, but as his methods were dismissed by Christy, it is unlikely that these patients were prescribed work. GPI was not nearly as common in rural asylums as in city establishments; syphilis was more prevalent in urban areas. GPI affected *c.*3% of male patients resident at the Asile de la Sarthe between 1921 and 1931, and an even smaller percentage of female patients. Nevertheless, it was a fatal condition before the (relatively late) introduction of malaria therapy at the asylum in 1933.[58] Before this date, GPI was the attributed cause of death in a quarter of male patients dying between 1921 and 1931. By 1937 GPI did not feature as a cause of death in the records.[59] It can be assumed that GPI patients would not have been considered capable of work before 1933, but thereafter, successful malaria treatment may have enabled convalescent patients, or those in remission, to work.

Although the Littlemore cared for curable, incurable and chronic cases, Dr. Good did not regard a mental hospital as an appropriate locus of care for the intellectually impaired, who comprised a growing proportion of admissions to the Littlemore each year. Good did not believe that these patients could benefit from the expensive treatment offered at a modern psychiatric hospital.[60] He acknowledged, however, the "extreme difficulty of obtaining vacancies for this class of case among the feeble-minded

[58] Rapport du Médecin en Chef, Asile de la Sarthe, Le Mans, 1933 [hand-written, no page nos.]. ADS-1X261.

[59] Rapport du Médecin en Chef, Asile de la Sarthe, 1937 (Le Mans, 1938), 23. ADS-1X964/5.

[60] Report of the Medical Superintendent, Annual Reports of the Littlemore Hospital, Oxford 1925/1926, 10; Ibid., 1932/3, 9. OHA-L1/A2/3; OHA-L1/A2/10.

institutions in this country which are now practically full".[61] As Kathleen Jones has pointed out, even in the 1950s there were still many patients in mental hospitals whose primary condition was intellectual impairment, rather than mental illness, due to lack of provision for the intellectually impaired in colonies.[62] In 1927, Good highlighted the number of "mental defectives" who had been admitted, and in 1930 he noted that the proportion of individuals suffering from "organic physical diseases and congenital mental defect" had increased.[63] The Board of Control inspectors also noticed the large number of "mental defectives" they encountered on their visit to the Littlemore in 1931 whom they felt would be better placed in a colony.[64] In 1932, the medical superintendent noted that there were 77 intellectually impaired male patients and 84 females in residence.[65] This was both costly for the local authority, and, "owing to their habits [the intellectually impaired] are detrimental to the acute cases".[66] "Occupational training", Good observed, was of "great value in constantly employing many feeble-minded patients who [were] otherwise mischievous and troublesome".[67] This suggests that many of the Littlemore's intellectually impaired patients were given some sort of work or occupation, if only to keep them out of mischief.

Senile dementia was suffered by a significant proportion of new patients during the interwar period (between 15% and 21%). This is in keeping with Good's comments regarding the high proportion of elderly admissions. Patients suffering from senile dementia would have been unlikely to work due to their age but they may have benefited from some form of occupational therapy. 'Confusion' was also a significant category (23%–41%); these patients would probably have been allocated simple taks, such as winding yarn, on the wards. GPI was also an important category amongst the male patients; work would not have been appropriate

[61] Ibid., 1925/6, 10.

[62] Kathleen Jones, *Asylums and After. A Revised History of the Mental Health Services: From the Early 18th Century to the 1990s* (London: Athlone Press, 1992), 123.

[63] Report of the Medical Superintendent, Annual Reports of the Littlemore Hospital, Oxford, 1927/1928, 8–9; Ibid., 1930/1, 9. OHA-L1/A2/5; OHA-L1/A2/8.

[64] Report of the Board of Control, Annual Reports of the Littlemore Hospital, Oxford, 1931/1932, 17. OHA-L1/A2/9.

[65] Report of the Medical Superintendent, Annual Reports of the Littlemore Hospital, Oxford, 1932/1933, 9. OHA-L1/A2/10.

[66] Ibid., 9.

[67] Ibid., 1927/8, 9. OHA-L1/A2/5.

for GPI patients until malaria therapy had been introduced. Good's 1926/1927 report mentions that research was being conducted into malaria therapy as a treatment for both encephalitis and GPI, although he does not refer to the results or whether patients were able to work after treatment.[68] Dr Robert Armstrong reported in 1938/1939 that malaria therapy was being continued with some successful results, but again, he did not mention patients' ability to work following treatment, nor whether they were prescribed occupational therapy.[69]

The numbers of acute cases, or patients exhibiting the first signs of mental disorder, admitted to the Littlemore gradually increased after the passing of the Mental Treatment Act in 1930. Voluntary admissions grew from seven patients in 1931 to 1947 in 1935 and 1991 in 1939. These acute-stage patients would have been given craft activities on the wards after an initial rest period. They were likely to have worked, or to have continued with occupational therapy during convalescence, prior to discharge, like patients at the Maudsley. Good expected all patients who were physically fit enough to be kept busy, whether on the wards or around the hospital. As one nurse put it, "We weren't allowed to let them do nothing!".[70] Between 50% and 100% of these voluntary patients were discharged within the same year of admission. This should have had the effect of increasing the annual rate of recovery or improvement, particularly towards the end of the 1930s when the proportion of voluntary admissions was *c*.30% of all admissions [see Table 8.6]. There was no significant change, however, probably because of the increasing number of incurable and elderly patients who were accumulating in the hospital. This is indicated by the increase in the number of patients who were discharged "not improved" (presumably to an alternative institution, such as a colony) from an average of 14 patients per year before 1931 to an average of 30 patients between 1932 and 1939.

[68] Ibid., 1926/1927, 11. OHA-L1/A2/4.

[69] Ibid., 1938/1939, 15. OHA-L1/A2/16.

[70] J. Goddard, *Mixed Feelings: Littlemore Hospital—An Oral History Project* (Oxford: Oxfordshire County Council, 1996), 31.

Table 8.6 Table to show rates of cure ('recovered') and improvement ('relieved') amongst male (M) and female (F) patients at the Littlemore Hospital, 1923–1939. [Source: Littlemore Hospital Annual Reports, 1923-39. OHA-L1/A2/1-17]

	1923	*1924*	*1925*	*1926*	*1927*	*1928*
Present on 1 January (M: F)	552 (223:329)	602 (276:326)	606 (266:340)	667 (222:445)	707 (254:453)	729 (268:461)
Admissions during year	182 (104:78)	156 (52:104)	245 (62:183)	182 (77:105)	146 (57:89)	142 (54:88)
Recovered (M: F)	73 = 40% (29:44)	70 = 12% (25:45)	44 = 7.3% (23:21)	26 = 14% (8:18)	14 = 2% (7:7)	37 = 26% (15:22)
Improved (M: F)	5 = 1% (2:3)	24 = 4% (11:13)	46 = 7.5% (14:32)	43 = 6.4% (11:31)	48 = 6.8% (14:34)	27 = 3.7% (10:17)

	1929	*1930*	*1931*	*1932*	*1933*	*1934*
Present on 1 January (M: F)	743 (273:470)	735 (282:453)	787 (295:492)	786 (285:501)	805 (299:506)	798 (296:502)
Admissions during year (M: F)	148 (70:78)	194 (66:128)	130 (55:75)	176 (78:98)	217 (88:129)	200 (79:121)
Recovered (M: F)	32 = 22% (17:15)	20 = 2.7% (12:8)	17 = 2.2% (10:7)	22 = 13% (14:8)	30 = 14% (13:17)	30 = 15% (15:15)
Improved (M: F)	52 = 7% (12:40)	47 = 6.4% (11:36)	36 = 4.6% (8:28)	59 = 7.5% (22:37)	68 = 8.4% (26:42)	66 = 8.3% (21:45)

	1935	*1936*	*1937*	*1938*	*1939*
Present on 1 January (M: F)	809 (299:510)	828 (301:527)	865 (321:544)	878 (329:549)	891 (333:558)
Admissions during year (M: F)	203 (82:121)	232 (110:122)	225 (88:137)	213 (83:130)	230 (89:141)
Recovered (M: F)	29 = 14% (13:16)	33 = 14% (21:12)	63 = 28% (21:42)	61 = 29% (22:39)	63 = 27% (32:31)
Improved (M: F)	60 = 7.4% (30:30)	69 = 8.3% (25:44)	46 = 5.3% (23:23)	43 = 5% (15:28)	51 = 5.7% (13:38)

Source: Littlemore Hospital Annual Reports, 1923–39. OHA-L1/A2/1-17

The Physical Condition of Patients

Patients in a weak physical condition would have been exempt from work. The poor physical health of patients admitted to the Asile Clinique, many of whom suffered from tuberculosis and other respiratory conditions, may help to explain the relatively small number of patient workers prior to the hospital's transformation to an acute service. As Dr. Marchand of the Men's First Section highlighted, many patients arrived at the Asile Clinique in a "very grave condition".[71] A healthier patient population at the Maudsley and Bethlem hospitals, in comparison to that of the Asile Clinique, is indicated by the relative patient death rates. These were much higher at the Asile Clinique and a high percentage of the deaths occurred within the first month of arrival. One explanation may lie in the class of patients. Many of the patients admitted to the Maudsley, and particularly to Bethlem, were middle-class or from amongst the employed working classes. Those admitted to the Asile Clinique were destitute; many had been attracted to Paris, "like nocturnal insects to a lamp", seeking a better life than that which they had endured elsewhere, but had been unable to find work. Destitute in an unfamiliar and unforgiving environment, many succumbed to alcoholism and ended up in one of the Seine asylums.[72] Poverty, as Parisian psychiatrist Édouard Toulouse emphasised, was one of the social scourges responsible for mental disorder.[73] In 1900, over half of those who died in Paris were buried in pauper graves, even though the city generated around a quarter of the nation's wealth.[74] After World War I, migrant labourers continued to be attracted to the impoverished working-class suburbs of Paris, where the population grew from *c.*1.5 million inhabitants in 1920 to two million in the late 1930s.[75] Accommodation in the suburbs was dirty, damp and cramped and those who found work were forced to spend most of their meagre wages on rent.[76]

Poverty was also a significant factor in the rural regions of Oxfordshire and La Sarthe, where wages were particularly low and work was often seasonal. Many Littlemore patients were admitted in a poor physical state, leading Good to comment that "so many of the admissions now are found to be suffering from severe physical illness and quite incapable of being employed in any hard or continuous occupation".[77] Oxford was not

[71] Dr. Marchand, Rapport de la 1re section des hommes, Asile Clinique (Paris: Préfecture de la Seine, 1936), 82. ADP/PER-566-6.

[72] Dr. Marie, Rapport sur la service des aliénés du département de la Seine pendant l'année (Paris: Préfecture de la Seine, 1908); 244. ADP/PER566.

[73] Huteau, *Psychologie, Psychiatrie et Société*, 8.

[74] Colin Jones, *Paris: Biography of City* (London: Penguin, 2004), 358.

[75] Ibid., 394.

[76] Ibid., 344.

[77] Report of the Medical Superintendent, Annual Reports of the Littlemore Hospital, Oxford, 1926/1927, 12.

severely affected by the General Strike of 1926, but the latter caused prices to rise, resulting in hardship for many.[78] The incidence of tuberculosis at the Littlemore was higher than the average for English mental hospitals. This often fatal disease was "overwhelmingly a scourge of the labouring poor".[79] In 1929, the Board of Control inspectors noted that the average number of notifications of new cases of tuberculosis for all mental hospitals was 8.5 per thousand, while for the Littlemore it was 27.4 per thousand.[80] Deaths from the disease were 12.3 per thousand at the Littlemore, compared with the average of 6.9 for all mental hospitals in England and Wales.[81] Pneumonia, heart and kidney disease and "organic brain disease" were other major causes of death among patients. Littlemore patients were also subject to epidemics of encephalitis (1924–1925) and German measles (1934) and regular outbreaks of influenza, such as during the minor epidemic of 1927.[82] All these conditions would have weakened patients and compromised their ability to perform work around the hospital, but they may have been given occupational therapy, an opportunity denied patients of the Asile de la Sarthe.

As indicated in the medical report of 1927, many patients at the Asile de la Sarthe suffered from physical conditions unrelated to their mental disorders. These conditions varied but included migraine; rheumatism; digestive problems such as diarrhoea and vomiting; respiratory conditions, such as tuberculosis, pneumonia, bronchitis, laryngitis and influenza; heart problems; and lumbago.[83] Tuberculosis, for example, was responsible for between 15% and 23% of female deaths between 1927 and 1931. The most common physical ailments, according to the chief medical officer, were "always" those involving the digestive tract.[84] In 1927, 214 men (or 74% of the male patient population) and 327 women (70% of the female patient population) suffered from one or more of these complaints. 282 (36% of the total patient population) patients required some sort of

[78] Phyl Surman, *An Oxford Childhood: Pride of the Morning* (Stroud: The History Press, 1992), 39–40.

[79] F. B. Smith, *The Retreat of Tuberculosis, 1850–1950* (London: Croom Helm, 1988), 20.

[80] Report of the Board of Control, Annual Reports of the Littlemore Hospital, Oxford, 1930/1931, 15.

[81] Ibid.

[82] Report of the Medical Superintendent, Annual Reports of the Littlemore Hospital, Oxford, 1924/1925, 8; Ibid., 1934/1935, 9.

[83] Rapport du Médecin en Chef, Asile de la Sarthe, 1927 (Le Mans, 1928), 25.

[84] Ibid., 27.

surgery, such as a setting a fractured bone, tooth extraction or treating an abscess.[85] These conditions indicate the poor general health of patients which would have prevented many from working, either inside the asylum or outside it. In some cases, physical and mental conditions were linked, such as in the case of circulatory problems leading to the onset of senile dementia, as highlighted by the chief medical officer in 1937.[86] Circulatory problems were particularly prevalent amongst female patients, accounting for 19% of female deaths in 1931 and 27% in 1937. Again, these conditions would have compromised a patient's ability to work.

Patient Age

Age, as well as mental and physical health, also influenced whether patients were ascribed work. Age was particularly relevant in the provincial asylums where a significant proportion of incurable patients remained for long periods, often for life, and whose ongoing presence limited the admission of younger patients. The numbers of young men aged under 35 years (and therefore of working age) being admitted to the Littlemore fell from 18 in 1923 to 12 in 1925.[87] Dr. Good attributed the decrease in admissions of the under-35 s in part to the successful early treatment of younger patients at the outpatient clinic.[88] The numbers of young people being admitted to the Littlemore continued to fall, leading Good to observe in 1935 that younger people tended to present themselves to the outpatient clinic more readily than the middle-aged.[89] The treatment received in the outpatients clinic helped to prevent patients' needing to be admitted to hospital. This was a positive for both the local authority budget and the individual, but it also meant that the average age of inpatients increased, and the numbers of capable workers decreased.

In 1924, the average age on admission was 49 for men and 47 for women, but already resident in the hospital at that time were 59 males and 112 females over the age of 60 years.[90] In 1926, 10 male and 26 female admissions were over 60 years; the average age of the male residents was

[85] Ibid., 29.

[86] Rapport du Médecin en Chef, Asile de la Sarthe, 1937 (Le Mans, 1938), 12.

[87] Report of the Medical Superintendent, Annual Reports of the Littlemore Hospital, Oxford, 1925/1926, 9.

[88] Ibid.

[89] Ibid., 1935/1936, 9. OHA-L1/A2/13.

[90] Ibid., 1924/1925, 9. OHA-L1/A2/2.

68 and that of the females 75 years.[91] In 1927, 40% of patient deaths were attributed to "senile decay".[92] The elderly, in Dr. Good's view, should be looked after in Public Assistance Institutions or Infirmaries, and the intellectually impaired should be sent to specialist institutions, where they could receive appropriate care.[93] The Board of Control noted in 1933 that 163 (or 28.6% of the total hospital population chargeable to the Oxford City and Oxford County local authorities) of the Littlemore's patients were over 60.[94] By 1936, 191 patients were over 65 years, most of whom were suffering from "gross organic disease".[95] The same year, eight patients aged over 75 years were admitted, few of whom required "the skilled nursing and medical care of an up-to-date mental hospital".[96] During 1937, another 17 men and 26 women over 65 were admitted, comprising 23% of admissions.[97] While these patients were unlikely to be able to perform useful work, they may have benefited from occupational therapy provided by the nurses on the wards.

A high proportion of elderly, particularly female, former workhouse inmates could be explained in part by the government's policy of encouraging the employment of younger workers in British enterprises, exposing many older people to unemployment and poverty.[98] Employers' reluctance to employ women over 55 on account of their alleged poor health led to older women being particularly vulnerable to pauperism. Many women gave up work early to care for elderly relatives and afterwards found themselves unable to re-join the labour market.[99] For these destitute older people, the only option during the interwar period was the workhouse until 1929, and thereafter Public Assistance Institutions.[100] If the latter were overcrowded, or if the older inmates were mentally disordered, they might

[91] Ibid., 1926/1927, 8. OHA-L1/A2/4.

[92] Ibid., 1927/1928, 9. OHA-L1/A2/5.

[93] Ibid., 1924/1925, 9. OHA-L1/A2/2.

[94] Report of the Board of Control, Annual Reports of the Littlemore Hospital, Oxford, 1933/1934, 16. OA-L1/A2/11.

[95] Report of the Medical Superintendent, Annual Reports of the Littlemore Hospital, Oxford, 1932/1933, 9; and Ibid., 1936/1937, 9. OHA-L1/A2/10; OHA-L1/A2/14.

[96] Ibid., 9.

[97] Ibid., 1937/1938, 11. OHA-L1/A2/15.

[98] Pat Thane, *Old Age in English History: Past Experiences, Present Issues* (Oxford and New York: Oxford University Press, 2000), 283.

[99] Ibid., 284–5.

[100] Ibid., 173.

be transferred to the local mental hospital. In the 1930s, the old were the largest group of inmates of Public Assistance Institutions.[101]

The profile of patients at the Asile de la Sarthe was also gradually ageing. Between 1921 and 1931, the proportion of male patients aged 35–44 years fell from 26% to 20%, while the proportion of males aged 55–65 years increased from 9% in 1921 to 17% in 1931. The proportion of women aged 55–64 increased from 20% in 1921 to 25% in 1931 and that of those aged 25–34 fell from 9% in 1921 to 3% in 1931. In 1937, 30% of male admissions, and 24% of female admissions, were over 65 years. Patients over 65 were unlikely to be expected to work, so this increasing age profile provides a plausible explanation for the decrease in the numbers of patient workers at the Asile de la Sarthe indicated in Table 8.4, above. The patient workforce was gradually contracting as older patients became too frail to work. The lack of movement in the patient population (that is, very few discharges as a result of cure or improvement) meant that new, potentially younger, patients were not being admitted in sufficient numbers to take the place of "retired" patients. The experience at the Asile de la Sarthe was therefore similar to that of the Littlemore in terms of the ageing patient profile and the diminishing numbers of patient workers. But at the Littlemore, less importance was attached to the financial contribution made by patient work. Although produce from the hospital farm continued to be evaluated, the precise numbers of working patients were no longer recorded and the priority in terms of patient occupation was on the provision of therapy.

The Hospital Environment and Facilities

Contributing to the overall experience of patients in institutions, as Jane Hamlett has demonstrated, were the hospital environment and facilities.[102] Since the early days of moral treatment, the asylum environment itself had been considered curative. Surroundings aimed to be cheerful and comforting and the décor to reflect that of a middle-class family home.[103] An impressive building surrounded by ample grounds and views of the

[101] M. A. Crowther, *The Workhouse System, 1834–1929: The History of an English Institution* (London: Routledge, 1983), 111.

[102] Jane Hamlett, *At Home In The Institution: Material Life in Asylums, Lodging Houses and Schools in Victorian and Edwardian England* (Basingstoke and New York: Palgrave Macmillan, 2015).

[103] Ibid., 19.

countryside were considered crucial to the healing process.[104] Opportunities for patients to spend time in the fresh air were emphasised in the late nineteenth and early twentieth centuries when open-air therapies were in vogue for a variety of conditions from mental illness to tuberculosis.[105] But by the 1920s, overcrowding, the necessity of building on some of the institutional land, and the neglect of essential building maintenance work during World War I had taken their toll on the older hospitals. Édouard Toulouse had campaigned for improvements to patient living conditions since the early 1900s and lamented the dilapidated state of Parisian asylums in 1918.[106]

Overcrowding was an ongoing problem at the Asile de la Sarthe in Le Mans. Patient numbers had increased from 604 in 1919 to 899 in 1939, compromising the quality of life enjoyed by patients in terms of physical space, the enforced proximity of other patients, whose behaviour could be threatening or antisocial, and less attention from staff. The existence of an outpatients' clinic, which meant patients could receive treatment without admission to hospital, explained the lack of overcrowding at the Littlemore, according to its medical superintendent, but there was no such facility at the Asile de la Sarthe.[107] The grounds at the Littlemore were extensive compared with the relatively cramped outdoor space at the Asile de la Sarthe, which was surrounded by Le Mans town, the river and railway line [see Fig. 2.3]. Bethlem's move in 1930, from urban South London to purpose-built premises in the relatively rural suburb of Monks Orchard, Kent, gave patients far more indoor and outdoor space, including landscaped gardens and an orchard where patients could enjoy occupations outside. The Maudsley was also purpose-built, with verandas enabling patients to be in the fresh air whilst confined to bed. When it was built in 1867, the Asile Clinique had been designed for 300 male and 300 female patients, but twenty years after its opening, the patient population had reached 900.[108] The interwar years

[104] Leonard D. Smith, *'Cure, Comfort and Safe Custody': Public Lunatic Asylums in Early Nineteenth-Century England* (London and New York: Leicester University Press, 1999).

[105] Clare Hickman, "Care in the Countryside: the theory and practice of therapeutic landscapes in the early twentieth century," in *Gardens and Green Spaces in the West Midlands since 1700,* eds Malcolm Dick and Elaine Mitchell (Hatfield: West Midlands Publications, 2018), 165–66.

[106] Louis Dausset, Rapport Général au nom de la 3e Commission sur les propositions budgétaires pour le Service des Aliénés (budget de 1919) (Paris: Conseil Général de la Seine, 1918), 26. ADP/D.10 K3/27/20.

[107] Thomas Saxty Good, "History and Progress of Littlemore Hospital," *Journal of Mental Science* October (1930): 602–21.

[108] Benoît Majerus, *Du Moyen Âge à Nos Jours, Expériences et Représentations de* la *Folie à Paris* (Paris: Parigramme, 2018), 59.

saw the population stabilise at around 1000 patients. The impressive buildings and grounds at Ste. Anne's (in which the Asile Clinique and the Henri Rousselle Hospital were situated) were designed by the architect, Charles Auguste Questel according to Baron Haussman's brief. The covered walkways, manicured lawns and elegant statues could not, however, compensate for overcrowded wards and malfunctioning baths.

At the Asile de la Sarthe, central heating and electric lighting were only installed in 1922 and 1923 respectively. Until 1924, when washbasins were provided in all sections, patients had to wash at the pump in the mornings.[109] Work to replace the existing "slopping out" buckets for the disposal of human waste with flushing water-closets began in 1929 and was completed in 1932.[110] The new chief medical officer appointed in 1933, Dr. Schutzenberger, was very critical of facilities at the Asile de la Sarthe. He was particularly appalled by the level of overcrowding, which denied patients the regulated amounts of space or air cubage, and by the lack of baths.[111] This was not only problematic from the perspective of personal hygiene (an average of only 10 baths were taken per year per patient) but meant that agitated patients were missing out on hydrotherapy, which was considered a valuable means of soothing them.[112] The financial crisis of 1931 limited expenditure on building maintenance and repairs, and new works had had to be postponed.[113] The asylum director remarked that "more than one project has had to be cancelled in these difficult circumstances".[114] Plans to extend the central heating system and modernise the kitchens (reportedly in a "dangerous" condition) were shelved. Nonetheless, the asylum director professed to be committed to making the hospital a more agreeable and happier place for the "*pauvres déshérités*" in their care. By 1937, patients were able to have weekly baths or showers, and the interiors had been repainted.[115]

[109] Rapport du Médecin en Chef, Asile de la Sarthe, 1924 (Le Mans, 1925), 35–6.

[110] Rapport du Médecin en Chef, Asile de la Sarthe, 1929, (Le Mans, 1930), 36; Rapport du Directeur, Asile de la Sarthe, 1932 (Le Mans, 1933), 437.

[111] Rapport du Médecin en Chef, Asile de la Sarthe, 1933 [handwritten, no page nos.].

[112] Ibid. Hermann Simon, the originator of MAT, used work to calm agitated patients, obviating the need for hydrotherapy equipment, but his methods were not recognised in provincial French asylums where work was the preserve of the tranquil, incurable and chronic patients and the convalescents.

[113] Rapport du Directeur, Asile de la Sarthe, 1932 (Le Mans, 1933), 435.

[114] Rapport du Directeur, Asile de la Sarthe, 1937 (Le Mans, 1938), 8.

[115] Ibid.; Rapport du Directeur, Asile de la Sarthe, 1938 (Le Mans, 1939), 10.

Standards of living for patients appeared to be higher at the Littlemore Hospital. The patients' quality of life, indicated by measures such as the decoration of the wards and communal spaces, a more varied diet and personalised clothing, gradually improved during the interwar period. Being allowed to stay up until 10 pm was an unusual (at most English mental hospitals patients retired at around 7.30 pm) and welcome privilege for patients.[116] The Board of Control noted that since women inspectors had been permitted to conduct "statutory visits" to mental hospitals, more attention was paid to the clothing of female patients. "Greater variety in dress helps to lessen the monotony which is the bane of institutional life," they maintained, "and is a step towards recovery when a patient can be induced to take some interest in her appearance".[117] Following pressure from the Board, "suitable" female patients at the Littlemore were able to choose the colour of the dresses made for them and to have "clothing marked for their own personal use".[118] The women's underclothing had shocked one nurse who joined the staff in 1936; she described the "calico drawstring drawers" as very old fashioned.[119] The underwear was modernised in 1937, and patients' boots began to be fitted individually.[120] Dentures "for the better types of patient" and glasses for those with defective vision were provided from 1937.[121] A fish fryer was installed in the kitchen in 1938 so that fish and chips could be served once a week; a measure that was "much appreciated" by both staff and patients.[122]

All these factors—the décor of patients' quarters, the facilities for personal hygiene, the quality of shoes and clothing, the numbers of patients occupying the same space, the provision of something as simple as a fish fryer—contributed to the patient's experience of asylum life. They gave context to the time spent working, engaging with occupational therapy or receiving some other form of therapy, taking exercise, enjoying a

[116] Report of the Board of Control, Annual Reports of the Littlemore Hospital, Oxford, 1923/1924, 14. OHA-L1/A2/1.

[117] Nineteenth Annual Report of the Board of Control, 1932, 8.

[118] Report of the Board of Control, Annual Reports of the Littlemore Hospital, Oxford 1936/1937, 20. OHA-L1/A2/14.

[119] J. Goddard, Oral History Interview with Former Littlemore Nurse, Littlemore History Project, 1994/1995. OHA/LH/OT233.

[120] Ibid.

[121] Report of the Medical Superintendent, Annual Reports of the Littlemore Hospital, Oxford, 1937/1938, 9. OHA-L1/A2/15.

[122] Ibid., 1938/1939, 9. OHA-L1/A2/16.

recreational activity, or being left with nothing to do. The ability to buy small luxuries with money earned from work (in French institutions), or to enjoy a few hard-earned privileges, also contributed to the patient experience. This study has not been able to ascertain what patients thought about the occupations they were given, but the potential of some form of reward may have encouraged the unwilling, or provided a boost to a patient worker's self-esteem. At the Asile de la Sarthe, an ability to work was one of the criteria for discharge, so this too may have encouraged patients to work. One patient, Jean S. wrote to the asylum director in 1919 asking for permission to work in the gardens as he believed that this might ultimately lead to his discharge [see Fig. 8.1].

Incentives to Work

In France, work around the hospital was incentivised with a nominal wage or "pécule" paid according to each day or half day worked. The legislation of 1839/57 stated that it was "the right" of pauper (but not private) patients to receive remuneration for their work.[123] In England, the Cobb Report of 1922 had recommended the monetary payment of English patients who worked, but this had not materialised.[124] Instead, patients were given an extra cup of tea with bread and butter or cheese, a custom that began in the nineteenth century. These additional workers' rations were itemised in the "dietary" section of Bethlem and the Littlemore's annual reports. In Victorian times, English patient workers had been rewarded with beer, but this practice had ended in the 1880s following growing awareness and concern over the harm to health and morals caused by alcohol.[125] Payment for work around the hospital (albeit just for food and drink in England), but not for occupational therapy, highlights the fact that this work was valuable to the hospital and may not have been perceived as beneficial to the patient by the patients themselves. Douglas Bennet's distinction between the two forms of activity is useful here. He maintains that with work, "one is generally doing something for other people" while in occupational therapy "one is doing something for

[123] Article 151, Bulletin Officiel du Ministère de l'Intérieur. Circulaire No. 7, 20/03/1857, 57.

[124] Cyril Cobb, Ministry of Health, Report of the Committee on Administration of Public Mental Hospitals (London: HMSO, 1922), 52.

[125] See: Niall McCrae, "The Beer Ration in Victorian Asylums," *History of Psychiatry* 15 no. 2 (2004): 155–175.

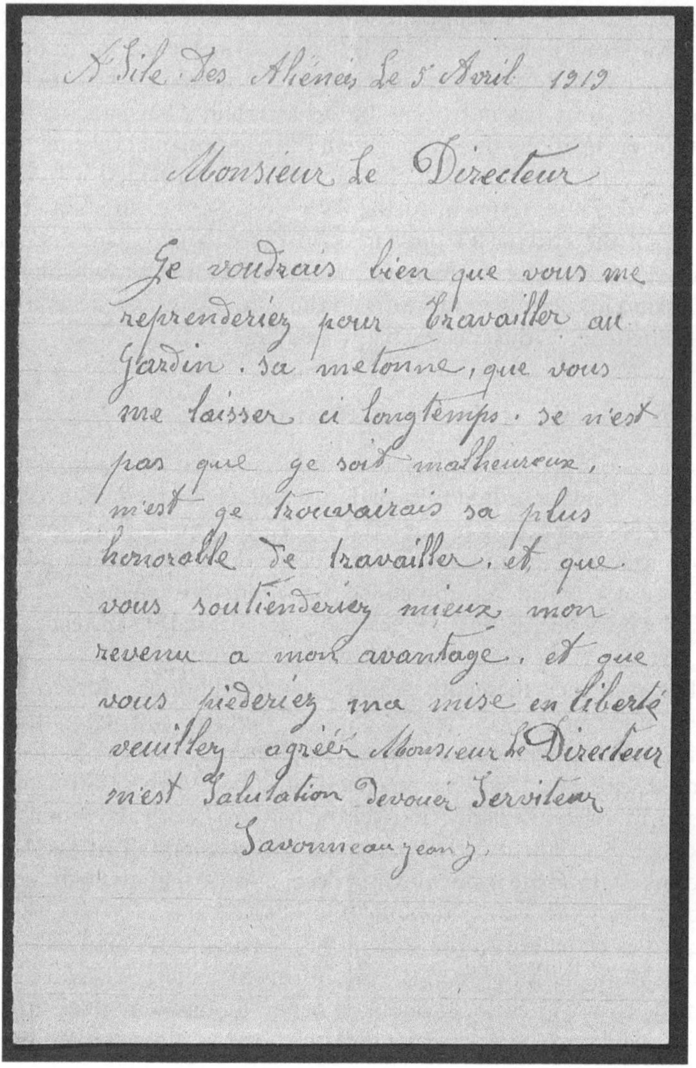

Asile Des Aliénés Le 5 Avril 1919

Monsieur Le Directeur

Ge voudrais bien que vous me
reprenderiez pour travailler au
Gardin. sa metonne, que vous
me laisser ci longtemps. se n'est
pas que ge soit malheureux,
m'est ge trouvairais sa plus
honorable De travailler. et que
vous soutienderiez mieux mon
revenu a mon avantage. et que
vous piederiez ma mise en liberté
veuillez agréer Monsieur Le Directeur
m'est Salutation Devouer Serviteur

Savonneau jean J.

Fig. 8.1 A letter dated 5 April 1919 from a patient to the asylum director of the Asile de la Sarthe requesting to work in the gardens, in the knowledge that work was a crucial step in his being allowed to leave the asylum. (© *Arch. dép. Sarthe, 1 × 633*)

oneself".[126] The aim of the arts and crafts activities that comprised occupational therapy (AOT) was to inspire creativity and to generate feelings of satisfaction and pride in the items created. These positive, self-enhancing emotions were considered curative. Patients were not paid for undergoing other forms of therapy such as shock treatments or hydrotherapy, so it seems logical that patients would not be paid for the arts and crafts aspects occupational therapy.

On the matter of patient choice, the Board of Control maintained in 1928 that employment had been restricted to "those patients whose readiness to work was spontaneous or needed only the urge of some small reward".[127] The Board regarded occupational therapy as a means of occupying the many patients who had been considered "unemployable" provided that the tasks were appealing to the patient, suggesting a degree of choice.[128] A later remark that, "An idle patient ought to be regarded as a reproach to the hospital", indicates an intolerance of idleness.[129] At the Ranchi asylum in British colonial India during the 1930s, patients were allowed to choose their occupation, as superintendent Dhunjibhoy was keen to point out, but they were not allowed to remain idle. Those who refused to work in the gardens or undertake domestic duties were given instruction in arts and crafts.[130] Nurses at the Littlemore hospital reported being very anxious to ensure that all patients were engaged in a task when the superintendent did his rounds; they got into trouble if patients were idle.[131] Edward Mapother also expected his patients at the Maudsley to do some form of work as soon as they were capable. He maintained that "every effort has to be made to prevent loafing among patients capable of occupation" to prevent "the boredom and deterioration" that would result from remaining unoccupied.[132]

[126] Douglas Bennett, "Work and Occupation for the Mentally Ill," in *150 Years of British Psychiatry, Vol II: The Aftermath*, eds Hugh Freeman and German E. Berrios (London and Atlantic Highlands, NJ: 1996), 193.

[127] Fifteenth Annual Report of the Board of Control (London: HMSO, 1928), 4.

[128] Ibid., 5.

[129] Nineteenth Annual Report of the Board of Control (London: HMSO, 1932), 5.

[130] Waltraud Ernst, "'Useful both to the patients as well as to the State': Patient work in colonial mental hospitals in South Asia, *c*.1818–1948," in *Work, Psychiatry and Society c.1750–2015*, ed. Waltraud Ernst (Manchester: Manchester University Press, 2016), 123.

[131] J. Goddard, Oral History Interview with Former Littlemore Nurse, Littlemore History Project, 1994/5. OHA/L1/OT232.

[132] Report of the Medical Superintendent, Maudsley Hospital, London, 1925/1926, 5.

A refusal to work to work at the Zwiefalten asylum in southern Germany incurred a loss of privileges and rewards during the nineteenth century, leading Thomas Mueller to question the allegedly 'voluntary' nature of patient work.[133] At the Hamburg-Langenhorn mental hospital during the Weimar period, the amount of food patients received depended on the work they did. Patients working in agriculture, the laundry and gardens, and as craftsmen, seamstresses or tailors were given more food than those engaged in mending clothes, housework or vegetable peeling.[134] This discrepancy was justified on the grounds of the energy requirements of patients engaged in certain activities, but the food allocation was also related to the perceived value of the patient's work to the institution.[135] An ability to work became a matter of life and death for mentally ill or disabled patients in Germany under the national-socialist "T4 Euthanasia Programme". Those who were unable to work were killed, while those whose work was perceived as valuable had a higher chance of survival.[136] These examples show that the boundaries between voluntary participation, incentivisation and coercion could be very fluid indeed.

Conclusion

This chapter has shown that although the treatment preferences of psychiatrists and the way certain institutions were managed held considerable sway over the nature of patient occupation, patient class, gender, physical and mental health, and age were also influential. The fact that it was considered inappropriate for the middle classes, and particularly middle-class women, to engage in any form of manual work, meant that middle-class patients had fewer options for keeping busy in France, where occupational therapy was not available. The arts and crafts activities that comprised the American style of occupational therapy (AOT) prescribed in English mental hospitals were well suited to the middle-class patient, particularly amongst the women, whose upbringing encouraged them to develop proficiency in arts and crafts as hobbies. When occupational therapy was

[133] Thomas Mueller, "Between therapeutic instrument and exploitation of labour force: Patient work in rural asylums in Wurttemberg, *c.*1810–1945," in *Work, Psychiatry and Society*, ed. Ernst, 225.

[134] Monika Ankele, "The Patient's View of Work Therapy: The Mental Hospital Hamburg-Langenhorn during the Weimar Republic," in *Work, Psychiatry and Society*, ed. Ernst, 247.

[135] Ibid.

[136] Ibid.

eventually introduced at Bethlem, it proved extremely popular with the middle-class patients. That said, at the Maudsley, where a quarter of patients paid fees, there did not seem to be any discrepancy in the allocation of occupation between the classes. The Maudsley was run as a modern hospital, where issues of class were less relevant than at the very traditional Bethlem, where nineteenth-century values and the practices of a traditional asylum persisted. The rigid segregation of the sexes, characteristic of traditional asylums, was also less evident at the Maudsley, where there was even a female psychiatrist.[137]

Whether a patient's mental condition was at the acute stage, and perceived curable, or incurable was fundamental. Acute-stage patients were not given any form of work or occupation in French institutions (apart from the Henri Rousselle Hospital) as this continued to be perceived as inappropriate for all but the calm, chronic and incurable cases and convalescent patients. Most French acute-stage patients were treated with "bed-rest" and sedated. This might be followed by "aggressive biological treatment" prescribed by the more progressive psychiatrists, such as Henry Christy at the Asile de la Sarthe and the doctors at the Asile Clinique. Sedation of acute-stage patients was also the policy at Bethlem, which appeared slow to embrace new ideas. At the Maudsley, Henri Rousselle and Littlemore Hospitals, occupation was considered beneficial for acute-stage patients, after a short rest period. The precise nature of a patient's condition (such as mania, melancholia, delirium or schizophrenia) was less influential than the stage of their illness, such as acute or chronic, and whether it was perceived as curable or incurable.

Poor physical health, that might prevent patients from working, was associated with the extremes of poverty found in Paris and the rural regions of Oxfordshire and La Sarthe. At the Asile Clinique, GPI (the result of tertiary syphilis) and alcoholism were common and debilitating problems, while tuberculosis and gastric diseases were characteristic of all three institutions. At the Maudsley and Bethlem Hospitals, where there was a higher proportion of middle-class and employed working-class individuals, patients were physically healthier and more likely to be able to undertake some form of occupation. The provincial institutions had a high, and increasing, proportion of elderly patients due to the long-term residency of incurable patients that limited the admission of younger, working-age

[137] Dr. Mary Barkas was one of four senior psychiatrists (and the only trained psycho-analyst) at the Maudsley between 1923 and 1928.

patients. This resulted in the gradual contraction of the patient workforce of the Asile de la Sarthe and the Littlemore. At these provincial establishments, preparing patients for work outside the asylum was arguably less important than at the institutions for acute patients, because so few were likely to be discharged.[138] Whether curable patients gained useful work experience whilst in hospital is considered in the next chapter.

Bibliography

Secondary Sources

Accampo, Elinor, Rachel Fuchs, and Mary Lynn Stewart. *Gender and the Politics of Social Reform in France, 1870–1914.* Baltimore and London: Johns Hopkins University Press, 1995.

Andrews, Jonathan, Asa Briggs, Roy Porter, Penny Tucker, and Keir Waddington. *The History of Bethlem.* Abingdon and New York: Routledge, 1997.

Ankele, Monika. "The Patient's View of Work Therapy: The Mental Hospital Hamburg-Langenhorn during the Weimar Republic." In *Work, Psychiatry and Society, c.1750–2015*, edited by Waltraud Ernst, 238-61. Manchester: Manchester University Press, 2016.

Bennett, Douglas. "Work and Occupation for the Mentally Ill." *150 Years of British Psychiatry, Volume II: The Aftermath*, edited by Hugh Freeman and German E. Berrios, 193–208. London and Atlantic Highlands, NJ: 1996.

Crowther, M. A. *The Workhouse System, 1834–1929: The History of an English Institution.* London: Routledge, 1983.

Ernst, Waltraud. "'Useful both to the patients as well as to the State': Patient work in colonial mental hospitals in South Asia, *c.*1818–1948." In *Work, Psychiatry and Society, c.1750–2015*, edited by Waltraud Ernst, 117-41. Manchester: Manchester University Press, 2016.

Gazeley, Ian. "Manual Work and Pay, 1900–70." In *Work and Pay in 20th Century Britain*, edited by Nicholas Crafts, Ian Gazeley, and Andrew Newell, 55–79. Oxford: Oxford University Press, 2007.

Goddard, J. *Mixed Feelings: Littlemore Hospital—An Oral History Project.* Oxford: Oxfordshire County Council, 1996.

[138] This was especially true of the Asile de la Sarthe, where voluntary admission was not an option during this period. At the Littlemore, there was more movement of the patient population following the Mental Treatment Act of 1930.

Hamlett, Jane. *At Home in the Institution: Material Life in Asylums, Lodging Houses and Schools in Victorian and Edwardian England*. Basingstoke and New York: Palgrave Macmillan, 2013.

Hickman, Clare. "Care in the Countryside: the theory and practice of therapeutic landscapes in the early twentieth century." In *Gardens and Green Spaces in the West Midlands since 1700,* edited by Malcolm Dick and Elaine Mitchell, 160–185. Hatfield: West Midlands Publications, 2018.

Huteau, Michel. *Psychologie, Psychiatrie et Société Sous La Troisième République, La biocratie d'Édouard Toulouse (1865–1947)*. Paris: L'Harmattan, 2002.

Jones, Colin. *Paris: Biography of a City*. London: Penguin, 2004.

Jones, Kathleen. *Asylums and After. A Revised History of the Mental Health Services: From the Early 18th Century to the 1990s*. London: Athlone Press, 1992.

Majerus, Benoît. *Du Moyen Âge à Nos Jours, Expériences et Représentations de* la *Folie à Paris*. Paris: Parigramme, 2018.

McCrae, Niall. "The Beer Ration in Victorian Asylums." *History of Psychiatry* 15 no. 2 (2004): 155–175.

Mueller, Thomas. "Between therapeutic instrument and exploitation of labour force: Patient work in rural asylums in Wurttemberg, *c.*1810–1945." In *Work, Psychiatry and Society, c.1750–2015*, edited by Waltraud Ernst, 220–37. Manchester: Manchester University Press, 2016.

Pederson, Susan. *Family, Dependence, and the Origins of the Welfare State: Britain and France 1914–1945*. Cambridge: Cambridge University Press, 1993.

Smith, F. B. *The Retreat of Tuberculosis, 1850–1950*. London: Croom Helm, 1988.

Smith, Leonard D. *'Cure, Comfort and Safe Custody': Public Lunatic Asylums in Early Nineteenth Century England*. London and New York: Leicester University Press, 1999.

Surman, Phyl. *An Oxford Childhood: Pride of the Morning*. Stroud: The History Press, 1992.

Thane, Pat. *Old Age in English History: Past Experiences, Present Issues*. Oxford and New York: Oxford University Press, 2000.

Warner, Richard. *Recovery from Schizophrenia: Psychiatry and Political Economy*. Third edition. Philadelphia: Routledge and Kegan Paul, 2004

Archival Sources

Bulletin Officiel du Ministère de l'Intérieur. Circulaire No. 7, 20/03/1857, 42–79. Paris, 1857. Accessed 12/02/2018. https://gallica.bnf.fr/ark:/12148/bpt6k5539701n/f2.item.zoom.

Capgras, Dr. Rapport de la 1re section des hommes, Asile Clinique. Paris: Le Préfecture de la Seine, 1929. ADP/PER-566-5.

Cobb, Cyril. Ministry of Health. Report of the Committee on Administration of Public Mental Hospitals. London: HMSO, 1922. ["The Cobb Report"].

Dausset, Louis. Rapport Général au nom de la 3e Commission sur les propositions budgétaires pour le Service des Aliénés (budget de 1919). Paris: Conseil Général de la Seine, 1918. ADP/D.10K3/27/20.

Fifteenth Annual Report of the Board of Control. London: HMSO, 1928.

Goddard, J. Oral History Interview with Former Littlemore Nurse. Littlemore History Project, 1994/1995. OHA/L1/OT232.

Good, Thomas Saxty. "History and Progress of Littlemore Hospital." *Journal of Mental Science* October (1930): 602–21.

Inspection Report, Eighteenth Report of the Board of Control. London: HMSO, 1931.

Inspection Report, Seventeenth Report of the Board of Control. London: HMSO, 1930.

Inspection Report, Thirteenth Report of the Board of Control. London: HMSO, 1926.

Lautier, Dr. "Les Exagérations de la Thérapie par le Travail." *L'Hygiène Mentale* no. 2 (1929): 191–2.

Leroy, Dr. Rapport de la 1re Section des femmes, Asile Clinique. Paris: Préfecture de la Seine, 1931. ADP/PER-566-6.

Macdonald, Mary and Norah Haworth. *Theory of Occupational Therapy for Students and Nurses.* London: Bailliere, Tindall and Cox, 1940.

Marchand, L. Rapport de la 1re Section (Hommes), Asile Clinique. Paris: Le Préfecture de la Seine, 1931, 1933 and 1936. ADP/PER-566-6.

Marie, A. Rapport sur la service des aliénés du département de la Seine pendant l'année. Paris: Préfecture de la Seine, 1908. ADP/PER566.

Navarre, M. Rapport au nom de la 3e Commission sur les propositions budgétaires pour le service des aliénés. Paris: Conseil Général de la Seine, 1908. ADP/D.10K3/18/35.

Nineteenth Annual Report of the Board of Control. London: HMSO, 1932.

Rapport du Directeur. Asile de la Sarthe, Le Mans, 1932; 1937 and 1938. ADS-1X964/5.

Rapport du Médecin en Chef. Asile de la Sarthe, Le Mans, 1924, 1927, 1929, 1933, 1935, 1937. ADS-1X964/5.

Report of the Board of Control. Annual Reports of the Bethlem Royal Hospital, London, 1920–1928; 1938. BMM/BAR-53.

Report of the Board of Control. Annual Reports of the Littlemore Hospital. Oxford, 1923/1924; 1931/1932; 1933/1934; 1936/1937. OHA-L1/A2/10; OHA-L1/A2/11; OHA-L1/A2/13.

Report of the Medical Superintendent. Annual Reports of the Littlemore Hospital. Oxford, 1924a/1925; 1925/1926; 1926/1927; 1927/1928; 1930/1931; 1932/1933; 1933/1934; 1936/1937; 1937/1938; 1938/1939. OHA-L1/A2/2; OHA-L1/A2/3; OHA-L1/A2/4; OHA-L1/A2/5; OHA-L1/A2/8; OHA-L1/A2/10; OHA-L1/A2/11; OHA-L1/A2/14; OHA-L1/A2/5; OHA-L1/A2/16.

Report of the Medical Superintendent. Maudsley Hospital, London, 1924b; 1925; 1926. BMM/MSR-01.

Report of the Physician Superintendent. Annual Reports of the Bethlem Royal Hospital, London, 1911; 1921; 1924; 1926; 1927; 1934; 1938. BMM/BAR-53.

Rousselle, Henri. Rapport au nom de la 3e Commission sur les budgets et comptes de l'Asile Clinique. Paris: Conseil Général de la Seine, 1901, 1903, 1911. ADP/D.10K3/14/52; ADP/D.10K3/15/38; ADP/D.10K3/21/42.

Twenty-first Annual Report of the Board of Control. London: HMSO, 1934.

CHAPTER 9

Work and Support for Patients Outside Hospital

This chapter examines whether the occupations undertaken by patients in mental hospitals and asylums prepared them for a return to the workplace after discharge, either by helping them to maintain their pre-admission skills or by teaching them new ones. Preparing patients for the workplace had been one of the objectives of patient work in the context of moral therapy, and in the early nineteenth century the type of work prescribed for asylum patients equipped them well for work outside the asylum. Working on the asylum farm was familiar to many, given the high proportion of the English and particularly French populations employed in agriculture at that time.[1] The artisanal workshops of the asylum, such as the shoemakers', stonemasons', or tailors' workshops provided employment similar to that which was available in most French and many English towns. One hundred years later, the economies and working methods of both countries had changed significantly (even more dramatically in England) due to industrialisation, raising the question of whether the work offered to mental hospital patients had kept pace with those changes. Were the new methods of scientific labour management and assembly line production introduced into asylums work schemes? Could the American style of occupational therapy, with its focus on arts and crafts, which encouraged creativity and involvement in all stages of the production

[1] Approximately 22% of the working adult population was employed in agriculture in England and Wales in 1821; in France the figure was closer to 75%.

J. Freebody, *Work and Occupation in French and English Mental Hospitals, c.1918–1939*, Mental Health in Historical Perspective,
https://doi.org/10.1007/978-3-031-13105-9_9

process, from design to finishing, claim to prepare patients for the modern workplace? Further questions arise concerning the support available to patients after leaving hospital, such as help with finding employment or somewhere to live, or access to ongoing treatment as an outpatient. Were patients left to fend for themselves after discharge or how was support organised? Record levels of unemployment in the immediate aftermath of war and during the Great Depression made the situation for recently discharged patients particularly precarious. The stress associated with unemployment threatened their already fragile mental health as well as their ability to support themselves.

Rehabilitation or Therapy?

New ways of working, by adopting the principles of "scientific labour management", were developed in the USA before World War I by the American mechanical engineer Frederick Winslow Taylor (1856–1915) and automobile manufacturer Henry Ford (1863–1947). Their methods generated interest amongst British and European manufacturers, who were keen to improve the efficiency of their own enterprises. This interest increased as the need for efficiency and productivity escalated during World War I and afterwards, as economies recovered. Scientific labour management encouraged business owners to undertake a systematic evaluation of all the labour processes involved in their business, with the aid of a stopwatch, and to maximise efficiency by subdividing tasks.[2] The resulting de-skilling of the workforce and a loss of worker discretion and autonomy was taken to extremes by Charles Bedaux, (1886–1944), a French-born American management consultant.[3] Bedaux's methods were particularly effective in industries using assembly lines, such as the automobile industry and were adopted in around 250 of Britain's largest manufacturing firms.[4] This quest for efficiency was unpopular with workers, however, who resented the loss of control, the close links between effort and earnings, and the stringent monitoring of performance.[5]

There was opposition to scientific management methods by skilled workers in France too. Patrick Fridenson highlights the issue of boredom,

[2] Arthur J. McIvor, *A History of Work in Britain, 1880–1950* (Basingstoke and New York: Palgrave, 2001), 54.

[3] Ibid., 54–5.

[4] Ibid., 96.

[5] Ibid., 97.

fatigue and resentment of managerial control amongst French automobile factory workers, leading to absenteeism, the slowing down or limiting of production, and the disruption of work schedules.[6] During the Depression, the French government intervened directly in the economy, introducing employment regulations and systems for industrial negotiation, and attempting to modernise production methods beyond the automotive industries.[7] The methods of Taylor and Bedaux, pioneered by Renault, Citroen and Peugeot, were promoted as a means of raising labour productivity and "rationalising" employment. Unpopular and perceived as a threat to traditional French working hierarchies, the imposition of such methods generated a wave of strike action in Paris in 1936.[8] The new methods struck at the very core of a French worker's identity, which was based on the established pattern of progress from journeyman to skilled *petit entrepreneur* and master of one's own destiny.[9] Automobile manufacture played an important role in the economies of all the towns and cities where the hospitals in this study were located, namely London (Ford), Paris (Renault and Citroen), Oxford (Morris Motors) and Le Mans (Bollé and later Renault). Bedaux's methods, whilst unpopular, could not be overlooked if patients were to be prepared for the local labour market.

On the other hand, work satisfaction in a hospital setting could not be ignored either, since one of the purposes of work for patients was to promote well-being and a sense of agency, as Waltraud Ernst has emphasised.[10] The deleterious effects of the division of labour and the "deskilling" of workers had already been recognised by political economists Adam Smith in the eighteenth, and Karl Marx in the nineteenth centuries. In *The Wealth of Nations* (1776) Smith argues that a man spending all his time performing the same tasks had no need to "exert his understanding or exercise his invention" to solve problems, rendering him "stupid and

[6] Patrick Fridenson, "Automobile Workers in France and Their Work, 1914–1983," in *Work in France: Representations, Meaning, Organisation and Practice*, ed. Stephen Laurence Kaplan and Cynthia J. Koepp (Ithaca and London: Cornell University Press, 1986), 514–64.

[7] Noel Whiteside, "Constructing Unemployment: Britain and France in Historical Perspective," *Social Policy and Administration* 48 no. 1 (2014): 77.

[8] Ibid.

[9] Ibid.

[10] Waltraud Ernst, "Therapy and Empowerment, Coercion and Punishment. Historical and contemporary perspectives on work, psychiatry and society," in *Work, Psychiatry and Society, c.1750–2015*, ed. Waltraud Ernst, 1–30. Manchester: Manchester University Press, 2016..

ignorant".[11] Karl Marx was also highly critical of the division of labour, which he claimed "attacks an individual at the very roots of his life" and could eventually lead to what he termed "industrial pathology".[12] Describing factories as "mitigated jails", he argued that the narrow, specialised functions associated with the division of labour reduced workers to a "fragment" of their former selves.[13] Marx claimed that work under capitalism, divided up into meaningless repetitive tasks, was the opposite of *purposeful* activity. Marx believed that purposeful activity, such as the work of the skilled watchmaker creating a complete watch, was necessary for the "realisation of the full humanity of the individual".[14] While Smith and Marx were referring to the new factory conditions introduced during the first industrial revolution, their remarks are perhaps even more relevant to the very exacting methods of scientific management. The loss of autonomy and control, and the sheer monotony of the work, could be damaging to a worker's mental health and self-esteem. But the factory and the assembly line—rather than the artisan's workshop—had become the new working environment for which patients left the asylum ill-prepared.[15]

It is hard to understand why, at a time when considerable research was being carried out into working practices by the very psychiatrists and psychologists who were treating patients with mental disorder, there appeared to be little crossover. Édouard Toulouse, medical director of the Henri Rousselle and co-founder of the French League for Mental Hygiene, was passionate about workforce productivity and measures to improve occupational health, writing extensively about the need to manage work scientifically.[16] He had established research laboratories investigating what he termed the "psychobiology of labour" in 1900 at the Villejuif Asylum,

[11] Adam Smith, *An Inquiry into the Nature and Causes of the Wealth of Nations* (Ware, Herts.: Wordsworth Editions, 2012), 777–8.

[12] Karl Marx, *Capital, Volume I: Capitalist Production* [1885], trans. Samuel Moore and Edward Aveling (Ware, Herts.: Wordsworth Editions, 2013), 253.

[13] Edward Royce, *Classical Social Theory and Modern Society: Marx, Durkheim and Weber* (London and New York: Rowman and Littlefield, 2015), 135.

[14] John Dupré and Regenia Gagnier, "A Brief History of Work," *Journal of Economic Issues* 30 no. 2 (1996): 554.

[15] Gerard Noiriel, *Workers in French Society in the 19th and 20th Centuries* (New York: Berg, 1989), 135–6. [See chapter 2:2:1 for discussion of the scientific management methods of Taylor and Bedaux]

[16] William Schneider, "The Scientific Study of Labour in Interwar France," *French Historical Studies* 17 no. 2 (1991): 416.

where he then worked as a chief medical officer, and these laboratories had moved with him to the Henri Rousselle Hospital in 1922.[17] These included the Laboratory of Physiology, which conducted research into muscular fatigue and reaction times, and the Laboratory of Experimental Psychology, which performed various intelligence, perception and memory tests on school children and machine operators to ascertain their suitability for various professions.[18]

Toulouse claimed that the laboratories fulfilled both clinical and social functions. On the clinical side, doctors from the Henri Rousselle Hospital sent patients to the Laboratories of Psychology and Physiology to obtain precise information on their mental functioning which aided the doctors in diagnosis and treatment.[19] The results of tests carried out on patients, when compared with the average behaviour of "normal" subjects, enabled doctors to identify diminished functionality. Yet it did not appear that his findings informed how work was organised for patients either at the Henri Rousselle Hospital or at the neighbouring Asile Clinique. One explanation may lie in the fact that experimental psychology was a relatively new area, only emerging from the long-established discipline of philosophy at the end of the nineteenth century; it did not impinge upon asylum medicine until much later in France. Furthermore, Toulouse's research was aimed at the prevention of mental disorder, in line with the principles of the Mental Hygiene Movement, and so the focus of his work in this area was on work outside, rather than inside, the mental hospital.

In England, research into how to minimise "industrial fatigue" and maximise output by workers began before World War I and escalated during the conflict as the war-time requirement of long hours and intense, sustained effort took their toll on the health of munitions workers.[20] This research was continued after the war by the Industrial Fatigue and Industrial Health Research Boards, established in 1918 and 1928 respectively.[21] Experts in fatigue, nutrition, psychology and physiology subjected the human body's movements and rhythms to detailed laboratory

[17] Ibid., 414.

[18] Henri Rousselle, Rapport au nom de la 3e Commission sur le Service et les services annexes (Paris: Conseil Général 1923), 5. ADP/D.10 K3/34/87.

[19] Édouard Toulouse, Rapport du Dr. Toulouse, médecin-directeur, sur le fonctionnement du Service de la Prophylaxie Mentale durant l'année 1926 (Paris: Conseil Générale de la Seine, 1926), 54. ADP/D.10 K3/42/91 Annexe.

[20] McIvor, *History of Work in Britain*, 131.

[21] Ibid., 134.

investigations in a quest to achieve greater productivity.[22] By rationalising the physiological and psychological performance of the worker's body, scientists sought to eliminate fatigue, overwork and wasteful motion, thereby improving the health, efficiency and productivity of workers. Their recommendations referred to the hours and physical conditions of work (such as temperature, humidity, ventilation and lighting) and personal factors such as vocational selection and guidance, as well as clothing and seating arrangements, which were all found to influence productivity.[23] But these recommendations did not appear to influence how work was organised in mental hospitals. That said, the Mental Hygiene principle that work should be purposeful and satisfying to perform was reflected by the introduction of occupational therapy in English institutions.

The activities comprising American-style occupational therapy were quite different to work on the factory floor in London's expanding industrial sectors. Instead of performing work for which little or no skill was required, patients were taught a craft and how to make aesthetically pleasing objects from scratch. The teaching of arts and crafts to patients inherent in American-style occupational therapy was influenced by the Arts and Crafts Movement, which encouraged the creation of hand-made goods in place of machine-made uniformity. Craftsmanship was considered to be spiritually uplifting quality and to have a "regenerative power".[24] In other words, arts and crafts were perceived as an antidote to modern factory production methods. Elizabeth Casson, who founded the Dorset House School of Occupational Therapy in 1930, maintained that crafts activities were designed to "arouse curiosity and the desire to achieve".[25] The therapeutic benefits of occupational therapy were related to the pride taken in the patient's work, the care and attention to detail invested, and the fact that the patient was responsible for the whole task, not just one tiny aspect of production. While occupational therapy developed concentration and self-esteem, it was not rehabilitative in an economic sense—except for the

[22] Anson Rabinbach, *The Human Motor: Energy, Fatigue and the Origins of Modernity* (Berkeley: University of California Press), 8.

[23] Steffan Blayney, "Industrial Fatigue and the Productive Body: the science of work in Britain," *Social History of Medicine* 32 no.2 (2019): 322.

[24] Rosalind Blakesley, *The Arts and Crafts Movement* (London and New York: Phaidon, 2006), 8.

[25] Elizabeth Casson, "Foreword," in *The Story of Dorset House School of Occupational Therapy 1930–1986*, ed. Betty Collins (Oxford: unpublished, 1986), 3. OBU Special Collections: Dorset House Collection.

few Maudsley patients who managed to turn arts and crafts into a paying hobby or became occupational therapists themselves.[26]

One of the criticisms of occupational therapy and institutional work made by contemporary psychiatrists after World War II, as Vicky Long has highlighted, was that they failed to prepare patients for the modern workplace.[27] This anomaly was eventually addressed in the 1950s with the introduction of "industrial therapy" (IT) into English mental hospitals. The IT units incorporated industrial features such as factory lighting and seating, conveyor belts and time sheets. Work was supplied by local firms and involved simple tasks, such as folding cardboard boxes, for which patients were paid a low hourly rate.[28] Whilst arguably less satisfying than arts and crafts, IT prepared patients more effectively for the type of work they would find outside the hospital. This "social readjustment" was important at a time when institutional care was gradually being replaced with care in the community.[29]

Welfare Measures in France and England

The rehabilitation of patients was an important rationale for the prescription of work to mental hospital patients in France and England since there were few "safety nets" for those unable to find work in the early days of the asylum system. In nineteenth-century England, ensuring that patients were self-sufficient on leaving hospital was accorded high priority, as the workhouse offered the only source of assistance for the destitute. Poor relief was provided by the New Poor Law of 1834, designed, according to Peter Bartlett, to "root out and dismantle a culture of poverty, perceived in terms of immorality, intemperance and promiscuity, and replace it with a culture of self-help, respectability, sobriety and hard work".[30] Unemployment was considered a moral failing and idleness frowned upon. English attitudes towards the unemployed began to change in the early

[26] Report of the Medical Superintendent, Maudsley Hospital, London, 1927–1931, 16.

[27] Vicky Long, "Work Is Therapy? The function of employment in British Psychiatric Care after 1959," in *Work, Psychiatry and Society, c.1750–2015*, ed. Waltraud Ernst (Manchester: Manchester University Press, 2016), 336.

[28] Ibid., 339.

[29] Ibid., 336.

[30] Peter Bartlett, "The Asylum and the Poor Law: The Productive Alliance," in *Insanity, Institutions and Society, 1800–1914: A Social History of Madness in Comparative Perspective*, ed. Joseph Melling and Bill Forsythe (London and New York: Routledge, 1999), 53.

twentieth century, with joblessness no longer viewed as an individual's "fault", or a matter of choice, but as a social problem that called for increased state intervention.[31]

Limited measures to alleviate poverty were introduced in England by the Liberal administrations of 1906–1914.[32] These measures included the Labour Exchanges Act of 1909 that created a national network through which the unemployed could search for work, and the National Insurance Act of 1911 that provided compulsory, contributory insurance for most employed individuals against the financial consequences of sickness, disablement, maternity, and short-term unemployment.[33] In France, widespread social reform legislation was not forthcoming until the late 1920s. Until then, "initiative and prudence" were supposed to protect the working man and his family from the consequences of accidents, illness, disability, old age and unemployment.[34] Between 1928 and 1932, social insurance laws providing similar protection to that afforded by the British legislation, were introduced.[35] The mass unemployment generated by the Great Depression resulted in an extension of the British welfare benefits, including the establishment of an Unemployment Assistance Board in 1934.[36] In France, further welfare measures were introduced under Léon Blum's Popular Front government of 1936–1938.[37]

These initiatives demonstrate that in both England and France, the state assumed increasing responsibility for citizens' welfare, and acknowledged that individuals were not to be blamed for unemployment and other social misfortunes. The evolution in the provision of state aid for the poor began some twenty years earlier in England. Late nineteenth-century English psychiatric texts made frequent references to the "creation of useful members of society" who could earn their own living after discharge and not pose a burden on society. But, as Sarah Chaney observes, this type

[31] Lynn Hollen Lees, *The Solidarities of Strangers. The English Poor Laws and the People, 1700–1948* (Cambridge: Cambridge University Press, 1998), 310.

[32] McIvor, *History of Work in Britain,* 156.

[33] Ibid.

[34] Ibid., 5–6.

[35] Ibid., 97.

[36] Edward Royle, *Modern Britain: A Social History, 1750–2011*, third ed. (London and New York: Bloomsbury, 2012), 239–40.

[37] Ibid., 183.

of rhetoric became far less common after World War I.[38] It is therefore plausible to assume a corresponding lowering of the priority accorded to ensuring self-sufficiency amongst patients. Indeed, the Macmillan Report of 1926 stressed that the purpose of occupation for the patients of mental hospitals (as opposed to the institutions caring for the "mentally deficient") was purely therapeutic, whether this comprised work around the hospital or arts and crafts activities in the occupational therapy workshop.[39] In France, where the social reforms came later, there did not appear to be any downgrading of the need to rehabilitate patients for the labour market. Here, an ability to work was one of the criteria for discharge.[40] Furthermore, France was in desperate need of capable workers after World War I. An acute labour shortage was the result of a very low birth rate and the death of so many working-age men during the conflict. The urgent need to rebuild France's workforce was emphasised by Édouard Toulouse who claimed in 1920 that "the nation's biggest problem was the re-establishment of its human capital".[41]

During the Great Depression, unemployment was rife in both countries and welfare measures were stretched to breaking point. The economic crisis led some patients to choose to remain in French asylums as workers, despite being eligible for discharge. This was an "unofficial" means of support offered by chief medical officers who were aware of the unemployment situation outside the hospital. Equally, impoverished families were slow to collect their cured or improved relatives from hospital, knowing that work was hard to find particularly for those bearing the stigma of internment.[42] In 1935, Dr. Marchand of the Men's First Section of the Asile Clinique, reported that an additional 10 patients could have been discharged that year, but had elected to remain at the Asile Clinique as workers and to continue their treatments as in-patients. These patients

[38] Sarah Chaney, "Useful Members of Society or Motiveless Malingerers? Occupation and malingering in British asylum psychiatry, 1870–1914," In *Work, Psychiatry and Society*, ed. Ernst," 282.

[39] Hugh Pattison Macmillan, Report of the Royal Commission on Lunacy and Mental Disorder (London: HMSO, 1926).

[40] Hervé Guillemain, *Chronique de la Psychiatrie Ordinaire: Patients, Soignants et Institutions en Sarthe du 19e au 21e Siècle* (Tours: Éditions de la Reinette, 2010), 45.

[41] Jean-Bernard Wojciechowski, *Hygiène Mentale et Hygiène Sociale: Contribution à l'Histoire de l'Hygiénisme, tome II* (Paris: L'Harmattan, 1997), 41.

[42] M.E. Chausse, Rapport Général au nom de la 3e Commission sur le compte du Service des aliénés, (Paris: Conseil Général de la Seine, 1932), 4. ADP/D.10 K3/48/95.

had been out of work when they were admitted and were unlikely to find work in the current economic climate.[43] Marchand remarked the following year that previously unemployed patients who had been allowed to remain made excellent workers.[44] A similar situation emerged at the Henri Rousselle Hospital. Former patients who had been unable to find work after discharge, returned to the hospital where they were employed as clerks, or in such tasks as locksmithing, plumbing, stone-masonry and painting.[45] A sum of 43,196F was spent on remunerating these patients in 1936.[46] This unofficial support for patients was no doubt a life-line for some, but successful longer term rehabilitation depended on whether the activities undertaken by patients in hospital prepared them for the type of work available in their local community.

The Local Relevance of Work within Institutions

The types of occupation offered to patients in both French and English institutions did not equip patients with the skills required in the modern workplace, although it was appropriate for those returning to more traditional trades. Patient occupation continued to be based on institutional requirements in France, and on a blend of institutional requirements and arts and crafts in England. The type of work performed within the Asile Clinique or the Asile de la Sarthe did not evolve to reflect developments in local employment during the interwar period, while the arts and crafts activities comprising occupational therapy at Bethlem, the Maudsley and the Littlemore hospitals were not vocational. These occupations may have equipped patients with the skills required in traditional sectors of employment, such as tailoring or furniture-making, but they did not prepare patients for work in new industrial sectors, such as the automotive sector or electrical goods' manufacture. John Burnett highlights the plight of an unemployed skilled woodcarver who had worked for thirty years on high-quality furniture. In the 1920s he found that "machine-turned furniture

[43] Dr. Marchand, Rapport de la 1re section des hommes, Asile Clinique (Paris: Préfecture de la Seine, 1935), 90. ADP/PER-566-6.

[44] Ibid., 1936; 72. ADP/PER-566-6.

[45] René Fiquet, Rapport au nom de la 3e Commission sur le Centre de psychiatrie et de prophylaxie mentale Henri-Rousselle et les services annexes (Paris: Conseil Général de la Seine, 1936), 3. ADP/D.10 K3/54/117.

[46] Ibid.

and machine-carved wood have put us out of business".[47] Furthermore, the occupations offered to female patients failed to reflect the increasingly diverse employment opportunities open to women after World War I.

While traditional industries persisted in the city centres, new industries mushroomed in the suburbs, many of which were based on modern management and production techniques, such as the assembly line. In Paris, for example, the traditional sectors of publishing, *haute couture* and jewellery remained in central Paris, while the suburbs became host to large-scale enterprises in new industrial sectors, such as automobiles (the Citroen plant was established in 1919) and chemicals.[48] In London, too, major industrial growth took place in the suburbs (the Ford Motor Company arrived in Dagenham in 1931) but almost 37,000 factories involved in clothing, furniture, food and drink, printing and engineering remained in London's old industrial centre.[49] Industries deploying new technology showed the most spectacular growth in interwar London. Between 1925 and 1937 the numbers employed in the manufacture of electrical goods rose from 48,000 to 113,500, an increase of 138%.[50] The other major growth area in both Paris and London was the service sector. In Paris, between 1906 and 1931, employment in banking and business grew by 52% and in administration and retail by 77%.[51]

The categories of asylum workshops (namely plumbing, stone-masonry, carpentry, locksmithing, painting, mechanical work, laundry and ironing services, shoe-making, tailoring and needlework) at the Asile Clinique had not changed since 1900, with the exception of an electricity workshop added in 1924.[52] The latter only employed two or three patients, and therefore did not permit the widespread acquisition of skills in this new growth area. Some of the occupations carried out in the traditional workshops mirrored the artisanal aspect of Paris' dual economy. Patients leaving the Asile Clinique having worked in the tailoring, needlework, carpentry or shoe-making workshops might still find work in these sectors, but they were not showing the rapid growth of the new industries. Many,

[47] John Burnett, *Idle Hands: The Experience of Unemployment, 1790–1990* (London and New York: Routledge, 1994), 218.

[48] Colin Jones, *Paris, Biography of a City* (New York and London: Penguin, 2004), 407.

[49] Stephen Inwood, *A History of London* (Basingstoke: Macmillan, 1998), 728.

[50] Ibid., 731.

[51] Jones, *Paris, Biography of a City*, 407.

[52] Henri Rousselle, Rapport au nom de la 3e Commission sur les comptes du directeur de l'Asile Clinique (Paris: Conseil Général de la Seine, 1924). ADP/D.10 K3/38/75.

such as shoe-making, were switching to modern machine production methods not used in hospital workshops.

Market-gardening was not an activity associated with the Parisian economy, but this occupation made an important contribution to the asylum's supply of food. Physical work in the open air had been considered particularly beneficial for patients since the early nineteenth century.[53] Farm work in particular, which gave patients the satisfying experience of growing their own food, had been recommended by Pinel in his *Traité* of 1800.[54] The restricted nature of this activity at the Asile Clinique (market gardening occupied between just seven and 12 patients each year) was due to the availability of land in a city-centre institution, rather than a reflection of the scope for employment in market-gardening in Paris. The large numbers of patients employed in the laundry reflected the fact that a hospital of over 1000 patients produced a significant amount of laundry, rather than size of the laundry sector outside the hospital. The workshops only occupied approximately one third of the Asile Clinique's working patients [see Table 9.1]. A breakdown of how the other patient workers were occupied was only indicated for the years 1937 and 1939 [see Table 9.2], but these figures show that roughly two thirds of patients (32% of male and 43% of female patient workers in 1937) were allocated housework or cleaning duties. This unskilled work was essential to institutional maintenance. In 1937, clerical work was allocated to 10% of male and 5% of female patient workers, while in 1939 these figures had decreased to 8% of males and 4% of females [see Table 9.2]. Hospital employment in clerical work was therefore contracting, whilst outside hospital it was expanding. What is not clear from the tables is which (convalescent) patients were from the treatment divisions, and therefore soon to be discharged, and which (incurable) patients were from the workers' pavilion, who were unlikely to leave the asylum.

The provincial institutions of the Littlemore and the Asile de la Sarthe catered for both curable and incurable patients. Rehabilitation for work outside hospital was particularly important for the curable who, on leaving the institution, would have to support themselves. Incurable patients were likely to spend the rest of their lives in hospital. Rehabilitation for the

[53] Louis Dausset, Rapport Général au nom de la 3e Commission sur les propositions budgétaires pour le Service des Aliénés (budget de 1919) (Paris: Conseil Général de la Seine, 1918), 45. ADP/D.10 K3/27/20.

[54] Philippe Pinel, *Traité médico-psychologique sur l'aliénation mentale, ou la manie* (Paris: chez Richard, Calille et Ravier, 1800), 225–6.

Table 9.1 Table to show a breakdown of the work performed by patients in the various workshops at the Asile Clinique, and the average daily wage ("pécule") paid to patients for each type of work, in 1925 and 1932

1925

Workshop	*Nos. Workshop Staff*	*Nos. Patient Workers*	*Average Daily Pécule (F=Francs)*
Market garden	1	12	0.45 F
Smoke house	3	1	0.40 F
Plumbing/stone masonry	3	4	0.50 F
Carpentry	2	6	0.47 F
Locksmiths	1	8	0.59 F
Electricity	1	3	0.30 F
Painting/decorating	3	3	0.37 F
Mechanical	4	1	1.5 F
Laundry	13	12	0.56 F
Ironing	3	4	0.37 F
Shoe-making	1	5	0.40 F
Tailors	2	8	0.39 F
Sewing group A	3	11	0.28 F
Sewing group B	2	20	0.30 F
No. patients employed in the workshops		98	

1932

Workshop	*Nos. Workshop Staff*	*Nos. Patient Workers*	*Average Daily Pécule*
Market garden	2	8	0.85 F
Plumbing	2	2	1.23 F
Stone-masonry	3	3	0.86 F
Carpentry	2	4	1.46 F
Locksmiths	1	7	1.56 F
Electricity	2	2	1.69 F
Painting/decorating	5	3	?
Mechanical	2	1	2.5 F
Laundry	10	17 men 2 women	1.10 F men 1.00 F women
Ironing	1	5 women	0.84 F
Shoe-making	1	3	1.20 F
Tailors	2	4	1.25 F
Sewing group A	1	14 women	0.75 F
Sewing group B	1	24 women	0.67 F
No. patients employed in the workshops		99	

Source: Reports to the General Council of the Seine, 1925 and 1932. ADP/D.10K3 33 & 40

Table 9.2 Table to show a breakdown of the work performed by patients at the Asile Clinique 1937–1940

Category /place of Work	*1937*	*1939*	*1940*
Administrative	9	10	10
Office work (men)	8	7	7
Office work (women)	7	6	6
Library—Patients	2	2	2
Library—Medical	3	3	2
Admissions service	7	9	8
Shoemaking workshop	2	4	3
Guard room (men)	1	3	3
Guard room (women)	1	1	1
Tennis courts	1	1	1
Pharmacy	2	3	3
Plumbing workshop	1	2	2
Decorating workshop	2	2	2
Lecture theatre	1	1	1
Transport	1	1	1
Cloakroom	2	2	2
Gardens	13	17	12
Mechanics' workshop	1	1	1
Construction/builders' yard	3	4	3
Electricity workshop	3	4	3
Tailoring workshop (men)	4	5	6
Tailoring workshop (women)	1	1	–
Locksmiths' workshop	4	5	4
Carpentry workshop	3	5	3
Heating workshop	2	2	2
Laundry (men)	16	24	21
Laundry (women)	2	2	2
Lingerie (men)	4	4	3
Lingerie (women)	1	1	1
Wine cellar	3	3	2
Surgery	5	5	6
Road mending	9	11	10
Kitchen (men)	9	14	10
Kitchen (women)	13	12	14
Mending/sewing	5	5	5
Ironing workshop	6	7	7
Couture A workshop	16	20	19
Couture B workshop	23	28	28
Housework	56	58	50
Floor scrubbers	56	58	50
Hairdresser	–	–	1
TOTALS	**308**	**349**	**320**

Source: Reports to the General Council of the Seine, 1937–40. ADP/D.10K3 57-60

labour market was less of an issue for these patients, but making a contribution to institutional running costs (and therefore to the costs of their care) was a priority for the authorities. In England, the Mental Treatment Act of 1930 meant that the movement of patients was greater, although the numbers of patients discharged "recovered" was still outnumbered by those remaining in hospital. An ability to work was considered one of the criteria for discharge at the Asile de la Sarthe, where a patient with no means of support (either by working or living with family) was to be retained (the phrase *à maintenir* appeared in their records).[55] A patient who was working well in the asylum was deemed capable of doing so outside it and therefore able to be self-sufficient and not a drain on society. The knowledge that his liberty depended on his capacity to work led one patient to write to the asylum director requesting employment in the asylum gardens [see Fig. 8.1].[56]

At the Asile de la Sarthe, the types of work offered to patients inside the asylum during the interwar period were identical to those offered before World War I [see Tables 9.3 and 9.4]. The tasks had been engineered to serve the needs of the asylum in the mid-nineteenth century and had not evolved to prepare patients for changes in the local economy. Farming, however, as the second largest category of employment within the Asile de la Sarthe for men, after cleaning, was useful preparation for patients seeking work in the local community. Although the numbers employed in agriculture were declining in the department of La Sarthe (22,000 Sarthois agricultural workers left the countryside in 1922 alone), 67% of the population still lived in rural areas in 1936.[57]

The third largest category of male employment was work in the laundry, but, as in Paris, this was indicative of institutional need. The only type of work available in the asylum that represented a growth area in the local economy (metallurgy) was tin-plate-making, but this involved very few patients. Metallurgy flourished in Le Mans, with the opening of La Maison Chappée, a new foundry producing radiators, which serviced the growing motor industry in the town. The latter employed 2000 workers in 1920 and 17,000 in 1933.[58] Another firm involved in metalwork was Carel et Fouché, which produced metal construction materials for the railways. Le Mans' artisanal activities, such as shoe-making, saddlery, brickmaking,

[55] Guillemain, *Chronique de la Psychiatrie Ordinaire*, 45.

[56] Ibid.

[57] André Lévy, ed. *La Sarthe des Origines à Nos Jours* (Saint-Jean-d'Angély: Éditions Bordessoules, 1983), 343–4.

[58] Ibid., 344.

Table 9.3 Table to show a breakdown of the work performed by male patients (paying and pauper) at the Asile de la Sarthe in 1925; 1929; 1934 and 1937. The figures represent the number of days worked in each category

MALES	*1925*	*1925*	*1925*	*1929*	*1929*	*1929*	*1934*	*1934*	*1934*	*1937*	*1937*	*1937*
Type of work	*Pauper*	*Paying*	***Total***	*Pauper*	*Paying*	***Total***	*Pauper*	*Paying*	***Total***	*Pauper*	*Paying*	***Total***
Farmyard	1457	–	**1457**	2026	77	**2103**	1832	102	**1934**	1629	–	**1629**
Bakery	317	–	**317**	299	–	**299**	614	–	**614**	612	–	**612**
Laundry	2398	710	**3108**	3208	668	**3876**	3284	614	**3898**	4285	152	**4437**
Wood yard	2282	584	**2866**	1972	450	**2422**	3519	–	**3519**	2271	–	**2271**
Shoe making	540	–	**540**	308	–	**308**	409	278	**687**	–	306	**306**
Kitchen	1551	–	**1551**	1540	205	**1745**	1535	307	**1842**	2090	–	**2090**
Farming	4161	1226	**5387**	4278	489	**4767**	10,485	1045	**11,530**	5139	410	**5549**
Clerical work	1076	–	**1076**	1175	–	**1175**	1186	230	**1416**	868	567	**1455**
Tin-plate making	459	–	**459**	500	–	**500**	614	–	**614**	306	–	**306**
Stone masonry	425	–	**425**	308	–	**308**	307	–	**307**	767	301	**1068**
Housework	7035	1131	**8166**	6943	1245	**3188**	307	–	**307**	10,733	684	**11,417**
Carpentry	521	–	**521**	206	46	**252**	564	178	**742**	437	229	**666**
Painting	352	–	**352**	–	–	–	331	–	**331**	561	-	**561**
Locksmiths	327	–	**327**	308	–	**308**	793	257	**1050**	343	–	**343**
Exterior maintenance	1202	152	**1354**	1217	415	**1632**	30,896	3510	**34,406**	612	–	**612**
TOTALS	24,103	3803	**27,906**	24,288	3595	**27,883**				30,653	2669	**33,322**

Source: Reports of the Asylum Director, Asile de la Sarthe, 1925, 1929, 1934 and 1937. ADS-1X964/5

Table 9.4 Table to show a breakdown of the work performed by female patients (paying and pauper) at the Asile de la Sarthe in 1925; 1929; 1934 and 1937. The figures represent the number of days worked in each category

FEMALES	*1925*	*1925*	***1925***	*1929*	*1929*	***1929***	*1934*	*1934*	***1934***	*1937*	*1937*	***1937***
Type of work	*Pauper*	*Paying*	***Total***	*Pauper*	*Paying*	***Total***	*Pauper*	*Paying*	***Total***	*Pauper*	*Paying*	***Total***
Farmyard	1220	–	**1220**	918	183	**1101**	307	236	**543**	308	–	**308**
Laundry	4252	26	**4278**	4220	–	**4220**	3129	–	**3129**	4443	153	**4596**
Lingerie	989	1037	**2.026**	1562	610	**2172**	461	833	**1294**	16,836	2039	**18,875**
Sewing	12,877	1831	**14,708**	12,974	756	**13,730**	14,786	1166	**15,952**	6209	56	**6265**
Kitchen	7707	305	**8012**	6499	305	**6804**	6202	410	**6612**	1232	616	**1848**
Housework	9025	1188	**10,213**	11,554	610	**12,671**	10,249	1354	**11,603**	11,735	614	**12,349**
Pharmacy	610	–	**610**	610	–	**610**	614	–	**614**	308	–	**308**
Mending	4225	301	**4526**	2907	123	**3030**	9122	780	**9902**	3571	580	**4151**
Ironing	1907	305	**2212**	1880	305	**2185**	1838	334	**2172**	2.947	–	**2947**
Knitting	1331	532	**1963**	301	647	**4046**	307	675	**982**	6279	837	**7116**
TOTALS	44,143	5625	**49,768**	43,925	4046	**47,971**	47,015	5788	**52,803**			

Source: Reports of the Asylum Director, Asile de la Sarthe, 1925, 1929, 1934 and 1937. ADS-1X964/5

ceramics, and tanning were in decline while food processing prospered.[59] It was the artisanal occupations, notably shoe-making, that employed patients within the Asile de la Sarthe.

For women, the largest employment categories in the asylum were housework and sewing. Whilst textiles had been an important aspect of the local Sarthois economy in the nineteenth century and earlier, this industry was in sharp decline during the interwar period. Needlework, housework, ironing, knitting or working in the kitchen, were tasks with which women were familiar in a domestic context. So, whilst the work available to women was not representative of the changing local economic profile, nor of the expansion in the range of work opportunities for women, it could be construed as rehabilitative for those returning to a domestic role. The employment of female patients in the farmyard also provided a useful and familiar task for those returning to work on the family small-holding. On the one hand, the work for women had changed little since the nineteenth century, while on the other, its emphasis on women's domestic role was reflective of contemporary measures to discourage women from working outside the home amid concerns regarding France's low birth-rate and high levels of infant mortality.[60]

At the Littlemore, the occupation undertaken by the largest group of patients prior to admission was labouring, which occupied *c*.20% of male patients between 1926 and 1939. "Labouring" could be interpreted in a variety of ways, and it is unclear whether patients were agricultural labourers, workers on a building site or casual labourers. The building industry in Oxford was expanding, stimulated by the requirement for new housing, factories and extensions to the university colleges.[61] Brickmaking at the Littlemore was therefore a useful skill to develop. Agriculture, on the other hand, was in sharp decline in Oxfordshire. Between 1921 and 1931, the percentage of agricultural workers in the county fell by 32%, compared to a fall of 17% nationally.[62] By 1931, just 6% of the nation's working

[59] Ibid., 350.

[60] Elinor Acampo et al., *Gender and the Politics of Social Reform in France, 1870–1914* (Baltimore and London: Johns Hopkins University Press, 1995), 9.

[61] Eleanor Chance, "Modern Oxford," in *A History of the County of Oxfordshire: Volume IV, the City of Oxford*, eds Alan Crossley and C.R. Elrington (London: Victoria County History, 1979), *British History Online*, http://www.british-history.ac.uk/vch/oxon/vol4 (accessed 02/03/2019).

[62] C. Whiting, *The View from Cowley: The Impact of Industrialisation upon Oxford, 1918–1939* (Oxford: Clarendon Press, 1983), 39.

population were involved in agriculture (compared to 36% in France).[63] Work on the hospital farm was not, therefore, particularly useful in terms of future employment prospects, but agricultural work had been considered a valuable therapy since the early nineteenth century, and was a useful means of producing food for the asylum.

Between *c.*3% and 13% of Littlemore patients were engaged in factory work prior to admission. This was a rapidly growing sector in Oxford. As in Le Mans, car manufacture and metallurgy developed rapidly in Oxford during the interwar period. In 1923, Morris Motors employed 1650 workers and by 1927 this number had grown to *c.*5000.[64] Pressed Steel's workforce grew from 546 in 1926 to 5250 in 1939. Morris Motors, Pressed Steel and Osberton Radiators between them employed 30% of Oxford's insured population in 1939.[65] Other major employers in Oxford included the many breweries, such as Morrells, print works, Cooper's Marmalade Factory, the Oxford and District Gas Company, the Oxford Electric Company and the railway.[66] There was no work within the Littlemore Hospital that would have prepared patients for the type of work involved in these industries. Between 4% and 9% of Littlemore patients were previously employed in traditional trades such shoe-making, carpentry, stone masonry and tailoring. Although these artisanal activities were in decline in Oxford, they were still viable, so work within the hospital in these areas could be conceived as rehabilitative. Domestic service was a large employer within Oxford due to the continued growth of the university. 23% of Oxford's workforce in 1931 were employed in domestic service, compared with the national average of 13%.[67] Between 4% and 14% of patients admitted to the Littlemore between 1926 and 1939 were employed in some form of service.[68] Work around the hospital, such as cleaning and helping in the kitchens or laundry, would have been useful preparation for this type of work. It is not clear whether Littlemore patients were offered any form of clerical work (as some patients were at the Asile

[63] Erik Buyst and Piotr Franaszek, "Sectoral Developments, 1870–1914," in *The Cambridge Economic History of Modern Europe, Vol. II: 1870 to the Present,* eds Stephen Broadberry and Kevin O'Rourke (Cambridge: Cambridge University Press, 2010), 210.

[64] Chance, "Modern Oxford."

[65] Ibid.

[66] See: Liz Woolley, "Industrial Architecture in Oxford, 1870 to 1914," *Oxonia* (2010): 67–96.

[67] Chance, "Modern Oxford."

[68] Littlemore Hospital Admissions Book 1922–1965 gives (partially complete) details of male patients' previous occupations. OHA/L7/A1/1/10.

de la Sarthe), but this was another expanding sector in Oxford. The numbers employed in local government and the civil service increased dramatically, from fewer than 500 in 1911 to over 2700 in 1931.[69] Approximately 6% of patients admitted to the Littlemore in 1926, and 11% in 1937, held a clerical or business role.

It appears that neither the work around the hospital nor occupational therapy were engineered towards preparing patients for work in the local economy. If the tasks allocated in hospital happened to be relevant to the type of work available locally, it was more by happy coincidence than by design. In France, the emphasis was on work to contribute to hospital maintenance, while in England, where psychiatrists placed a greater premium on the therapeutic properties of occupation, the emphasis was on therapeutic occupation. Whilst both work around the hospital and occupational therapy developed discipline and concentration, and some of the workshop activities promoted artisanal skills still in existence in local economies, patients did not leave hospital with experience of modern workplace methods. For the many incurable patients in the provincial asylums, this was not so important since they were unlikely ever to leave the institution, but curable patients would have to support themselves outside hospital. Their ability to cope with living and working independently, particularly in the busy cities of London and Paris, was aided by the support provided for patients after discharge, where this was available. The Board of Control in England emphasised the need for "after-care" for patients in 1928, observing that the transition from "the ordered and sheltered life of the hospital to the stress and competition of the work-a-day world is too sudden and too severe for the convalescent".[70] Psychiatrists, such as Aubrey Lewis (medical superintendent of the Maudsley from 1937), recognised that patients who left hospital without the prospect of employment or support were more likely to suffer a relapse of their mental symptoms as a result of anxiety and loss of self-esteem.[71] John Burnett highlights how unemployed men during the early 1930s spoke of "feeling 'lost' without work", and of the "hopelessness" of being unable to find a job. Research revealed that psychoneuroses increased with the duration of unemployment.[72]

[69] Chance, "Modern Oxford."
[70] Fifteenth Annual Report of the Board of Control (London: HMSO, 1928), 6.
[71] See: Aubrey Lewis, "Neurosis and Unemployment," *The Lancet* no. 2 (1935): 297.
[72] Burnett, *Idle Hands*, 229–30.

Support for Patients Outside Hospital

Whilst welfare measures to support the poor had been introduced in the 1910s in England and the late 1920s in France, the legislation provided uneven cover and individuals who had not been in a position to pay National Insurance contributions were excluded. There was, however, limited support for recently discharged mental hospital patients in both countries. Before World War I, this support was charitable, provided by a network of Lady Almoners and the Mental After Care Association (MACA) in England and the *Sociétés de patronage des aliénés guéris* (Patronage Societies for Cured Mental Patients) in France. Provision by these organisations was patchy in both France and England, but particularly so in France. French patients, however, had the benefit of leaving the asylum with a small sum of money, earned whilst in the asylum.

English patients who worked in the various hospital departments were not paid; they merely received additional food rations, such as an extra portion of bread and cheese, or cup of tea for their trouble.[73] French patients, in contrast, were paid a nominal daily wage, or *pécule*, for their work in the asylum, a principle established by the law of 1857.[74] Their wages accumulated during their stay and enabled them to leave with a small "nest-egg" or *pécule de sortie*, given to them on discharge. A minimum amount was set by law (and reviewed periodically) for the leavers' nest-egg. It was only after this amount had been earned and put aside that a patient was allowed to use their earnings to buy small luxuries such as chocolate, tobacco, or soap from the asylum shop. If the patient had not managed to earn the full amount before departure, this was made up by the asylum. During the interwar period, the value of the nest-egg varied between 15FF and 30FF. For the completely destitute among patients from the Seine, there was also recourse to the *Fondation André*, a legacy that had been invested for the purpose of giving small lump sums to the poorest asylum patients on discharge. It was divided between the institutions of the Seine; 300FF per annum was received by the Asile Clinique for distribution to its most needy patients.[75]

[73] These additional items appeared in the "Dietary" section of the annual reports of the Littlemore and Bethlem Royal hospitals.

[74] Bulletin Officiel du Ministère de l'Intérieur. Circulaire No. 7 du 20 mars 1857, 42–79. Paris, 1857. https://gallica.bnf.fr/ark:/12148/bpt6k5539701n/f2.item.zoom (accessed 12/02/2018).

Articles 154–163 referred to the payment of patients.

[75] Indicated in the annual accounts for the Asile Clinique.

Every French department was encouraged to create its own patronage society, subsidised by the prefecture, although many departments did not have one. Where they existed, the societies fulfilled a similar function to that of MACA in England. A patronage society was only established in La Sarthe in 1933, after Dr. Bourdin retired from the Asile de la Sarthe and beseeched his successor, Dr. Schultzenberger, to make it a priority.[76] The department of the Seine's patronage society, founded in 1886 and subsidised by an 8000FF annual grant from the General Council, aimed to assist indigent patients who had recently left a public asylum or hospice. Its function, however, was described by the Council as "incomplete and insufficient".[77] It was agreed that more comprehensive support for discharged patients was required, to avoid their return to the asylum. Many patients failed to secure work because employers were often reluctant to employ ex-asylum patients. They soon fell into poverty, causing their symptoms of mental illness to return. Families frequently rejected discharged relatives, owing to the stigma attached to mental disorder.[78]

In England, the Mental After Care Association (MACA) had been founded in 1879 by Rev. Henry Hawkins, chaplain of the Colney Hatch Asylum, to provide an alternative to the workhouse for pauper patients discharged from asylums. MACA was a national organisation, but most of its services were concentrated in the south-east. Former patients were assisted with lodging, money and clothing and helped to find suitable work. In the 1920s and 1930s, MACA provided cottage homes for convalescent patients, who stayed for a short period of between a fortnight and three months.[79] The Board of Control applauded the work of MACA, but felt that the larger hospitals should do more themselves in respect of providing after care.[80] The Board recommended the appointment of a full- or part-time psychiatric social worker.[81] Some English hospitals had a Lady Almoner, a voluntary role whose remit included after-care and visiting

[76] Rapport du Médecin en Chef, Asile de la Sarthe, Le Mans 1931 & 1933, [no page nos.]. ADS-1X965 and ADS-1X261.

[77] Dausset, "Rapport Général," 1918, 41.

[78] Ibid.

[79] The provision of interwar support for convalescent patients by MACA is discussed in: Stephen Soanes, *Rest and Restitution: Convalescence and the Public Mental Hospital in England, 1919–1939,* Thesis (Warwick: University of Warwick, 2011).

[80] Nineteenth Annual Report of the Board of Control (London: HMSO, 1932), 43.

[81] Ibid.

patients' homes.[82] When the Maudsley first opened, a Lady Almoner collected "useful and important knowledge" for the medical staff, supported patients whilst in hospital and helped them find work after discharge.[83] The medical superintendent, Edward Mapother, reported that a "great deal of time and thought" went into "re-establishing former patients in normal life again" in relatively stress-free situations that were unlikely to cause a relapse of their symptoms.[84] During 1925, he reported that 98 Maudsley patients had been assisted by the Lady Almoner.[85]

During the interwar period, support for recently discharged patients started to become professionalised, with the provision of paid psychiatric social workers taking over the role of the volunteer Lady Almoners in the more "progressive" city institutions.[86] The importance of the role of the psychiatric social worker (or PSW) to modern psychiatry had been recognised by American psychiatrists, notably Adolf Meyer, since the early twentieth century. By 1920, knowledge of a patient's social environment was considered vital to an understanding of individual behaviour and the PSW was regarded as an indispensable partner to the psychiatrist, gathering information on patients' family and community environment.[87] In 1929, a psychiatric social worker (PSW) replaced the Lady Almoner at the Maudsley Hospital, pre-empting the Board of Control's call for "well-trained social workers" to be attached to mental hospitals and clinics, as they were at US hospitals.[88] Edward Mapother cited as one of the most important developments during the Maudsley's first five years of operation, the increasing co-operation between PSWs and the hospital.[89] The investigations into a patient's social situation made by a PSW and the PSW's ongoing support of discharged patients were seen by Mapother as

[82] The emergence and role of hospital almoners, widely considered as the forerunners of modern-day hospital social workers, is discussed by Lynsey T. Cullen, "The First Lady Almoner: The Appointment, Position, and Findings of Miss Mary Stewart at the Royal Free Hospital, 1895–1899," *Journal of the History of Medicine and Allied Sciences* 68 no. 4 (2012): 551–82.

[83] Report of the Medical Superintendent, Maudsley Hospital, London, 1925, 8.

[84] Ibid., 1926, 6.

[85] Ibid., 1925, 8.

[86] See: Noel Timms, *Psychiatric Social Work in Great Britain, 1939–1962* (London: Routledge & Kegan Paul, 1964).

[87] Gerald Grob, *Mental Illness and American Society, 1875–1940* (Princeton: Princeton University Press, 1983), 250.

[88] Eighteenth Annual Report of the Board of Control (London: HMSO, 1931), 3.

[89] Report of the Medical Superintendent, Maudsley Hospital, London, 1927–1931, 17.

an "indispensable condition of ... progress in tackling mental illness".[90] "For modern institutional psychiatry" to function effectively, he maintained, the co-operation of specially trained PSWs was essential.[91]

The employment of a PSW was a luxury that not many English provincial mental hospitals could afford, despite the Board of Control's advocacy of such a service. At the Littlemore, a volunteer social worker assisted the medical team for two months during 1937, obtaining full case histories from patients on admission and making arrangements for patients who were about to be discharged. Her services were much appreciated by staff, leading Dr. Armstrong to observe that "a permanent psychiatric social worker will eventually be considered a necessity at this as well as many other mental hospitals".[92] By 1938, in the absence of a permanent social worker, the Littlemore relied upon the City Mental Health Visitor and her assistants for help with obtaining information regarding new patients and for arranging after-care for discharged patients.[93] A temporary part-time social worker, Miss Leslie, was appointed to the hospital staff on 1 January 1940.[94]

In France, the services of a PSW were regarded as essential by Édouard Toulouse, who ensured that a social services unit was attached to the Henri Rousselle Hospital.[95] The PSW, directed by the doctor, acted as a liaison between the hospital and the patient's family, providing advice, ensuring that treatment guidelines were being followed, and researching information about the patient's circumstances that might aid the doctor treating a patient in an outpatient facility or in hospital.[96] Mental crises, or relapses, were often caused by irregular life-styles, overwork, lack of work, domestic difficulties or the breakdown of family relations; the PSW could investigate these issues.[97] PSWs also helped patients find work that was appropriate for their condition, and to identify suitable employers.[98] To

[90] Ibid.

[91] Ibid., 1932–1935, 38.

[92] Report of the Medical Superintendent, Annual Reports of the Littlemore Hospital, Oxford, 1937/8, 8.

[93] Ibid., 1938/9, 8.

[94] Ibid., 1939/40, 8.

[95] Rousselle, Rapport au nom de la 3e Commission sur le Service départemental de prophylaxie mental les services annexes, 1923; 9.

[96] Ibid.

[97] Ibid.

[98] Ibid.

facilitate the "best use" of the individual, the social worker could call upon the Service of Professional Orientation (also part of the Henry Rousselle Hospital) which identified the most suitable (and unsuitable) types of work according to an individual's characteristics.[99] For those who were unable to work, the PSW found alternative means of support, such as placement in a family colony, hospice or retirement home.[100] Toulouse described the social assistants as "indispensable collaborators" from the doctors' perspective and as "precious guides" and "advisors" from that of patients.[101] A PSW service was not provided outside Paris during the interwar period (despite its recommendation to all departmental Prefects by the Ministry of Health in 1937), emphasising the difference in approach between the Henri Rousselle Hospital, which was heavily influenced by the American model of psychiatric care and the mental hygiene movement, and the traditional "closed" asylums, which remained custodial institutions. The Asile de la Sarthe was no exception to this rule and did not have a PSW service.

Outpatient clinics were another service for the mentally ill that emerged during the interwar period. They were of benefit to recently discharged patients, who could continue their treatment as an outpatient, as well as for those with mild symptoms who did not require admission. Although they had been recommended by the Medico-Psychological Association in their 1911–1914 report, and by Édouard Toulouse in his report to the Prefecture of 1913, outpatient services for the mentally ill were extremely rare in England and France before the end of World War I. In England, outpatient facilities were established in 1918 in Oxford and in London in 1919, as part of the Bethlem Royal Hospital. The Oxford clinic for nervous disorders, while established and managed by Thomas Saxty Good of the Littlemore Asylum, was attached to the Radcliffe Infirmary, a general hospital. This arrangement helped circumnavigate the law regarding the certification of mental patients that pertained until 1930. Good was also responsible for establishing Oxford's City Education Clinic for children with learning difficulties or behavioural problems.

[99] Ibid.

[100] Ibid.

[101] Édouard Toulouse, Rapport sur le fonctionnement et le développement de l'hôpital Henri-Rousselle (Paris: Conseil Général de la Seine, 1926–1927), 60. ADP/D.10 K3/42/91 Annexe.

A steady increase in the number of outpatients attending both the clinics during the interwar period corresponded with a reduction in the numbers of admissions to the Littlemore, particularly amongst young adults. The latter were able to remain living, and working, in the community while receiving treatment, mainly in the form of psychotherapy. The Board of Control noted in 1936 that the two outpatient clinics were "providing for a large section of the mentally sick, who, under other circumstances, would be mental hospital patients" and that many children were being "saved from becoming psychotics in later life".[102] Both clinics continued "to serve a useful function" after Good's retirement in 1936, and attendances remained high.[103] Dr. Armstrong, Good's successor at the Littlemore, remarked on recent praise for "Out-patient Clinics in the treatment of early mental conditions of both children and adults", noting that Oxford had "long been amply served by both types of Clinic" thanks to the enlightened policy of Dr. Good.[104] By treating so many cases of nervous disorders as outpatients, before symptoms became entrenched, the Littlemore avoided the overcrowding suffered by many provincial mental hospitals.

The outpatient department at Bethlem was the first to be established at a London mental hospital.[105] Dr. Porter-Phillips, Bethlem's physician superintendent, felt that an outpatient department was an important addition to the hospital because, according to Jonathan Andrews, it was a "progressive policy" expected of "high status" institutions.[106] When the department opened in November 1919, the treatments offered included massage, X-ray and electrical treatment.[107] Speech therapy and the services of a Lady Almoner were added later. Porter-Phillips also saw the clinic as a means of identifying cases of "mental deficiency", an area in which Dr. A. F. Tredgold, appointed in 1920, specialised.[108] Although the clinic was considered a success in terms of the numbers of patients treated and added

[102] Report of the Board of Control, Annual Reports of the Littlemore Hospital, Oxford, 1936/7, 17. OHA-L1/A2/14.

[103] Report of the Medical Superintendent, Annual Reports of the Littlemore Hospital, Oxford, 1937–1938, 11. OHA-L1/A2/15.

[104] Ibid., 1938–1939, 14. OHA-L1/A2/16.

[105] Andrews et al., *History of Bethlem,* 556.

[106] Ibid.

[107] Report of the Physician Superintendent, Annual Reports of Bethlem Royal Hospital, London, 1919, 10.

[108] Ibid., 1921, 8.

"great value" to the education of students (the University of London, for example, was impressed with its performance), the clinic closed in 1927.[109] Porter-Phillips felt that too many patients were of "a chronic order with a history of attendance elsewhere", rather than the incipient cases for which the service had been established.[110] Many of the war veterans referred by the Ministry of Pensions were found to be incurable and "quite unsuitable for treatment".[111] The closure of the clinic was perceived as a backwards step for Bethlem, since as Dr. Tredgold emphasised, "the whole tendency of modern medicine is to remove the stigma of Insanity, by breaking down the artificial barriers which has so long existed between the different forms of disease and disorders of the nervous system".[112] But the patients treated at Bethlem's clinic did not appear to be well enough to carry on with their daily lives and work whilst receiving treatment, and in this they differed from the Maudsley's outpatients.

Outpatient departments were incorporated into the Maudsley and Henri Rousselle hospitals from the outset and fared rather more successfully. The outpatient services worked in tandem with the main hospitals, referring cases that warranted in-patient treatment, but also treating some patients for long periods as outpatients. This enabled patients to continue with their regular work outside hospital, and thus to remain productive, whilst receiving treatment. At the Maudsley, the main treatment offered was psychotherapy, while at the Henri Rousselle patients a range of specialist treatments, including hydrotherapy, electrotherapy, UV-ray treatment, physiotherapy, organotherapy and vaccinations, as well as psychotherapy, were offered. These treatments could be accessed without admission or following discharge from either the Henri Rousselle or the Asile Clinique. At both the Maudsley and the Henri Rousselle hospitals, demand for outpatient services increased rapidly. Between 1923 and 1935, the number of new outpatients registered at the Maudsley rose three-fold, the total number treated four-fold and the number of attendances 13-fold.[113] By 1935, the outpatient facilities were deemed "grossly inadequate" for the numbers of patients wanting to use them and were

[109] Ibid., 1927, 8.
[110] Ibid.
[111] Ibid., 1924, 8.
[112] Andrews et al., History of Bethlem, 557.
[113] Report of the Medical Superintendent, Maudsley Hospital, London, 1932–1935, 8.

therefore extended in 1936.[114] Between the opening of the Henri Rousselle outpatient facilities in 1922 and 1934, the number of consultations given annually at the outpatient department rose tenfold from 3289 to 31,817.[115]

The success of the outpatient facilities at the Henri Rousselle prompted the Minister of Health, Marc Rucart, to encourage their provision nationwide in 1937. The Minister issued a Circular in which he proposed the reorganisation of care for the mentally ill, along the lines already established by the Henri Rousselle Hospital in Paris.[116] Acknowledging that measures to combat mental illness had not been pursued as vigorously as those taken to fight other social scourges (such as tuberculosis), Rucart set out similar proposals for reform to those put forward by the English authorities over a decade earlier. He stressed the "therapeutic, economic and social importance" of early treatment and maintained that pauper mental patients, "easily curable" at the start of their illness, became a danger to themselves and others if their condition was left untreated.[117] By the time their condition had deteriorated to the point where the law intervened, they faced the prospect of long-term internment in an asylum, perhaps for life, for which there was a heavy cost to society, both in terms of the patients' care and their lack of productivity.[118] Rucart wanted to see the provision of "open services" for the voluntary admission of those with mild symptoms, outpatient facilities, social services and assistance for "abnormal children" in all departments.[119] These recommendations were followed by another ministerial Circular, issued in 1938, which attempted to modify (but did not supplant) the regulations set out in 1857. They sought to re-orientate the care of patients in "closed" asylums, such as the Asile de la Sarthe, towards a focus on treatment rather than custodial

[114] Ibid.; Twenty-third Annual Report of the Board of Control (London: HMSO, 1936), 490.

[115] René Fiquet, Rapport au nom de la 3e Commission sur le Centre de psychiatrie et de prophylaxie mentale Henri-Rousselle et les services annexes (Paris: Conseil Général de la Seine, 1934), 3. ADP/D.10 K3/51/88.

[116] Marc Rucart, Le Ministre de la Santé publique à Messieurs les Préfets, Circulaire du 13/10/1937, Relative à la réorganisation de l'Assistance psychiatrique dans le cadre départemental, 35. https://www.ascodocpsy.org/wp-content/uploads/textes_officiels/Circulaire_13octobre1937.pdf (accessed 20/11/2018).

[117] Ibid., 34.

[118] Ibid.

[119] Ibid., 35–6.

care.[120] Henceforth asylums were to be known as "psychiatric hospitals" to emphasise this new focus, although as psychiatrist Paul Balvet observed, the change was in name only.[121]

Most French provincial institutions lacked either the finances or the will to instigate the proposals set out by Rucart before the outbreak of World War II in 1939. This lack of reform within provincial institutions led historians Postel and Quetel to observe that the further French asylums were from cities the more they remained locked into psychiatric conservatism. They point to a cleavage between the Parisian Henri Rousselle Hospital where the principles of mental hygiene were adopted, under Toulouse's influence (including the establishment of "open" services, outpatient clinics, social services and research facilities), and provincial institutions that were effectively left behind.[122] Their distance from the capital and the reformist agenda of Toulouse and his colleagues, together with their isolation from the rest of medicine, meant that many psychiatrists in French provincial asylums continued to run their asylums on a custodial basis.

There was no outpatient department at the Asile de la Sarthe, nor was it able to open its doors to voluntary patients. That said, as Dr. Christy (chief medical officer of the Asile de la Sarthe from 1935 to 1938) observed, "open" services might not have been appropriate in the department of La Sarthe whose high proportion of rural inhabitants rarely sought medical advice. The occupation and life-style of peasant farmers, in Christy's opinion, meant that they were less likely to seek help for mental illness voluntarily than urban dwellers.[123] Christy supported the idea of early treatment, particularly for cases of GPI and dementia praecox, which would benefit from the early administration of malaria therapy and shock treatments respectively, but he recognised that the delivery of any form of treatment was almost impossible at the Asile de la Sarthe with only one

[120] Marc Rucart, Ministre de la Santé publique, Circulaire du 5/02/1938, Règlement modèle du Service Intérieur des Hôpitaux psychiatriques, 39. https://www.ascodocpsy.org/wp-content/uploads/textes_officiels/Circulaire_5fevrier1938.pdf (accessed 03/12/2017).

[121] Jacques Postel and Claude Quétel, *Nouvelle Histoire de la Psychiatrie* (Paris: Dunod, 2012), 351. The Asile Clinique became a "treatment hospital" in 1927, but only officially changed its name in 1938 to the Hôpital Psychiatrique de Ste. Anne.

[122] Ibid., 352. See also: Aude Fauvel, "Aliénistes contre psychiatres: la médecine en crise (1890–1914)," *Psychologie Clinique* no. 17 (2004): 61–76.

[123] Rapport du Médecin-en-Chef sur l'Étude prescrite par M. le Ministre de la Santé Publique pour l'extension de la prophylaxie mentale, Le Mans, 18/02/1938. ADS-1X952.

doctor for 900 patients.[124] Despite Rucart's recommendations, the Asile de la Sarthe remained a "closed" institution, overcrowded and dominated by incurable and chronic cases, until after World War II. This had consequences for patient occupation, which also remained unchanged throughout the interwar period. Routinised work, organised to fulfil institutional requirements, remained the main means of occupying calm, chronic and incurable inmates and the few convalescent patients who made a recovery. The latter could not benefit from the services of a psychiatric social worker to help them secure employment, nor could they continue receiving treatment as an outpatient while pursuing their regular employment.

Had Rucart's recommendations been acted upon before the outbreak of World War II, the French system of mental health provision might have appeared altogether more uniform and more in line with the American system, but the American model remained the preserve of the Henri Rousselle Hospital. Although the English mental healthcare system might have been considered more "advanced" than the French in terms of the national distribution of outpatient facilities, educational and preventative work, psychiatric social work and support for discharged patients, provision was nonetheless piecemeal. Services varied across the country and were provided by a mixture of charitable and public organisations whose activities were poorly co-ordinated. While at one end of the scale, the Maudsley Hospital provided a model of best practice, there were parts of the country where there was no support at all for patients outside hospital. Mindful of these shortcomings, the English Ministry of Health set up a committee, led by Lord Feversham, to investigate. The Feversham Committee reported in 1939, recommending that in the interests of efficiency all the different bodies providing mental healthcare should be amalgamated.[125] The committee believed that "Mental Health should be recognised as a *single concept*" and this belief dictated the tone of the report.[126] The report also highlighted the "encouragement of community care" resulting from the Mental Treatment Act (1930), an early indication of how mental healthcare provision would develop.[127] The report recommended that every Local Authority should appoint a Mental Health

[124] Ibid.

[125] Kathleen Jones, *Asylums and After. A Revised History of the Mental Health Services: From the Early 18th Century to the 1990s* (London: Athlone Press, 1993), 139.

[126] Henry Herd, "The Voluntary Mental Health Services: Report of the Feversham Committee," *Mental Welfare* 20 no. 4 (1939), 99.

[127] Ibid.

Committee to deal with all matters relating to mental welfare, thereby bringing the control of institutions, early treatment of mental disorders, establishment of clinics, community care and education of the public under one co-ordinating body that would operate like the Public Health Committees.[128] The Feversham Report, like Rucart's Circular, was not acted upon before World War II, but its recommendations had considerable influence after the conflict.[129]

Conclusion

French patients, unlike their English counterparts, left the asylum with a small "nest-egg", money earned whilst working in the hospital, but in general support for patients outside hospital was more limited in France than in England. The Henri Rousselle was the only institution of its type in the capital, and its social service and outpatient facility could not provide support for all those who needed it. The increasing levels of demand for the Maudsley's outpatient facility were met by an expansion of the service and the opening of additional facilities in north London. This did not happen in Paris, where the Henri Rousselle remained the only provider. Édouard Toulouse and his associates in Paris were unique in their support for the holistic, American model of psychiatric care, which took longer to attract support from French psychiatrists. It was not until 1937 that the Health Minister, Marc Rucart, saw fit to recommend the services comprising the Henri Rousselle Hospital to other institutions in the capital and the provinces. This was not forthcoming until after World War II, however. In England, on the other hand, support for the American model was more widespread. Its principles had been recommended by the Macmillan Report of 1926. The Mental Treatment Act of 1930, which provided for the widespread introduction of the outpatient system and the relaxation of the laws regarding admission, paved the way for greater fluidity between hospital and community in England. Psychiatric social work services were embryonic during this period, but their emergence demonstrated the beginnings of a transition from voluntary to professional service provision. Psychiatric social workers were regarded as an "essential" aspect of modern psychiatry by English psychiatrists and by Toulouse in Paris. Their role in helping discharged patients find work outside hospital

[128] Ibid.
[129] Jones, *Asylums and After*, 140.

was a lifeline for some, particularly since the occupations allocated to patients in hospital did not prepare them for local employment opportunities or modern working practices in either England or France. Work around the hospital was anachronistic, remaining similar in character to the work provided in the early days of the asylum system. Occupational therapy, on the other hand, provided patients with an experience that was the opposite to that which they would encounter in a modern factory. This anomaly would eventually be addressed by the introduction of industrial therapy in England in the late 1950s, the same decade that occupational therapy was being introduced in France. Patient occupation could thus be regarded as a "barometer" of developments within psychiatry.

Bibliography

Secondary Sources

Accampo, Elinor, Rachel Fuchs and Mary Lynn Stewart. *Gender and the Politics of Social Reform in France, 1870–1914.* Baltimore and London: Johns Hopkins University Press, 1995.

Andrews, Jonathan, Asa Briggs, Roy Porter, Penny Tucker, and Keir Waddington. *The History of Bethlem.* Abingdon and New York: Routledge, 1997.

Bartlett, Peter. "The Asylum and the Poor Law: The Productive Alliance." In *Insanity, Institutions and Society, 1800–1914: A Social History of Madness in Comparative Perspective*, edited by Joseph Melling and Bill Forsythe, 48–67. London and New York: Routledge, 1999.

Blakesley, Rosalind. *The Arts and Crafts Movement.* London and New York : Phaidon, 2006.

Blayney, Steffan. "Industrial Fatigue and the Productive Body: the science of work in Britain, 1900–1918." *Social History of Medicine* 32 no. 2 (2019): 310–28.

Burnett, John. *Idle Hands: The Experience of Unemployment, 1790–1990.* London and New York: Routledge, 1994.

Buyst, Erik and Piotr Franaszek. "Sectoral Developments, 1914–1945." In *The Cambridge Economic History of Modern Europe, Volume II: 1870 to the Present*, edited by Stephen Broadberry and Kevin O'Rourke, 208–231. Cambridge: Cambridge University Press, 2010.

Chance, Eleanor. "Modern Oxford." In *A History of the County of Oxford: Volume IV, the City of Oxford,* edited by Alan Crossley and C. R. Elrington. London: Victoria County History 1979. *British History Online.* Accessed 02/03/2019. http://www.british-history.ac.uk/vch/oxon/vol4.

Chaney, Sarah. "Useful members of society, or motiveless malingerers? Occupation and malingering in British asylum psychiatry, 1870–1914." In *Work, Psychiatry and Society, c.1750–2015* edited by Waltraud Ernst, 277–97. Manchester: Manchester University Press, 2016.

Cullen, Lynsey T. "The First Lady Almoner: The Appointment, Position, and Findings of Miss Mary Stewart at the Royal Free Hospital, 1895–99." *Journal of the History of Medicine and Allied Sciences* 68 no. 4 (2012): 551–82.

Dupré, John and Regenia Gagnier. "A Brief History of Work." *Journal of Economic Issues*, 30 no. 2 (1996): 553–9.

Ernst, Waltraud. "Introduction: Therapy and Empowerment, Coercion and Punishment. Historical and contemporary perspectives on work, psychiatry and society." In *Work, Psychiatry and Society, c.1750–2015*, edited by Waltraud Ernst, 1–30. Manchester: Manchester University Press, 2016.

Fauvel, Aude. "Aliénistes contre psychiatres: la médecine en crise (1890–1914)." *Psychologie Clinique* no. 17 (2004): 61–76.

Fridenson, Patrick. "Automobile Workers in France and Their Work, 1914–1983," in *Work in France: Representations, Meaning, Organisation and Practice*, edited by Stephen Laurence Kaplan and Cynthia J. Koepp, 54–64. Ithaca and London: Cornell University Press, 1986.

Grob, Gerald. *Mental Illness and American Society, 1875–1940.* Princeton: Princeton University Press, 1983.

Guillemain, Hervé. *Chronique de la Psychiatrie Ordinaire: Patients, Soignants et Institutions en Sarthe du 19e au 21e Siècle.* Tours: Éditions de la Reinette, 2010.

Inwood, Stephen. *A History of London.* Basingstoke, Macmillan, 1998.

Jones, Colin. *Paris, Biography of a City.* New York and London: Penguin, 2004.

Jones, Kathleen. *Asylums and After. A Revised History of the Mental Health Services: From the Early 18th Century to the 1990s.* London: Athlone Press, 1993.

Lees, Lynn Hollen. *The Solidarity of Strangers. The English Poor Laws and the People, 1700–1948.* Cambridge: Cambridge University Press, 1998.

Lévy, André, ed. *La Sarthe Des Origines À Nos Jours.* Saint-Jean-d'Angély: Éditions Bordessoules, 1983.

Long, Vicky. "Work is therapy? The function of employment in British Psychiatric Care after 1959." In *Work, Psychiatry and Society, c.1750–2015,* edited by Waltraud Ernst, 334–50. Manchester: Manchester University Press, 2016.

Marx, Karl. *Capital, Volume I: Capitalist Production* [1885]. Tanslated by Samuel Moore and Edward Aveling. Ware, Herts: Woodsworth Editions, 2013.

McIvor, Arthur J. *A History of Work in Britain, 1880–1950.* Basingstoke and New York: Palgrave, 2001.

Noiriel, Gerard. *Workers in French Society in the 19th and 20th Centuries.* New York: Berg, 1989.

Postel, Jacques and Claude Quétel. *Nouvelle Histoire de la Psychiatrie.* Paris: Dunod, 2012.

Rabinbach, Anson. *The Human Motor: Energy, Fatigue and the Origins of Modernity.* Berkeley: University of California Press, 1992.
Royce, Edward. *Classical Theory and Modern Society: Marx, Durkheim and Weber.* London and New York: Rowman and Littlefield, 2015.
Royle, Edward. *Modern Britain: A Social History 1750–2011.* Third edition. London and New York: Bloomsbury, 2012.
Schneider, William. "The Scientific Study of Labour in Interwar France." *French Historical Studies* 17 no. 2 (1991): 410–46.
Smith, Adam. *An Inquiry into the Nature and Causes of the Wealth of Nations* [1776]. Ware, Herts.: Wordsworth Editions, 2012.
Soanes, Stephen. *Rest and Restitution: Convalescence and the Public Mental Hospital in England, 1919–1939.* Thesis. Warwick: University of Warwick, 2011.
Timms, Noel. *Social Work in Great Britain, 1939–1962.* London: Routledge & Kegan Paul, 1964.
Whiteside, Noel. "Constructing Unemployment: Britain and France in Historical Perspective," *Social Policy and Administration* 48, no. 1 (2014): 67–85.
Whiting, C. *The View from Cowley: The Impact of Industrialisation upon Oxford, 1918–1939.* Oxford: Clarendon Press, 1983.
Wojciechowski, Jean-Bernard. *Hygiène Mentale et Hygiène Sociale: Contribution à l'Histoire de l'Hygiénisme. Tome I.* Paris: L'Harmattan, 1997.
Woolley, Liz. "Industrial Architecture in Oxford, 1870 to 1914." *Oxonia* (2010): 67–96.

Primary Sources

Bulletin Officiel du Ministère de l'Intérieur. Circulaire No. 7 du 20 mars 1857, 42–79. Paris, 1857. Accessed 12/02/2018. https://gallica.bnf.fr/ark:/12148/bpt6k5539701n/f2.item.zoom [Articles 154–163]
Casson, Elizabeth. "Foreword." In *The Story of Dorset House School of Occupational Therapy 1930–1986,* edited by Betty Collins, 3–6. Oxford: unpublished, 1986. Oxford Brookes University Library, Special Collections: Dorset House Collection.
Chausse, M.E. Rapport Général au nom de la 3e Commission sur le compte du Service des aliénés. Paris : Conseil Général de la Seine, 1932. ADP/D.10K3/48/95.
Dausset, Louis. Rapport Général au nom de la 3e Commission sur les propositions budgétaires pour le Service des Aliénés (budget de 1919). Paris : Conseil Général de la Seine, 1918. ADP/D.10K3/27/20.
Eighteenth Annual Report of the Board of Control. London: HMSO, 1931.
Fifteenth Annual Report of the Board of Control. London: HMSO, 1928.
Fiquet, René. Rapport au nom de la 3e Commission sur le Centre de psychiatrie et de prophylaxie mentale Henri-Rousselle et les services annexes. Paris: Conseil Général de la Seine, 1936. ADP/D.10K3/54.

Fiquet, René. Rapport au nom de la 3e Commission sur le Centre de psychiatrie et de prophylaxie mentale Henri-Rousselle et les services annexes. Paris: Conseil Général de la Seine, 1934. ADP/D.10K3/51/88.

Herd, Henry. "The Voluntary Mental Health Services: Report of the Feversham Committee." *Mental Welfare* 20 no. 4 (1939), 98–103.

Lewis, Aubrey. "Neurosis and Unemployment." *The Lancet* no. 2 (1935): 293–97.

Littlemore Hospital Admissions Book 1922–1965. OHA/L7/A1/1/10.

Macmillan, Hugh Pattison. Report of the Royal Commission on Lunacy and Mental Disorder. London: HMSO, 1926. ["The Macmillan Report"]

Marchand, Dr, Rapport de la 1re Section (Hommes), Asile Clinique. Paris: Préfecture de la Seine, 1935 et 1936. PDS. ADP/PER-566-6.

Nineteenth Annual Report of the Board of Control. London: HMSO, 1932.

Pinel, Philippe. *Traité médico-philosophique sur l'aliénation mentale, ou la manie*. Paris: chez Richard, Calille et Ravier, 1800.

Rapport du Médecin en Chef, 1931. Asile de la Sarthe. Le Mans, 1932. [archive ref]

Rapport du Médecin en Chef, 1933. Asile de la Sarthe. Le Mans, 1933.

Rapport du Médecin-en-Chef sur l'Étude prescrite par M. le Ministre de la Santé Publique pour l'extension de la prophylaxie mentale. Le Mans 18/02/1938. ADS-1X952.

Report of the Board of Control. Annual Reports of the Littlemore Hospital. Oxford, 1936/1937. OHA-L1/A2/14.

Report of the Medical Superintendent. Annual Reports of the Littlemore Hospital. Oxford, 1937/1938; 1938/1939; 1939/19340. OHA-L1/A2/15; OHA-L1/A2/16; OHA-L1/A2/17.

Report of the Medical Superintendent. Maudsley Hospital, London, 1927–1931.

Report of the Medical Superintendent. Maudsley Hospital. London, 1925, 1926, 1927–1931, 1932–35.

Report of the Physician Superintendent. Annual Reports of Bethlem Royal Hospital. London, 1919; 1921; 924; 1927. BMM/BAR-53.

Rousselle, Henri. Rapport au nom de la 3e Commission sur le Service départemental et les services annexes. Paris: Conseil Général de la Seine, 1923. ADP/D.10K3/34/87.

Rousselle, Henri. Rapport au nom de la 3e Commission sur les comptes du directeur de l'Asile Clinique. Paris: Conseil Général de la Seine, 1924. ADP/D.10K3/38/75.

Rucart, Marc. Le Ministre de la Santé publique à Messieurs les Préfets. Circulaire du 13/10/1937, Relative à la réorganisation de l'Assistance psychiatrique dans le cadre départemental. Accessed 03/12/2017. https://www.ascodocpsy.org/wp-content/uploads/textes_officiels/Circulaire_13octobre1937.pdf

Toulouse, E. Rapport sur le fonctionnement et le développement de l'hôpital Henri-Rousselle. CGS, 1926a–1927 ADP/D.10K3/42/91 Annexe.

Toulouse, Édouard. Rapport du Dr Toulouse, médecin-directeur, sur le fonctionnement du Service de la Prophylaxie Mentale durant l'année 1926. Paris: Conseil Général de la Seine, 1926b. ADP/D.10K3/42/53 Annexe.

Toulouse, Édouard. Rapport sur le fonctionnement et le développement de l'hôpital Henri-Rousselle. Paris: Conseil Général de la Seine, 1926c–1927. ADP/D.10K3/42/91 Annexe.

Twenty-third Annual Report of the Board of Control. London: HMSO, 1936.

CHAPTER 10

Conclusions

Occupation has been a feature of asylum or mental hospital regimes since the early nineteenth century. The rationales for its prescription to patients have varied in time and space, but occupation itself has been a constant. The comparison of patient occupation in French and English institutions during the interwar period has revealed very different attitudes towards the therapeutic value of occupation not only between the two countries, but also between metropolitan and provincial institutions. In addition, the value attributed to therapeutic occupation in France and England has shone a light on the different perceptions of mental disorder held by psychiatrists, and the different models of care associated with each, in the two countries at that time. Comparing how patients were occupied has also highlighted the varied influences at play, such as attitudes towards labour, poverty and welfare, national competitiveness, professional pride, and the effects of war on both the psyche and the economy. The comparison accentuates the importance of personal relationships, professional networks, and the beliefs, training and aptitudes of those prescribing and supervising patient occupation. It draws attention to the different levels of resources available to metropolitan and rural institutions and the effect this had on the adoption of innovations in treatment. It has demonstrated that medical theories do not necessarily develop in a linear direction and that treatment methods discarded decades previously can be re-imagined and brought back into use.

J. Freebody, *Work and Occupation in French and English Mental Hospitals, c.1918–1939*, Mental Health in Historical Perspective, https://doi.org/10.1007/978-3-031-13105-9_10

Belief in the curative potential of occupation followed the acceptance and rejection of psychological interpretations of mental disorder. As this study has demonstrated, psychiatry has been torn between by two different and competing explanatory models, one focusing on the mind (the psychological model) and one on the body (the somatic or organicist model) throughout the nineteenth and early twentieth centuries. The Anglo-French moral therapists understood mental disorder in terms of the mind. For them, occupation in the form of work and recreational activities was curative. The moral therapists believed that taking a patient away from their usual environment, engaging them in a regular daily routine and providing carefully selected activities that distracted their minds and tired their bodies, would lead to a recovery. As the psychological model lost favour in the mid-nineteenth century, and the organicist model gained primacy, psychiatrists lost faith in the ability of occupation to "cure" mental disorder. Their efforts were focused on finding a somatic explanation and a biological cure for mental disease. Patient work had become an established part of the asylum regime by this time. It had proved to be a useful means of managing the now large numbers of patients packed into the public asylums and of off-setting institutional running costs. Occupation was still considered beneficial for patients, but not curative. Individually designed and supervised work programmes were replaced by a routinised bureaucratic system of work allocation that catered as much for the needs of the institution as the patient. Work was no longer considered suitable for acute-stage patients, and was restricted to calm, chronic and incurable cases and convalescent patients who required little supervision. The records indicate that between 30% and 55% of French and English patients worked as cleaners, or on the asylum farm, or in the grounds, workshops, laundry, and kitchen. The work was described as "mere drudgery" by psychiatrist David Henderson. Patient work continued to be organised in much the same way in France and England until the outbreak of war in 1914.

Acceptance of a "psychobiological" approach, which framed mental illness as a failure of adaptation to the human environment, led psychiatrists to regard mental disorders as "correctible maladjustments", rather than incurable diseases.[1] This approach restored psychiatrists' faith in the curative powers of occupation. Psychobiological theory was advanced by the influential American psychiatrist Adolf Meyer and informed the principles

[1] S. D. Lamb, *Pathologist of the Mind: Adolf Meyer and the Origins of American Psychiatry* (Baltimore: Johns Hopkins University Press, 2014), 4.

of the Mental Hygiene Movement (co-founded by Meyer). It struck a chord amongst progressive English psychiatrists, and a small group of Parisian psychiatrists, after World War I. These psychiatrists, who returned to the idea that mental disorder could have a psycho-social origin, sought to "re-educate" their patients to help them adapt more effectively to their environment. Their methods included psychotherapy and occupation. The (allegedly) new ideas regarding occupational therapy that emerged just before and during World War I in Germany and the USA therefore fell on fertile ground amongst these psychiatrists.

The transfer of these new ideas concerning occupational therapy demonstrates the importance of international personal connections and of the role played by political and scientific rivalries between nations. Clifford Beers' autobiographical account of his experiences led to a personal friendship developing between American former patient Beers and French psychiatrist Édouard Toulouse, and to Toulouse's introduction to the Mental Hygiene Movement. The personal connections developed between the co-founder of the Mental Hygiene Movement, Adolf Meyer, and the young British doctors who worked with him in Baltimore during their training, both before and after World War I, were responsible for the transfer of American theories of occupational therapy to Britain. Scottish psychiatrist David Henderson's enthusiasm for the American style of occupational therapy inspired his English colleagues, as did study visits to asylums at Gutersloh and Santpoort organised by the Royal Medico-Psychological Association in the late 1920s and early 1930s to see Hermann Simon's more active therapy in action. The latter also impressed a small group of French psychiatrists who visited Santpoort, but their positive reaction was not shared by their colleagues. The reception of German ideas in France was fraught with on-going feelings of *revanchisme* following defeat of France by Germany in the Franco-Prussian War of 1870–1871 and by the devastation of French territory caused by German forces during World War I. There was also resentment regarding the German usurpation of France's pole position in the ranks of continental psychiatry. Promotion of Hermann Simon's methods in France fell on deaf ears. "Traditional hostility towards anything German" felt by the French goes some way to explaining this, but another major factor was the on-going organicism of most French psychiatrists.[2]

[2] Jacques Postel and Claude Quétel, *Nouvelle Histoire de la psychiatrie* (Paris: Dunod, 2004), 350.

In France, most psychiatrists outside Paris rejected the psychobiological approach and remained wedded to the somatic model. This insistence on the exclusively physiological nature of mental disorder is explained by the way psychiatry had developed as a profession in France. French psychiatry had always been closely aligned with neurology, which enjoyed a high profile and great respect within medical circles. While in England, neurologists did not have much involvement in asylum medicine, in France, the opposite applied. Psychiatry had yet to develop independently of neurology and many chief medical officers of asylums were neurologists, who saw mental disorder solely in terms of physiology, and sought biological methods of treating it. They were sceptical about treatment methods that could be perceived as "unscientific" by their medical colleagues. Biological and physical treatments, that were being used with increasing success in other areas of medicine, were more likely to enhance the psychiatrists' scientific credentials and professional standing. The biological treatment of acute patients was preferred by the chief medical officers at the Asile Clinique and Asile de la Sarthe. Here, only chronic, incurable and convalescent patients were given work, the nature of which was unchanged since the late nineteenth century.

The profession of psychiatry in England had evolved rather differently, becoming an independent discipline far sooner than in France. This allowed English psychiatrists to break free of the rigidly organicist interpretation of mental disorder far more readily than their French colleagues. The theories of Freud, Janet and James were known—although not widely circulated—by English psychiatrists before World War I. This knowledge was crucial to the development of psychological treatments for the huge number of soldiers suffering from war neuroses during the conflict. The efficacy of these psychological methods broadened English psychiatrists' understanding of mental disorder and paved the way for acceptance of the psychobiological approach developed by Meyer. The French response to war neuroses relied on neurological methods of treatment since most psychiatrists were trained neurologists. The French experience of war neuroses reinforced, rather than challenged, organicist thinking and led to the divergence between French and English approaches to psychiatry during the interwar period. This divergence was highlighted by the acceptance of new thinking about the curative use of occupation in England, and its rejection in most of France.

Paris proved an exception to this rule. The Parisian psychiatrist Édouard Toulouse denounced the rigid organicism of his French colleagues and developed his own version of psychobiology, which he called *la biocratie*. Toulouse supported the preventative agenda of the Mental Hygiene Movement and founded the French League of Mental Hygiene in 1920. Toulouse, an active socialist, was convinced that mental disorder was often linked to poverty and the psychological stress caused by deprivation. His acceptance of psychosocial factors as potential causes of mental disorder put him at odds with French psychiatrists outside the capital and many within it. The Asile Clinique, for example, continued to be led by neurologically oriented psychiatrists, although younger members of staff who joined in the 1930s, such as Jacques Lacan (1901–1981), were more psychologically inclined. In contrast to the mainly biological treatments dispensed at the Asile Clinique, the Henri Rousselle Hospital established by Toulouse, offered psychotherapy and occupational therapy including art classes. All patients were encouraged to keep busy at the Henri Rousselle, but work was voluntary and was not evaluated in the same way as at the Asile Clinique or the Asile de la Sarthe, where the financial contribution made by patient labour was calculated and appeared in the asylum accounts. Toulouse's attitude to occupation marked a departure from the late nineteenth-century asylum system of patient work that remained in place in provincial asylums and highlighted the difference in his overall approach to mental disorder. The Henri Rousselle Hospital had much more in common with London's Maudsley Hospital, where psychotherapy and occupational therapy were also offered, than with other establishments in France.

Both the Maudsley and the Henri Rousselle were unique in their respective countries in providing a combination of voluntary admission, both inpatient and outpatient facilities, research laboratories and psychiatric social services, and for their focus on acute, incipient cases of mental disorder. These public hospitals represented a new model of care, based on that of the general hospital and informed by the principles of the Mental Hygiene Movement. Inpatients were admitted voluntarily, enabling them to avoid the often-lengthy process of certification and to start active treatment at the onset of their symptoms. Outpatient clinics allowed patients to seek treatment without ever being admitted to a mental institution, or to continue their treatment after discharge. Patients could carry on with their regular lives and work whilst still receiving treatment. The employment of professional psychiatric social workers gradually began to replace an uneven patchwork of voluntary service provision in England, while in France the concept was quite new. As well as providing vital information to psychiatrists about their patients' domestic situation, the psychiatric social worker helped patients adjust to life outside hospital, and assisted with finding employment, training, or access to welfare support. The care at the Maudsley and Henri Rousselle was considered "state of the art" and it is noteworthy that the active treatment recommended for acute patients included occupational therapy. Occupational therapy had become the hallmark of a modern hospital.

The Board of Control chastised Bethlem for its lack of occupational therapy provision, maintaining that Bethlem did not deserve its reputation as a progressive institution without making occupational therapy available to patients. The delay in establishing an occupational therapy department was in part due to the enduring organicism of Bethlem's physician superintendent, John Porter-Phillips, who remained committed to identifying the physical causes of mental disorder. Occupational therapy was eventually provided at Bethlem in 1932 following complaints of boredom by patients.[3] Porter-Phillips' conservatism was in marked contrast to the progressive views of Thomas Saxty Good of the Littlemore Hospital. Good, who became convinced of the efficacy of psychotherapy during World War I while treating soldiers suffering from war neuroses, was an early adopter of occupational therapy. He rarely prescribed sedative drugs, while these were routinely used at Bethlem. Good's patients were trusted to take trips into the village or into Oxford for shopping, and to wander freely in the grounds. Very few of the Littlemore wards were locked, a measure applauded by the Board of Control. This "open door" policy represented a significant step towards the opening up of asylums that followed the Mental Treatment Act of 1930. Such measures were unknown in provincial France where asylums remained "closed" institutions. At the Asile de la Sarthe security was tight and work in the open air was limited for fear that patients might escape. Neither psychotherapy nor occupational therapy were offered at the Asile de la Sarthe. Chief medical officer Henry Christy's self-professed enthusiasm for neuropsychiatry and biological methods of treatment for acute patients meant that work remained confined to the chronic, incurable and convalescent patients. The preferences of the individuals in charge of mental hospitals were thus of fundamental importance to the nature of treatment provided. Even within hospitals, there could be differences of approach. At the Maudsley, for example, Edward Mapother's foremost concern was that psychiatry should "do no

[3] Andrews et al., *History of Bethlem* (Abingdon and New York: Routledge, 1997), 689.

harm". He responded far more cautiously to the emergence of the shock treatments than some of his colleagues, such as Eliot Slater.[4]

As well as support from those in charge of mental hospitals, the successful adoption of occupational therapy depended upon well-trained, dedicated nursing staff or occupational therapists. As Good remarked, "the greatest essential to any hospital's success is the human element, *i.e.* the staff and the way it fits in and pulls together".[5] Hermann Simon's more active therapy (MAT) required a cultural change within the hospital and the involvement and support of all members of staff. Nurses played an essential role in supervising patients in their various tasks; they needed to be well trained and fully committed. The American style of occupational therapy required practitioners to have expert skills in arts and crafts. Mental nurse training was available nationwide in England, but in France its provision outside the capital was uneven. Even in Paris, mental nursing failed to attract high quality candidates. While pay and conditions were slightly better than those of provincial asylums, they compared unfavourably with other types of work available in the capital. Inadequate educational attainment (many French mental nurses were illiterate even in the 1930s) compromised the ability of nurses to take advantage of training. The few French psychiatrists who were in favour of occupational therapy, such as Paul Courbon, had little confidence in the ability of their nurses to deliver it.[6] In England, becoming qualified in mental nursing was not obligatory in most hospitals (although it was for senior staff at the Maudsley), but all hospitals provided training and encouraged nurses to sit the mental nursing examinations set by the Royal Medico-Psychological Association or the General Nursing Council.

At the Littlemore, nurses enhanced their mental nursing skills by spending three months training at the Radcliffe Infirmary, to gain experience of general nursing. Mental nurses were expected to learn a craft that they could then teach to patients, enabling them to deliver occupational therapy, thereby avoiding the expense of employing an occupational therapist. Good ensured that nurses were relieved of administrative duties so they could spend more time developing relationships with patients and

[4] Edgar Jones, "Aubrey Lewis, Edward Mapother and the Maudsley," *Medical History Supplement* 47 no. 22 (2003): 13.

[5] Thomas Saxty Good, "The History and Progress of Littlemore Hospital," *Journal of Mental Science* October (1930): 613.

[6] Paul Courbon, "Un voyage d'étude dans les asiles de Hollande," *Annales Médico-psychologiques* no. 2 (1928): 399.

engaging them in activities. The calibre of nurses at the Littlemore was very different to that of nurses at the Asile de la Sarthe. In Le Mans, male nurses had little experience of any form of nursing. Their ineptitude was a source of anxiety for the chief medical officer. Whilst the sisters who provided nursing care on the female side were diligent, the fact that nursing was still being delivered by nuns, so long after the secularisation of most public institutions in France, suggests that methods were old-fashioned. The delivery of occupational therapy by nursing staff therefore offers an insight into the different levels of professionalisation of mental nursing in France and England, and between metropolitan and provincial institutions.

As Good recognised, nurses played a crucial role in the happiness and general wellbeing of patients. Clifford Beers had experienced brutality and a total lack of empathy from his nurses in the 1900s. It can be assumed that where nurse training was prioritised, as at the Littlemore, the patient experience was far better than where nurses had only rudimentary skills and little interest in their work, as on the male side of the Asile de la Sarthe. The damning remarks made by Lomax had prompted an enquiry into nursing standards in England, and fed into the Macmillan Report of 1926, both of which helped to raise standards in England. The implementation of occupational therapy, whether based on Hermann Simon's method or the American style, would have improved patients' experience of work from the grim description given by Lomax, quoted at the beginning of this book. The individualised programmes of activity that characterised occupational therapy—where they were implemented—were a far cry from the conditions endured by Lomax's "closet barrow gang". For the French patients at the Asile de la Sarthe, however, the nature of work for the incurable and convalescent patients did not change a great deal. The lack of occupation for acute-stage patients here and at the Asile Clinique contrasts with the activities devised for patients at this early stage of their illness in English mental hospitals. In England, the middle-class patient's experience of occupation at Bethlem, where sports and leisure activities, and arts and crafts in the occupational therapy department (from 1932) comprised the main forms of occupation, was very different from that of the private patient at the Asile de la Sarthe, where the only option was manual labour. The latter, of course, was considered unsuitable for the middle classes, and made occupation less accessible for them. Patients experienced occupation differently according to the rationale for its prescription.

The varied rationales for occupying patients have provided insights into the financial circumstances of institutions, national and local labour requirements, and attitudes towards individuals with different types of mental disorder. As we have seen, the economic rationale for patient work was bound up with the need for asylums to offset their running costs. Budgets were tight, particularly during the challenging economic climate of the interwar period. Employing patients to perform many of the tasks that would otherwise have incurred significant expense made financial sense, particularly since engaging patients in work could be justified as therapy. The necessity of using patient labour was particularly apparent in France, where plans to transform the Asile Clinique to a hospital for acute patients were delayed due to concerns regarding the loss of patient workers. Here, and even more so at the Asile de la Sarthe, patient work contributed significantly to the asylum budgets, both in terms of labour costs and in terms of the products made or grown by patients, throughout the interwar period. Pressure exerted by the asylum director to meet institutional labour requirements affected decision-making regarding the allocation of patient work. The asylum director's role has been compared to that of a company Chief Executive Officer, managing the workforce to maximise profits.[7] It was for this reason that Édouard Toulouse was keen to become medical director of the Henri Rousselle Hospital (and not chief medical officer), as this dual role put him in charge of both administrative and medical matters.

The economic contribution of patient work was viewed positively by the French authorities who saw it as helping to reduce the burden on the public purse of caring for the mentally disordered. In English mental hospitals, on the other hand, the economic contribution made by occupation was downplayed. The emphasis here, expressly stated in the Macmillan Report of 1926, was on providing active treatment, including occupational therapy, for curable patients. The Board of Control's *Memorandum* also spelled out that occupation was for the purpose of therapy and not for saving money on the production of commodities for the hospital. This was justified on the grounds that focusing on active therapy would enable curable patients to recover faster and to leave the hospital sooner, which would be cheaper in the long term. The attitude towards the work undertaken by incurable, or "mentally deficient" patients was quite different.

[7] Pierre Delion, "Preface" in François Tosquelles, *Le Travail Thérapeutique En Psychiatrie* (Paris: Éditions du Scarabée, 1967), 10–11.

The work carried out by incurable patients who remained in a mental hospital (as many did at the Littlemore), or who had been placed in specialist institutions, *was* supposed to make a financial contribution to institutional running costs. So, if a patient's condition was curable, the rationale for occupation was therapy, while if incurable, the primary rationale was economic. Examined from this perspective, the rationale for work for incurable patients was similar in France and England.

A rehabilitative rationale for patient work was evident in the English asylum system in the early nineteenth century. Asylum work programmes, whilst modelled on the teaching of the moral therapists, were created in the context of debates concerning the Old Poor Law and its ability to provide for an ever-increasing number of paupers. An ethos of productivity and self-sufficiency, inherent in the New Poor Law of 1834, informed the asylum system. Medical superintendents like Sir William Charles Ellis were keen for patients to have a professional skill that they could use to secure employment after leaving the asylum. The jobs around the asylum, from shoemaking to gardening, were similar to the type of work that could be found outside the institution. In France too, where being a productive citizen was *de rigueur* after the Revolution of 1789, asylum work programmes reflected the work available in the community. Work and working practices in asylums failed to evolve with the changing profile of local and national industry. The type of work offered in mental institutions after World War I did not equip patients either for the emerging new industries, such as the manufacture of automobiles or electrical goods, nor for the modern working practices that were associated with them, such as assembly-line production. Occupational therapy, with its focus on arts and crafts, represented the antithesis of working practices in the modern factory associated with scientific labour management. This failure to provide vocational training, or work designed to replicate local labour market conditions, can be explained in terms of the emphasis on active therapy for curable patients. It was no longer incumbent upon mental hospitals to prepare patients for the workplace. The goal was to cure and expedite discharge. This was particularly true of the Maudsley and Henri Rousselle hospitals, which specialised in treating acute, curable patients and where support with finding employment was available from a psychiatric social worker. For incurable patients who were destined to remain in an institution for the rest of their lives, such as the majority of patients at the Asile de la Sarthe, the question of rehabilitation did not arise as they were unlikely to have to find work outside the asylum.

The need for occupation to fulfil a more rehabilitative remit in English mental institutions was expressed in the late 1950s as deinstitutionalisation loomed. During the interwar period, the main goal of occupational therapy had been to restore mental health, rather than to provide vocational training. As the locus of patient care began to shift from the hospital to the community after 1959, former in-patients were required to support themselves and needed the skills to do so. Despite the alleged potential of modern working practices to compromise mental health, "industrial therapy" (IT) was introduced in the late 1950s. Factory-style workshops were recreated in many mental hospitals, as Vicky Long has outlined, either in addition to, or in many cases, replacing occupational therapy workshops.[8] Work on the assembly line was monotonous, mundane and offered little satisfaction. It was "alienating" in the Marxist sense and could be psychologically damaging. While industrial therapy prepared patients for the working environments they were likely to encounter outside hospital, whether it could accurately be described as "therapy" is debatable. A therapeutic rationale for occupation had ceded "pole position" to a rehabilitative one in response to changing mental health policies rather than new psychiatric ideology.

As occupational therapy was being replaced by industrial therapy in many English mental hospitals in the 1950s, occupational therapy was emerging as a new profession in France. A major re-assessment of care for the mentally disordered took place after 1945, prompted by revelations of the neglect and ill-treatment of asylum patients during World War II.[9] The re-assessment led to the introduction of "institutional psychotherapy", promoted by a group of young, militant psychiatrists known as *L'information psychiatrique*. This group included the influential post-war psychiatrist François Tosquelles, who was inspired by the work of both Sigmund Freud and Hermann Simon.[10] Institutional psychotherapy was based on a return to the founding principles of modern psychiatry, as laid down by moral therapist and creator of French asylum legislation,

[8] Vicky Long, "Work Is Therapy? The function of employment in British Psychiatric Care after 1959." in *Work, Psychiatry and Society, c.1750–2015*, ed. Waltraud Ernst (Manchester: Manchester University Press, 2016), 342.

[9] Isabelle von Bueltzingsloewen, *L'Hécatombe des Fous: La Famine dans les hôpitaux psychiatriques français sous l'occupation* (Paris: Éditions Flammarion, 2007), 403.

[10] Tosquelles produced his own guide to occupational therapy; François Tosquelles, *Le Travail Thérapeutique En Psychiatrie* (Paris: Éditions du Scarabée, 1967).

Jean-Étienne Dominique Esquirol, pupil of the revered Philippe Pinel.[11] This marked a renewed acceptance of a more psychological interpretation of mental disorder by French psychiatrists and an attempt to "rehumanise" psychiatry. Tosquelles sought to transform patient work from the way it was practised in most French asylums (which he regarded as having nothing to do with therapy) to occupational therapy. The latter became a key aspect of "institutional psychotherapy". Occupational therapy services were provided by those who had trained in Anglo-Saxon schools until two French schools of occupational therapy were established in 1954, one in Paris and one in Nancy.[12] A national association of occupational therapy was created in 1961 and a diploma in 1970.[13]

French psychiatry had operated in the shadow of neurology for so long that it had lost sight of the "mind". Psychiatry's return to the founding principles of the profession after World War II saw the "mind" return. As it did so, the benefits of using occupation therapeutically were re-evaluated and its curative potential recognised once more. A similar transformation occurred in English psychiatry after World War I, as psychobiology took the place of organicism, re-introducing the psychological perspective to psychiatry. In the wake of this ideological shift, occupational therapy replaced the routinised patient work of the late nineteenth century for curable patients. Occupational therapy was not really new, however. It was a re-imagined, more sophisticated version of moral treatment. It represented a return to prioritising a therapeutic agenda for patient occupation, whether this involved work around the hospital, or the production of arts and crafts. Its remit was formalised and professionalised, but the basic principles were the same as those of moral treatment. Both occupational therapy and work in the context of moral treatment provided a distraction from a patient's troubles, instilled regular habits and boosted self-esteem. Both encouraged patients to control their symptoms and behave in a way that corresponded to contemporary social norms. This in turn helped patients to adapt more effectively to their environment. Both were based on an individualised approach designed to suit the needs, interests and condition of the patient. These transformations in theory and practice

[11] Nicolas Henckes, "Réformer et Soigner. L'Émergence de la Psychothérapie Institutionnelle en France, 1944–1955," in *Psychiatries dans L'Histoire*, ed. Jacques Arveiller (Caen: PUC, 2008), 278.

[12] Lisbeth Charret and Sarah Thiébaut Samson, "Histoire, fondements et enjeux actuels de l'ergothérapie," *Contraste* 1 no. 45 (2017): 21.

[13] Ibid., 22.

emphasise the cyclical nature of the paradigms underpinning psychiatry and occupational therapy.

English psychiatrists began to question the efficacy of activities based on arts and crafts in the 1960s, considering them too unscientific and insufficiently evidence based.[14] While arts and crafts still comprised a significant aspect of occupational therapy training in the 1960s and 1970s, they were used less and less in hospitals. Ironically, and a further indication of the cyclical nature of treatments, art therapy, music therapy and drama therapy developed as an independent disciplines after World War II, filling the gap left in these areas by occupational therapy.[15] Most recently, music therapy has been found to be particularly helpful in the treatment of dementia and schizophrenia.[16] "Social prescribing" is another recent non-clinical intervention aimed at improving mental health and wellbeing.[17] Patients can be referred by their general practitioner to local community based organisations for access to support and advice on such topics as loneliness, social networking, volunteering and opportunities to engage in arts and crafts, and other creative activities, as well as healthy eating, legal advice and debt counselling.[18] Estimates that around 20% of patients consult their general practitioner for social issues have prompted a proliferation of social prescribing and have led to its description as the "topic of the moment" by one group of researchers.[19] The aim of social prescribing to apply the "common knowledge that people's health is largely determined by socioeconomic factors" resonates with the preventative agenda of the

[14] Annie Turner, "The Elizabeth Casson Memorial Lecture 2011: Occupational Therapy—a profession in adolescence?" *British Journal of Occupational Therapy* 74 no.7 (2011): 318.

[15] See: David Edwards, *Art Therapy*, 2nd edition (Los Angeles and London: SAGE, 2014); Judith A. Rubin, *Introduction to Art Therapy*, 2nd edition (Hoboken: Taylor and Francis, 2009); Leslie Bunt and Brynjulf Stige, *Music Therapy: An Art Beyond Words*, 2nd edition (London: Routledge, 2014); A. Lewis and R.W. Alley, *Music Therapy* (Newburyport: Abbey Press, 2014); Phil Jones, *Drama as Therapy, Theatre as Living* (East Sussex and New York: Brunner-Routledge, 2002); Roger Grainger, *Drama and Healing: the roots of drama therapy* (London: J. Kingsley, 1990).

[16] Melissa Owens, "Remembering Through Music: Music therapy and Dementia," *Age in Action* 29 no. 3 (2014): 1–5.

[17] M. Mofizul Islam, "Social Prescribing—An Effort to Apply a Common Knowledge: Impelling Forces and Challenges," *Frontiers in Public Health* 8 no. 515469 (2020): 1–7.

[18] Ibid., 2.

[19] Kerryn Husk et al., "Social prescribing: where is the evidence?" *British Journal of General Practice* 69 no.678 (2019): 6–7.

Mental Hygiene Movement.[20] Art, music, drama, and social activities have all featured in therapeutic occupation since the early days of moral treatment, resurfacing at intervals right up to the present day.

Occupation of some form, or perhaps the lack of it, has been at the heart of the mental patient's experience in England and France since the early nineteenth century. Activities that patients enjoyed, such as arts and crafts, or work that was satisfying (rather than "mere drudgery"), such as gardening, were more likely to be provided to patients who were deemed curable in English mental hospitals during the interwar period than in French institutions. Curable French patients, other than those at the Henri Rousselle Hospital, were either unoccupied whilst undergoing biological treatments, or, if convalescent, obliged to perform work around the hospital about which there appeared to be little choice, but for which they were paid. Occupation did not appear to reflect the individual tastes or aptitudes of patients in French institutions; rather, it was allocated according to the needs of the institution. Whilst occupation (or idleness) was fundamental to the patient experience, so too were the institution's urban or rural location, and whether it was publicly or charitably funded. The living conditions, the recreational facilities, the number and quality of staff were all superior at the charitably funded Bethlem which catered for a middle-class clientele. In France, provincial institutions enjoyed significantly fewer resources that metropolitan hospitals; innovations that occurred at the centre rarely reached the periphery. Patients in French provincial institutions, where the model of care remained custodial during the interwar period, did not enjoy the same degree of liberty as their counterparts in England. The Littlemore patient who could wander into town to do some shopping must have experienced hospital life very differently to the patient kept under lock and key at the Asile de la Sarthe. The knowledge that someone could leave the institution at will—as patients could at the Maudsley and Henri Rousselle hospitals—must have made an enormous difference to the overall experience. This freedom (also available to voluntary patients after 1930 in England) must also have changed a patient's attitude towards work. Even if a patient disliked the occupation prescribed for them, it was not something that had to be endured forever, or for as long as a doctor decreed. The voluntary patient had far more agency, knowing they could leave the institution at any time.

[20] Islam, "Social Prescribing," 2.

This study has revealed the interwar years as a dynamic period for the profession of psychiatry, a period that witnessed the beginnings of a diversification of services for the mentally ill and an "opening up" of institutions in England and Paris. The period was characterised by experimentation with a range of treatments, from malaria therapy to organotherapy, from the shock treatments to occupational therapy. Patients began to be treated as individuals with complex needs. But however innovative the interwar years were, they also revealed that "even in medicine, history repeats itself" as David Henderson observed in 1939.[21] He was talking about putting the "mind" back in psychiatry, which occurred in England and Paris after World War I but not until after World War II in the rest of France. Henderson was keen to stress the "close correlation" between anatomical, physiological and psychological factors.[22] It was acceptance of the psychological dimension of mental disorder that enabled the moral therapists of the early nineteenth century and psychiatrists of the mid-twentieth to recognise the curative potential of occupation. The use of occupation as a medical tool may have waxed and waned over the past two hundred years, but the centrality of work and occupation to an individual's humanity and wellbeing have never been in doubt.

Bibliography

Andrews, Jonathan, Asa Briggs, Roy Porter, Penny Tucker, and Keir Waddington. *The History of Bethlem*. Abingdon and New York: Routledge, 1997.

Bunt, Leslie and Brynjulf Stige. *Music Therapy: An Art Beyond Words*. Second edition. London: Routledge, 2014.

Charret, Lisbeth and Sarah Thiébaut Samson, "Histoire, fondements et enjeux actuels de l'ergothérapie," *Contraste* 1 no. 45 (2017): 17–36.

Courbon, Paul. "Un voyage d'étude dans les asiles de Hollande." *Annales Médico-psychologiques* no. 2 (1928): 289–306; 385–405.

Delion, Pierre. "Preface." In *Le Travail Thérapeutique En Psychiatrie* by François Tosquelles, 7–15. Paris: Éditions du Scarabée, 1967.

Edwards, David. *Art Therapy*. Second edition. Los Angeles and London: SAGE, 2014.

Good, Thomas Saxty. "The History and Progress of Littlemore Hospital." *Journal of Mental Science* October (1930): 602–21.

[21] D.K. Henderson, "The Nineteenth Maudsley Lecture: A Revaluation of Psychiatry," *Journal of Mental Science* 85 no. 354 (1939): 17.

[22] Ibid.

Grainger, Roger. *Drama and Healing: the roots of drama therapy.* London: J. Kingsley, 1990.

Henckes, Nicholas. "Réformer et Soigner. L'Émergence de la Psychothérapie Institutionnelle en France, 1944–1955." In *Psychiatries dans L'Histoire*, edited by Jacques Arveiller, 277–88. Caen: PUC, 2008.

Henderson, D. K. "The Nineteenth Maudsley Lecture: A Revaluation of Psychiatry," *Journal of Mental Science* 85 no. 354 (1939): 1–21.

Husk, K., J. Elston, F. Gradinger, L. Callaghan and S. Asthana. "Social prescribing: where is the evidence?" *British Journal of General Practice* 69 no.678 (2019): 6–7.

Islam, M. Mofizul. "Social Prescribing—An Effort to Apply a Common Knowledge: Impelling Forces and Challenges," *Frontiers in Public Health* 8 no. 515469 (2020): 1–7.

Jones, Edgar. "Aubrey Lewis, Edward Mapother and the Maudsley." *Medical History Supplement* 47 no. 22 (2003): 3–38.

Jones, Phil. *Drama as Therapy, Theatre as Living.* East Sussex and New York: Brunner-Routledge, 2002.

Lamb, S. D. *Pathologist of the Mind: Adolf Meyer and the Origins of American Psychiatry.* Baltimore: Johns Hopkins University Press, 2014.

Lewis, A. and R. W. Alley. *Music Therapy.* Newburyport: Abbey Press, 2014.

Long, Vicky. "Work is therapy? The function of employment in British Psychiatric Care after 1959." In *Work, Psychiatry and Society, c.1750–2015*, edited by Waltraud Ernst, 334–50. Manchester: Manchester University Press, 2016.

Owens, Melissa. "Remembering Through Music: Music therapy and Dementia," *Age in Action* 29 no. 3 (2014): 1–5.

Postel, Jacques and Claude Quétel. *Nouvelle Histoire de la psychiatrie.* Paris: Dunod, 2004.

Rubin, Judith A. *Introduction to Art Therapy*, Second edition. Hoboken: Taylor and Francis, 2009.

Turner, Annie. "The Elizabeth Casson Memorial Lecture 2011: Occupational Therapy—a profession in adolescence?" *British Journal of Occupational Therapy* 74 no.7 (2011): 315–22.

Von Bueltzingsloewen, Isabelle. *L'Hécatombe des Fous: La Famine dans les hôpitaux psychiatriques français sous l'occupation.* Paris: Éditions Flammarion, 2007.

BY

Appendix

Table A.1 Table to show the male and female patient population of the Asile Clinique (comprising the male and female divisions, and the Faculty Clinic) and the population of all six asylums of the Seine department, 1919–1938, as recorded on 31 December of each year

	Males—Asile Clinique	*Females—Asile Clinique*	*Total—Asile Clinique*	*Males—Seine*	*Females—Seine*	*Total—Seine*
1919	568–51%	546	1114	3060	4723	7783
1921	600	509	1109	3752	5584	9338
1922	595	499	1094	3364	5286	8650
1923	522	533	1055	3414	5411	8825
1924	499	495	994	3486	5596	9082
1925	506	512	1018	3478	5754	9232
1926	507	500	1007	3416	5524	8940
1927	466	457	923	3564	5617	9181
1928	542	501	1043	3523	5694	9217
1929	571 = 53%	506	1077	3638	5766	9404
1930	581	502	1083	3835	5847	9682
1931	573	501	1074	3814	6028	9842
1932	915	716	1631	4531	6575	11,106
1933	574	511	1085	4235	6677	10,912
1934	582	510	1092	4519	6647	11,166
1935	585	509	1094	4931	7185	12,116
1936	577	513	1090	5351	7550	12,901
1937	588 = 54%	496	1084	5653	6900	12,553
1938						

Source for Tables A.1-8: Reports of the Chief Medical Officers to the Prefecture, Asile Clinique, 1919–1938. ADP/PER-566-5/6

J. Freebody, *Work and Occupation in French and English Mental Hospitals, c.1918–1939*, Mental Health in Historical Perspective,
https://doi.org/10.1007/978-3-031-13105-9

Table A.2 Table to show the population of the Asile Clinique men's division, 1919–1925 (prior to transformation of the service in 1927)

	1919	*1921*	*1922*	*1923*	*1924*	*1925*
Present 1 January	266	380	405	394	401	404
Admitted during year	502	285	304	341	334	423
Total treated	768	665	709	735	735	827
Discharged	220	161	234	228	228	324
Deaths	175	99	81	106	103	101
Remaining	373	405	394	401	404	402

Table A.3 Table to show the population of the Asile Clinique women's division, 1919–1926 (prior to transformation of the service in 1927)

	1919	*1921*	*1922*	*1923*	*1924*	*1925*	*1926*
Present 1 January	361	382	413	343	372	365	362
Admitted during year	189	175	259	215	238	220	117
Total treated	550	557	672	558	610	585	479
Discharged	128	142	260	134	190	163	123
Deaths	40	42	69	52	55	60	38
Remaining	382	373	343	372	365	362	318

Table A.4 Table to show the number of patients resident at the Asile Clinique men's first section on 1 January; those admitted during the year; the total treated during the year and the numbers of patients who were discharged, or died, 1928–1938 (after transformation of the service in 1927)

	1927	*1928*	*1929*	*1930*	*1931*	*1932*	*1933*	*1934*	*1935*	*1936*	*1937*	*1938*
Present 1 January	336	163	218	220	221	220	215	219	225	221	223	222
Admitted during year	148	190	275	257	189	190	217	259	248	221	203	386
Total treated	484[a]	353	493	477	410	410	432	478	473	442	426	608
Discharged	112	109	196	212	154	171	171	219	210	164	147	302
Deaths	21	26	77	44	36	23	42	34	42	55	57	60
Remaining	163	218	220	221	220	216	219	225	221	223	222	246

[a]*Of whom 188 were passed to the 2nd Section following the separation of the service on 1 April*

Table A.5 Table to show the number of patients resident at the Asile Clinique men's second section on 1 January; those admitted during the year; the total treated during the year and the numbers of patients who were discharged, or died, 1928–1938 (after transformation of the service in 1927)

	1927	*1928*	*1929*	*1930*	*1931*	*1932*	*1933*	*1934*	*1935*	*1936*	*1938*
Present 1 January		180	217	243	244	245	252	262	245	249	245
Admitted during year	225	276	316	298	320	242	284	236	225	192	469
Total treated	225	456	533	541	564	487	536	498	470	441	714
Discharged		183	226	239	270	200	230	203	186	150	404
Deaths	33	39	64	58	49	35	44	27	35	43	81
Remaining		234	243	244	245	252	262	268	249	248	229

Table A.6 Table to show the number of patients resident at the Asile Clinique women's first section on 1 January; those admitted during the year; the total treated during the year and the numbers of patients who were discharged, or died, 1928–1938 (after transformation of the service in 1927)

	1927	*1928*	*1929*	*1931*	*1932*	*1933*	*1934*	*1935*	*1936*	*1937*	*1938*
Present 1 January	318	152	177	178	175	178	172	179	178	183	179
Admitted during year	146	96	107	166	138	172	221	213	243	260	253
Total treated	464	248	284	344	313	350	393	392	421	443	432
Discharged	288	60	95	149	114	124	188	170	197	211	197
Deaths	24	11	12	20	21	54	26	44	41	53	54
Remaining	152	177	177	175	178	172	179	178	183	179	181

Table A.7 Table to show the number of patients resident at the Asile Clinique women's second section on 1 January; those admitted during the year; the total treated during the year and the numbers of patients who were discharged, or died, 1928–1938 (after transformation of the service in 1927)

	1927	*1928*	*1929*	*1930*	*1931*	*1932*	*1933*	*1934*	*1935*	*1936*	*1938*
Present 1 January	110	148 + 29[a]	177	178	169	178	175	187	182	181	179
Admitted during year	245	252	233	247	289	240	214	223	207	236	347
Total treated	405	400 + 29[a]	474	425	449	418	389	410	389	417	526
Discharged	224	190	177	111	194	192	179	202	181	192	266
Deaths	33	38	55	59	77	50	23	27	26	41	86
Remaining	148	177	178	169	178	176	187	181	182	184	174

[a] *These patients were sent to the Pavilion for workers which opened on 24 Sept 1928*

Table A.8 Table to show the number of male (M) and female (F) patients resident at the Asile Clinique's Faculty Clinic on 1 January; those admitted during the year; the total treated during the year and the numbers of patients who were discharged, or died, 1919–1938

	1919	*1921*	*1922*	*1923*	*1924*	*1925*	*1926*
Present 1 January	161	190	194	203	181	82	70
(M:F)	(80:81)	(117:73)	(118:76)	(127:76)	(72:109)	(37:45)	(38:32)
Admitted during year	308	226	505	495	460	3824	3575
(M:F)	(191:117)	(114:112)	(244:261)	(193:302)	(183:277)	(1885:1939)	(1690:1885)
Total treated (M:F)	469	416	699	698	641	3906	3645
	(271:198)	(231:185)	(362:337)	(320:378)	(255:386)	(1922:1984)	(1728:1917)
Discharged/died	272	222	495	517	466	3836	3567
(M:F)	(150:122)	(113:109)	(234:261)	(248:269)	(185:281)	(1884:1952)	(1686:1881)
Remaining	197	194	204	181	175	70	78
(M:F)	(121:76)	(118:76)	(128:76)	(72:109)	(70:105)	(38:32)	(42:36)

	1927	*1928*	*1930*	*1932*	*1933*	*1934*	*1936*	*1937*	*1938*
Present 1 January	177	180	181	185	183	182	184	184	175
(M:F)	(68:109)	(69:111)	(69:112)	(70:115)	(67:116)	(69:113)	(70:114)	(67:117)	(67:108)
Admitted during year	447	410	413	313	354	367	347	320	356
(M:F)	(219:228)	(208:202)	(227:186)	(170:143)	(199:155)	(178:189)	(172:175)	(182:138)	(125:231)
Total treated (M:F)	624	590	594	498	537	549	531	504	531
	(287:337)	(277:313)	(296:298)	(240:258)	(266:271)	(247:302)	(242:289)	(249:255)	(192:339)
Discharged/died	444	420	414	315	355	364	347	329	351
(M:F)	(218:226)	(209:211)	(227:187)	(173:142)	(197:158)	(175:189)	(175:172)	(182:147)	(123:228)
Remaining	180	170	180	183	182	185	184	175	180
(M:F)	(69:111)	(68:102)	(69:111)	(67:116)	(69:113)	(72:113)	(67:117)	(67:108)	(69:111)

Table A.9 Table to show the number of patients admitted to the Henri Rousselle Hospital; the number of days' treatment received by in-patients; and the number of consultations conducted at the outpatients' clinic, 1924–1938

Henri Rousselle Hospital	*1924*	*1925*	*1928*	*1929*	*1930*	*1931*	*1932*
No. of days present—Men	1501	1582	1665	1580	1455	1558	1598
No. of days present—Women	4721	4504	4218	4549	4607	4807	3765
Total no. days—Men and Women	6222	6086	5883	6129	6062	6365	5363
No. of men admitted during year	62	54	69	70	84	72	47
No. of women admitted during year	99	92	82	116	127	105	94
Total no. admissions—Men and Women	161	146	151	186	211	177	141
Outpatients dept.							
No. of consultations during year—Men and Women	c.800	c.3000	c.5000	c.6800	c.7500	c.?	c.8200

Henri Rousselle Hospital	*1933*	*1934*	*1935*	*1936*	*1937*	*1938*
No. of days present—Men	1270	1498	1554	1604	1561	1336
No. of days present—Women	2716	1815	3142	2879	3160	3115
Total no. days—Men and Women	3986	3313	4696	4483	4721	4451
Total no. existing on 1/1 (Men: Women)		8 (3:5)	9 (6:3)	16 (6:10)	12 (4:8)	14 (3:11)
No. of men admitted/ discharged during year	83 / 78	84 / 81	83 / 84	62 /64	79 /80	63 / 59
No. of women admitted/ discharged during year	81 / 83	55 / 56	98 /85	73 / 75	78 /75	73 / 78
Deaths (Men: Women)	2 (2:0)	1 (0:1)	0	0	0	1 (1:0)
Total no. remaining 31/12 (M:W)	8 (3:5)	9 (6:3)	18 (5:13)	12 (4:8)	14 (3:11)	12 (6:6)
Outpatients dept.						
No. of consultations during year—Men and women	c.7800	c.8200	c.8700	c.7500	c.8200	c.7800

Source: Reports of the Medical Director, Henri Rousselle Hospital, 1924–1938. ADP/D.10K3/40-55

Table A.10 Table to show the number of male (M) and female (F) patients resident at the Maudsley Hospital on 1 January (or 1 February); those admitted during the year and the total number of patients treated during the year

Year No.	*Year*	*Resident at start of period—M*	*Resident at start of period—F*	*Admitted during year—M*	*Admitted during year—F*	*Total treated during year—M*	*Total treated during year—F*	*Total No. patients treated*
1	To 31/1/1924					189	263	452
2	1/2/1924–31/1/25	65	77	190	258	233	335	590
3	1/2/25–31/1/26	58	78	197	355	255	433	688
4	1/2/26–31/12/26	52	96	189	360	241	456	697
5	1/1/1927–31/12/27	57	96	260	385	317	480	797
6	1/1/1928–31/12/28	52	91	310	343	362	434	796
7	1/1/1929–12/31/29	64	105	269	309	333	414	747
8	1/1/1930–31/12/30	58	103	283	325	341	428	769
9	1/1/1931–31/12/31	75	109	264	348	339	457	796
10	1/1/1932–31/12/32	77	111	299	403	376	514	890
11	1/1/1933–31/12/33	93	133	271	423	364	556	920
12	1/1/1934–31/12/34	94	136	300	413	394	549	943
13	1/1/1935–31/12/35	85	131	322	422	407	553	960

Source: Maudsley Hospital Annual Reports, 1924–1935. BMM/MSR-01

Table A.11 Table to show the number of male (M) and female (F) patients resident at the Bethlem Royal Hospital on 1 January; those admitted during the year; the total number of patients treated during the year; the number of patients discharged due to recovery, improvement, other reasons (such as not responding to treatment within one year), or death, 1919–1939

	1919	*1920*	*1921*	*1922*	*1923*	*1924*	*1925*	*1926*	*1927*	*1928*	*1929*
Present on 1 January (M:F)	210[a] (75:135)	226 (85:141)	213 (87:126)	243 (105:138)	234 (103:131)	224 (103:121)	223 (103:120)	218 (97:121)	208 (20:118)	202 (89:113)	196 (83:113)
Admissions during year	240 (89:151)	261 (110:151)	298 (125:178)	187 (84:103)	174 (73:101)	228 (87:141)	272 (106:166)	298 (124:174)	248 (104:144)	235 (103:132)	220 (91:129)
Total treated during year	450 (164:286)	487 (195:232)	511 (212:299)	430 (189:241)	408 (176:232)	452 (190:262)	495 (209:286)	516 (221:295)	456 (194:262)	437 (192:245)	416 (174:242)
Recoveries (M:F)	97 (32:65)	124 (46:78)	132 (50:82)	83 (39:44)	74 (26:48)	107 (43:64)	106 (49:57)	124 (51:73)	82 (28:54)	82 (43:39)	84 (35:49)
Relieved/ improved (M:F)	31 (12:19)	49 (19:30)	25 (11:14)	24 (9:15)	22 (11:11)	17 (4:13)	29 (11:18)	20 (9:11)	19 (10:9)	30 (12:18)	24 (8:16)
Not improved (M:F)	76 (21:55)	77 (29:48)	91 (34:57)	64 (19:45)	66 (25:41)	79 (32:47)	121 (41:80)	137 (61:76)	125 (53:72)	93 (30:63)	73 (31:42)
Deaths (M:F)	20 (14:6)	26 (14:12)	20 (12:8)	25 (19:6)	22 (11:11)	26 (8:18)	21 (11:10)	27 (10:17)	28 (14:14)	32 (19:13)	29 (16:13)
Remaining 31 December (M:F)	226 (85:141)	211 (87:124)	243 (105:138)	234 (103:131)	224 (103:121)	223 (103:120)	218 (97:121)	208 (90:118)	202 (89:113)	196 (83:113)	178 (76:102)
Average daily no. resident	262[b] (94:168)	210 (85:125)	219 (89:130)	213 (91:122)	209 (97:112)	205 (91:114)	206 (93:113)	214 (95:119)	209 (92:117)	198 (90:108)	196 (83:113)

(continued)

Table A.11 (continued)

	1930	*1931*	*1932*	*1933*	*1934*	*1935*	*1936*	*1937*	*1938*	*1939*
Present on 1 January (M:F)	177 (76:84)	77 (33:44)	174 (65:109)	207 (88:119)	206 (82:124)	224 (99:125)	228 (103:125)	218 (97:121)	224 (96:128)	215 (90:125)
Admissions during year	122 (52:70)	241 (100:141)	259 (122:137)	211 (96:115)	235 (103:132)	234 (76:158)	229 (81:148)	233 (76:157)	229 (82:147)	262 (98:164)
Total treated during year	299 (128:171)	318 (133:185)	433 (187:256)	418 (184:234)	441 (185:256)	458 (175:283)	457 (184:273)	451 (173:278)	453 (178:275)	477 (188:289)
Recoveries (M:F)	34 (19:15)	20 (11:9)	94 (40:54)	82 (43:39)	92 (38:54)	92 (31:61)	104 (33:71)	104 (32:72)	117 (40:77)	124 (50:74)
Relieved/ improved (M:F)	40 (11:29)	21 (10:11)	25 (18:7)	29 (13:16)	26 (11:15)	29 (8:21)	26 (14:12)	23 (9:14)	38 (12:26)	40 (12:27)
Not improved (M:F)	142 (63:79)	35 (13:22)	83 (30:53)	75 (27:48)	85 (31:54)	90 (27:63)	87 (33:54)	76 (51:25)	70 (31:39)	82 (28:54)
Deaths (M:F)	6 (2:4)	26 (13:13)	24 (11:13)	26 (19:7)	14 (6:8)	19 (6:13)	22 (7:15)	24 (11:13)	13 (5:8)	8 (2:6)
Remaining 31 December (M:F)	77 (33:44)	174 (65:109)	207 (88:119(	206 (82:124)	224 (99:125)	228 (103:125)	218 (97:121)	227 (77:150)	215 (90:125)	223 (95:128)
Average daily no. resident	97 (42:55)	137 (54:83)	194 (77:117)	208 (88:120)	211 (89:122)	225 (100:125)	215 (93:122)	224 (96:128)	220 (99:121)	207 (87:120)
% Recoveries on admissions	27.86% (36.5%: 21.4%)	8.3% (11%: 6.4%)	36.29% (32.78%: 39.42%)	38.86% (44.79%: 33.91%)	39.15% (36.89%: 40.91%)	39.31% (40.79%: 38.61%)	45.41% (40.74%: 47.97%)	44.63% (42.1%: 45.86%)	51.09% (48.79%: 52.38%)	47.32% (51.02%: 45.12%)

Source: Bethlem Royal Hospital Annual Reports, 1921–1938. BMM/BAR-53

[a] Included in these figures were fee-paying and non-fee paying patients and both certified and voluntary patients

[b] of whom were 211 certified (77 men/134 women) and 51 voluntary (17 men/34 women) patients

Table A.12 Table to show the number of male (M) and female (F) patients resident at the Asile de la Sarthe on 31 December; those admitted during the year; the total number of patients treated during the year; the number of patients discharged due to recovery, improvement, other reasons, or death, 1919–1939

	1919	*1920*	*1921*	*1922*	*1923*	*1924*	*1925*	*1926*	*1927*	*1928*	*1929*
Present on 31 December (M:F)	721	588[a]	604 195:409	614 198:416	621 205:416	648 216:432	683 235:448	707 266:441	757 290:467	742 289:453	777 297:480
Admissions during year (M:F)	189 75:114	177 71:106	134	123	138 57:81	154 64:90	146 81:65	153 67:86	106 41:65	139 49:90	108 39:69
Total treated during year	910	565	738	737	759	802	829	860	863	881	885
Leaving the asylum:	**121**	**112**	**74**	**60**	**58**	**68**	72	**58**	**63**	**53**	**45**
Cured			17	19	24	22	23	20	15	11	5
Improved			34	26	16	30	33	21	34	28	24
Other			23	15	18	16	16	17	15	14	16
Deaths	201#	49	50	56	53	51	50	45	58	51	58
Average no. residents	609	594	615	618	642	658	692	745	746	761	777

	1930	*1931*	*1932*	*1933*	*1934*	*1935*	*1936*	*1937*	*1938*	*1939*
Present on 31 December	782	792	799	812	831	837	845	869	879	899
(M:F)	297:485	298:494	305:494	312:500	335:496	347:490	331:514	343:526	362:517	373:526
Admissions during year	122	118	142	142	162	171	157	172	165	231
(M:F)	46:76	42:76	49:93	68:74	76:86	71:100	67:90	82:90	71:94	127:104
Total treated during year	904	910	941	954	993	1008	1002	1041	1044	1130
Leaving the asylum	**60**	**47**	**54**	**60**	**91**	**89**	**81**	**94**	**72**	**117**
Cured	16	15	5	1	–	–	8	10	6	29
Improved	26	14	28	29	39	46	47	55	38	54
Other	18	18	21	30	59	43	26	29	28	34
Deaths	52	64	75	63	65	74	52	68	73	65
Average no. residents	791	796	812	822	843	857	854	863	883	918

Source: Reports of the Chief Medical Officer, Asile de la Sarthe, 1919–1938. ADS-1X964/5

[a] *High death rate in 1919/fall in patient nos.1920 due to influenza epidemic*

Table A.13 Table to show the number of male (M) and female (F) patients resident at the Littlemore Hospital on 1 January; those admitted during the year; the total number of patients treated during the year; the number of patients discharged due to recovery, improvement, other reasons, or death, 1923–1939

	1923	*1924*	*1925*	*1926*	*1927*	*1928*	*1929*	*1930*	*1931*
Present on 1 January	552	602	606	667	707	729	743	735	787
(M:F)	(223:329)	(276:326)	(266:340)	(222:445)	(254:453)	(268:461)	(273:470)	(282:453)	(295:492)
Admissions during year	182	156	245	182	146	142	148	194	130
	(104:78)	(52:104)	(62:183)	(77:105)	(57:89)	(54:88)	(70:78)	(66:128)	(55:75)
Total treated during year	734	758	851	849	853	871	891	929	917
	(327:407)	(328:430)	(328:523)	(299:550)	(311:542)	(322:549)	(343:548)	(348:581)	(350:567)
Recoveries	73 = 40%	70	44	26 = 14%	14	37 = 26%	32 = 22%	20	17
(M:F)	(29:44)	(25:45)	(23:21)	(8:18)	(7:7)	(15:22)	(17:15)	(12:8)	(10:7)
Relieved	5	24	46	43	48	27	52	47	36
(M:F)	(2:3)	(11:13)	(14:32)	(11:31)	(14:34)	(10:17)	(12:40)	(11:36)	(8:28)
Not improved	1	14	46	15	5	12	13	12	11
	(1:0)	(9:5)	(45:1)	(4:11)	(2:3)	(5:7)	(11:2)	(7:5)	(8:3)
Deaths	42	44	48	59	57	52	59	63	67
		(17:27)	(24:24)	(22:37)	(20:37)	(19:33)	(21:38)	(23:40)	(39:28)
Total no. leaving	132	152	184	142	124	128	156	142	131
(M:F)	(51:81)	(62:90)	(106:78)	(45:97)	(43:81)	(49:79)	(61:95)	(53:89)	(65:66)
Average daily no. resident	556		598	704	711	721	731	759	782
	(239:317)		(229:369)	(250:454)	(256:455)	(266:455)	(275:456)	(292:467)	(290:492)
No. days resident	207,836	215,597	229,306	258,772	261,965	265,228	267,344	281,244	286,240

	1932	*1933*	*1934*	*1935*	*1936*	*1937*	*1938*	*1939*
Present on 1 January (M:F)	786	805	798	809	828	865	878	891
	(285:501)	(299:506)	(296:502)	(299:510)	(301:527)	(321:544)	(329:549)	(333:558)
Admissions during year (M:F)	176	217	200	203	232	225	213	230
	(78:98)	(88:129)	(79:121)	(82:121)	(110:122)	(88:137)	(83:130)	(89:141)
Total cases treated (M:F)	962	1022	998	1012	1060	1090	1091	1121
	(363:599)	(387:635)	(375:623)	(381:631)	(411:649)	(409:681)	(412:679)	(422:699)
Recoveries	22 = 13%	30 = 14%	30 = 15%	29 = 14%	33 = 14%	63 = 28%	61 = 29%	63 = 27%
(M:F)	(14:8)	(13:17)	(15:15)	(13:16)	(21:12)	(21:42)	(22:39)	(32:31)
Relieved	59	68	66	60	69	46	43	51
(M:F)	(22:37)	(26:42)	(21:45)	(30:30)	(25:44)	(23:23)	(15:28)	(13:38)
Not improved	7	43	36	23	17	37	34	39
(M:F)	(3:4)	(19:24)	(15:21)	(7:16)	(10:7)	(9:28)	(17:17)	(16:23)
Regraded	5	8	13	7	12	8	6	14
(M:F)	(2:3)	(5:3)	(3:10)	(2:5)	(6:6)	(3:5)	(1:5)	(11:3)
Deaths	64	75	44	65	64	66	57	53
(M:F)	(23:41)	(28:47)	(22:22)	(28:37)	(28:36)	(27:39)	(24:33)	(13:40)
Total nos. Leaving (M:F)	157	224	189	184	195	220	201	220
	(64:93)	(91:133)	(76:510)	(80:104)	(90:105)	(83:137)	(79:122)	(85:135)
Average daily no. resident	737	791	800	807	841	869	877	892
	(290:497)	(294:497)	(296:504)	(298:509)	(311:530)	(321:548)	(330:547)	(338:554)
No. days resident	238,404	289,873	292,986	297,614	310,351	317,898	322,153	326,792

Source for Tables A.13-14: Littlemore Hospital Annual Reports, 1923/4-1939/40. OHA-L1/A2/1-17

Table A.14 Table to show the origin of Littlemore patients (indicated by the local authorities paying for their care) in 1923, 1929, 1935 and 1939

Local authority	*1923—Men*	*1923—Women*	*1923—Total*	*1929—Men*	*1929—Women*	*1929—Total*	*1935—Men*	*1935—Women*	*1935—Total*	*1939—Men*	*1939—Women*	*1939—Total*
Oxfordshire *County:*							155	223	378	180	251	431
Abingdon	3	5	8	2	4	6						
Banbury	29	40	69	28	37	65						
Bicester	5	15	20	12	22	34						
Chipping Norton	21	30	51	24	28	52						
Faringdon	1	5	6	1	3	6						
Headington	42	85	127	44	85	129						
Henley	9	24	33	10	17	27						
Oxford	41	39	80 = 13%	47	37	84 = 11%	83	112	195 = 24%	102	136	238 = 26%
Thame	13	10	23	12	14	26						
Wallingford	4	6	10	5	6	11						
Witney	24	29	53	18	35	53						
Woodstock	11	17	28	13	19	32						
Out-county:												
Various	1	2	2	–	3	3	3	9	12	5	4	9
Contracts:												
London	49	20	69	–	20	20	–	20	20	10	10	20
Middlesex				27	91	118	30	98	128	30	100	130
Croydon				19	20	39						
Nottingham				–	9	9	–	32	32	–	32	32
Buckinghamshire							10	20	30	–	30	30
East Ham							–	12	12			
Service patients	23	–	23	20	–	20	20	–	20			
Private patients							1	1	2			
TOTALS	276	326	602	282	453	735	301	527	828	337	564	901

Index[1]

[1] Note: Page numbers followed by 'n' refer to notes.

J. Freebody, *Work and Occupation in French and English Mental Hospitals, c.1918–1939*, Mental Health in Historical Perspective, https://doi.org/10.1007/978-3-031-13105-9

B

E

F

G

P